Producing Fuels and Fine Chemicals from Biomass Using Nanomaterials

Edited by
Rafael Luque
Alina Mariana Balu

CRC Press
Taylor & Francis Group
Boca Raton London New York

CRC Press is an imprint of the
Taylor & Francis Group, an **informa** business

CRC Press
Taylor & Francis Group
6000 Broken Sound Parkway NW, Suite 300
Boca Raton, FL 33487-2742

First issued in paperback 2019

ISBN-13: 978-1-4665-5339-2 (hbk)
ISBN-13: 978-0-367-37926-1 (pbk)

Library of Congress Cataloging-in-Publication Data

Producing fuels and fine chemicals from biomass using nanomaterials / editors, Rafael Luque, Alina Mariana Balu.
    pages cm
    Summary: "This book explores the available technologies for the preparation of fuels and chemicals from biomass using nanomaterials. This focus bridges the gap between three hot topics: nanomaterials, energy, and the environment. The book also deals with other important topics related to nanomaterials including toxicity and sustainability and environmental aspects. "-- Provided by publisher.
    Includes bibliographical references and index.
    ISBN 978-1-4665-5339-2 (hardback)
    1. Biomass energy. 2. Nanotechnology. 3. Nanostructured materials--Industrial applications. I. Luque, Rafael. II. Balu, Alina Mariana.

TP339.P755 2013
620'.5--dc23
            2013032516

Visit the Taylor & Francis Web site at
http://www.taylorandfrancis.com

and the CRC Press Web site at
http://www.crcpress.com

Contents

SECTION I   Nanomaterials for Energy Storage and Conversion

SECTION II   Biofuels from Biomass Valorization Using Nanomaterials

SECTION III   Production of High–Added-Value Chemicals from Biomass Using Nanomaterials

Preface

The design of novel and innovative methodologies to maximize current available resources without compromising the future of coming generations is one of the most important challenges of the 21st century. The scarcity of resources and the expected increase in population and energy demands are two of the most important issues to be addressed. In this regard, green chemical and low environmental-impact technologies combined with renewable resources through innovation will be able to offer alternatives to potentially useful processes for a more sustainable bio-based society in which we will move away from the petrol-based economy we have relied upon the past 50-plus years. Biomass is one of the most promising and widely available renewable feedstocks that has a significant potential to offer a number of alternatives to be converted to materials, fuels, and chemicals. Waste residues can also partially help in contributing to this aim, leading to advanced valorization technologies for energy and fuels production. On the other hand, nanotechnology and nanomaterials development have experienced a staggering evolution in recent decades to the point that scientists are currently able to design and optimize a surprisingly significant number of nanomaterials for an extensive range of applications including energy storage, fuels production, biomedicine, and nanocatalysis, which are also taken to industrial and commercial practices. Combining the design of nanomaterials for the valorization of biomass and waste feedstocks to energy and chemicals as well as to (solar) energy storage can constitute a major step forward to further advancing current society in a scientific understanding of properties and alternative applications of nanomaterials in our goal toward a sustainable biorefinery-based economy.

In light of these premises, this book aims to provide a comprehensive and most varied approach to the development of innovative nanomaterials for the production of energy and high added-value chemicals through the use of nanomaterials via energy storage, solar conversion, and biomass and waste valorization practices. The book contains a range of topics bringing together disciplines such as photocatalysis, energy storage, biofuels, cellulose conversion, carbonaceous nanomaterials, supported nanoparticles, among others, aiming to provide a multidisciplinary approach to the fascinating area of nanomaterials properties and applications. We hope this timely book will be able to inspire scientists to devise ever innovative and low environmental-impact alternatives to contribute to the advancement of mankind and we look forward to enjoying further advances in the field in the next few years.

With best wishes for an enjoyable reading.

Acknowledgments

Rafael Luque gratefully acknowledges support from the Spanish MICINN via the concession of a RyC contract (ref. RYC-2009-04199) and funding under projects P10-FQM-6711 (Consejeria de Ciencia e Innovacion, Junta de Andalucia) and CTQ2011 28954-C02-02 (MICINN). He is also indebted to his co-editor, Alina Mariana Balu, for her hard work, dedication, support, and patience in the process of developing the book.

Both editors would like to dedicate this book to the memory of their former mentor, colleague, and friend Professor Juan Manuel Campelo, who unexpectedly passed away in October 2012.

Editors

Rafael Luque earned his PhD in 2005 from Universidad de Cordoba, Spain. He has amassed significant experience on biomass and waste valorization practices to materials, fuels, and chemicals over the past 10 years. He has also published over 150 research articles, filed 3 patent applications, and edited 5 books as well as made numerous contributions to book chapters. He has been an invited guest, and keynote lecturer worldwide in the areas of (nano)materials science, heterogeneous (nano)catalysis, microwave and flow chemistry, biofuels, and green chemical methods in synthetic organic chemistry. Dr. Luque is also a member of the editorial advisory boards of several journals including *Chemical Society Reviews, Catalysis Communications, Current Organic Synthesis, Current Green Chemistry, Journal of Technology Innovations in Renewable Energy, Sustainable Chemical Processes*, and has recently been appointed as editor-in-chief of the section "Porous Materials" of the journal *Materials*.

Among recent awards and recognition to his scientific career, Dr. Luque was awarded the Marie Curie Prize from Instituto Andaluz de Quimica Fina in Spain (2011), the Green Talents award from the Federal Ministry of Education and Research in Germany (2011), and the TR35 Spain 2012 from MIT (USA) as one of the top 10 innovative young entrepreneurs of Spain last year. He has also been recently honored as Distinguished Engineering Fellow 2013 from the Hong Kong University of Science and Technology (HKUST). Dr. Luque also combines his academic duties with his activities as a young entrepreneur after co-founding the spinoff companies Starbon® Technologies at York, UK (2011), and Green Applied Solutions S.L. in Cordoba, Spain (2012).

Alina Mariana Balu earned her PhD in 2012 from Universidad de Cordoba, Spain. Her work focuses on alternative and greener methodologies in chemistry, including novel technologies for materials preparation, the production of biofuels, and photocatalysis—a greener alternative for the production of chemicals and energy. As part of her PhD thesis, Balu brought together three important disciplines—nanomaterials, energy, and environment—and led a multidisciplinary team to develop more sustainable processes for the preparation of supported nanoparticles for various energy and environmental applications. This included the production of advanced second-generation biofuels and high added-value chemicals as well as biomass valorization and environmental remediation. Dr. Balu has coauthored 35 scientific papers and presented at more than 30 international conferences and has recently been awarded a Marie Currie Intra-European Fellowship (2013) and the Green Talents award for high potential in sustainable development from the Federal Ministry of Education and Research in Germany (2012).

Contributors

Muthupandian Ashokkumar
The University of Melbourne
Parkville, Victoria, Australia

Ramazan Asmatulu
Wichita State University
Wichita, Kansas

Alina M. Balu
Aalto University
Espoo, Finland

Claudia L. Bianchi
University of Milan
Milan, Italy

Daria C. Boffito
Department of Chemical Engineering
Polytechnique Montreal
University of Montreal
and
Polytechnic
Montreal, Quebec, Canada

Giuseppina Cerrato
University of Turin
Turin, Italy

Simona Coman
University of Bucharest
Bucharest, Romania

Ilja Gasan Osojnik Črnivec
National Institute of Chemistry
Ljubljana, Slovenia

Yuliya Demidova
Boreskov Institute of Catalysis
Novosibirsk, Russia

Petar Djinović
National Institute of Chemistry
and
Center of Excellence "Low Carbon
    Technologies"
Ljubljana, Slovenia

Boštjan Erjavec
National Institute of Chemistry
and
Center of Excellence "Low Carbon
    Technologies"
Ljubljana, Slovenia

Bastian Etzold
Friedrich-Alexander University of
    Erlangen-Nuremberg
Erlangen, Germany

Benjamin Hasse
Friedrich-Alexander Universität
    Erlangen-Nürnberg
Erlangen, Germany

Peter J. C. Hausoul
RWTH Aachen University
Aachen, Germany

Denisa Hulicova-Jurcakova
The University of Queensland
St. Lucia, Queensland, Australia

Ed de Jong
Avantium Chemicals
Amsterdam, the Netherlands

Sarkyt Kudaibergenov
Kazakh National Technical University
and
Institute of Polymer Materials and
    Technology
Almaty, Kazakhstan

Ji Liang
The University of Adelaide
Adelaide, South Australia, Australia

Rafael Luque
University of Cordoba
Cordoba, Spain

Sara Morandi
University of Turin
Turin, Italy

Dmitry Yu. Murzin
Åbo Akademi University
Turku/Åbo, Finland

Nurxat Nuraje
Massachusetts Institute of Technology
Cambridge, Massachusetts

Regina Palkovits
RWTH Aachen University
Aachen, Germany

Vasile I. Parvulescu
University of Bucharest
Bucharest, Romania

Albin Pintar
National Institute of Chemistry
and
Center of Excellence "Low Carbon
    Technologies"
Ljubljana, Slovenia

Carlo Pirola
University of Milan
Milan, Italy

Shi Zhang Qiao
The University of Adelaide
Adelaide, South Australia, Australia

Juan Carlos Colmenares Quintero
Polish Academy of Sciences
Warsaw, Poland

Marcus Rose
RWTH Aachen University
Aachen, Germany

Irina L. Simakova
Boreskov Institute of Catalysis
Novosibirsk, Russia

Maria-Magdalena Titirici
Queen Mary University of London
London, UK

Madalina Tudorache
University of Bucharest
Bucharest, Romania

Jan C. van der Waal
Avantium Chemicals
Amsterdam, the Netherlands

Ruifeng Zhou
The University of Queensland
St. Lucia, Queensland, Australia

and

The University of Adelaide
Adelaide, South Australia, Australia

1 Introduction to Production of Valuable Compounds from Biomass and Waste Valorization Using Nanomaterials

Alina M. Balu and Rafael Luque

CONTENTS

Society faces a daunting future in terms of water, food, and resource scarcity. This has been clearly evidenced by recent studies showing a continuous decrease in fossil fuel resources, the increasing generation of waste, and the expected increase in population in future years [1–4]. Facing these challenges is not an easy task. A multidisciplinary team effort from many disciplines is needed to come up with suitable alternatives for a future more sustainable society able to deal with these important issues. Renewable resources for the production of alternative energies are believed to be key to a future scenario in the absence of crude oil. The paradigm shift from petroleum hydrocarbons to biobased feedstock provides remarkable opportunities for the chemical processing industry and production of sustainable materials able to replace petrol, strongly linked with a more sustainable future.

Biomass emerged in recent years as a promising alternative and renewable feedstock to be converted into chemicals, materials, and fuels using a range of transformations that should encompass green chemical methods and low environmental impact processes [5,6]. An approximate estimation of terrestrial biomass growth amounts to 118 billion tons/year, dried. About 14 billion tons of biomass/year are produced in agricultural cycles, and out of this, about 12 billion tons/year are essentially discharged as waste. Importantly, the high structural and chemical complexity of most biomass feedstocks, from the simplest polysaccharides to more complex lignocellulosics and lignin, is a key factor to consider the design and development

1

of novel processes aimed to facilitate their conversion to simple fractions able to be (bio)chemically processed to high–added-value products [7].

A series of (bio)chemical as well as thermochemical processing strategies have been developed to process biomass feedstocks with different high–added-value products obtained from processing of such feedstocks [7,8]. Some of these include gasification, pyrolysis, fermentation, anaerobic digestion, catalytic transformations, etc. [7–9]. In any case, advances in the field have provided multiple and most promising possibilities to deal with this renewable, abundant, and widely available type of feedstock.

A recent emerging area in biomass valorization to high–added-value compounds including biochemicals, biomaterials, and biofuels entails the use of designer nanomaterials for biomass conversion. This interesting concept is driven by the enormous possibilities in the design of nanoentities in terms of particle size, nanostructuration, and catalytic sites that are able to provide accessible and highly active sites for the conversion of complex biomass feedstocks [7]. These include (but are not restricted to) nanostructured solid acids, nanoparticles (as such or supported on nanoporous supports), and related nanoentities that hold significant potential for biomass valorization [7,10].

In the light of these premises, this monograph aims to provide a highly multidisciplinary and multitopic viewpoint on the utilization of nanostructured and nanocatalytic materials for the production of high–added-value end products. The whole manuscript has been essentially structured to cover the most important aspects of topics ranging from the design and development of innovative carbonaceous nanomaterials from biomass to related biomass valorization to biofuels and biochemicals using novel (bio)nanocatalytic materials as well as topics on solar energy storage and photocatalysis that provide a highly heterogeneous approach to the use of nanomaterials for various applications. In this regard, the book has been structured into three different parts, mostly related to the applications of the described nanomaterials. The first part is related to the design of nanomaterials for energy storage and conversion-related applications as compared with the second part, which describes the development of nanocatalytic materials for biomass valorization to biofuels, and the third part, which deals with biomass conversion to high–added-value chemicals. Commencing from part I (Nanomaterials for Energy Storage and Conversion), Chapter 2 from Titirici et al. from the Max Planck Institute of Colloids and Interfaces in Golm (Germany), recently moved to Queen Mary University of London in London (United Kingdom), provides a general overview on the state of the art for the production of environmentally sound carbonaceous materials with key applications as heterogeneous catalysts for the production of biofuels and their integration into future biorefineries, an interesting topic that will be subsequently covered in Chapters 6 and 7.

Complementing this chapter, Qiao et al. from the University of Adelaide (Australia) describes in Chapter 3 the design of tailored-made nanomaterial–based carbons for a range of energy-related applications including fuel and solar cells, lithium-ion batteries, and electrochemical capacitors. This chapter highlights in a very comprehensive way the potential of nanomaterials for energy conversion applications. Chapter 4 by Nuraje et al. from MIT (USA) details the utilization of nanomaterials for solar energy storage.

Part II (Biofuels from Biomass Valorization Using Nanomaterials) comprises three highly comprehensive contributions on catalytic reforming of biogas into syngas using a range of supported noble-metal and transition-metal nanoparticles by Pintar et al. from the National Institute of Environmental Sciences and Engineering (Slovenia, Chapter 5) as well as two more specific contributions on biofuels production from waste oils and fats using chemical methods based on nanostructured sulfonated zirconia solid acid catalysts with enhanced acidity synthesized by means of novel ultrasound methodologies (Chapter 6, Boffito et al., University of Milano, Italy) and different supported and immobilized biocatalysts on nanoporous supports for biodiesel, bioethanol, and related high–added-value compounds from crude glycerol (Chapter 7, Parvulescu et al., University of Bucharest, Romania). These chapters highlight the importance of a careful and well-understood design of nanomaterials for biofuels production.

The last part of the book (part III, Production of High–Added-Value Chemicals from Biomass Using Nanomaterials) is probably the most detailed and extensive part of the book, with a number of contributions covering the design and development of biocatalytic, solid acid, photocatalytic, and nanostructured materials for the conversion of a wide range of biomass feedstocks into a wide variety of valuable chemicals (e.g., platform molecules) as intermediates to end products such as biopolymers, bioplastics, biofuels, agrochemicals, and pharmaceutical products. Chapter 8 from Palkovits et al. at RWTH Aachen (Germany) discusses in detail the design of nanostructured solid acid catalysts in the conversion of cellulose as well as nanocatalytic materials for the subsequent processing of platform molecules obtained in cellulose deconstruction to valuable chemicals (e.g., oxidations and hydrogenations of 5-hydroxymethylfurfural and furfural). To complete this emerging and highly trendy topic, Chapter 9 from De Jong et al. from Avantium Chemicals (The Netherlands) provides an interesting industrial viewpoint contribution on key chemocatalytic industrial and pilot-plant processes related to sugars valorization to high-added-value products including furfural, dicarboxylic acids, aromatics, and isosorbides, as well as related aqueous-phase reforming of lignocellulosics.

In continuation with the topic on biomass valorization, Chapter 10 from Murzin et al. at Turku University (Finland) completes the overview on fine chemical production using catalytic nanomaterials by providing useful insights into structure sensitive chemistries of two case studies, namely the isomerization of α-pinene to camphene and hydrogenation of thymol to menthol. These examples highlight the importance of metal nanoparticle sizes in activity, selectivity, and stability of catalysts.

The book is finally completed with Chapter 11 from Colmenares et al. at the Polish Academy of Sciences in Warsaw (Poland) who discloses the basics in the design of nanotitania photocatalytic systems for biomass valorization to valuable chemicals via photocatalytic processing, with examples on hydrogen production via water splitting, photocatalytic reforming, and transformations of platform molecules to valuable acids and important intermediates (e.g., photocatalytic oxidation of glucose to gluconic and glucaric acids avoiding full mineralization to CO_2 and H_2O).

With this full overview of the monograph as a whole, readers will identify several key areas for further development in which we hope the included contributions can provide a good source of information to stimulate more advances in the field of designer nanomaterials for biomass valorization to materials, fuels, and chemicals as well as energy conversion and storage.

ACKNOWLEDGMENTS

Rafael Luque gratefully acknowledges MICINN for the concession of a Ramon y Cajal Contract (ref. RYC-2009-04199) and funding from projects CTQ2011-28954-C02-02 (MEC) and P10-FQM-6711 (Consejeria de Ciencia e Innovacion, Junta de Andalucía).

REFERENCES

1. The Royal Society, Sustainable biofuels: prospects and challenges, 2008, ISBN 978 0 85403 662 2.
2. IEA, *World energy outlook* 2007. International Energy Agency: Paris.
3. C. Ford Runge, B. Senauer, *Foreign Affairs* 2007, 86.
4. P. Hazell, R. K. Pachauri (eds.), *Bioenergy and agriculture: promises and challenges*, International Food Policy Research Institute 2020 Focus No. 14, 2006.
5. A. Corma, S. Iborra, A. Velty, *Chem. Rev.* 2007, 107, 2411–2502.
6. A. J. Ragauskas, C. K. Williams, B. H. Davison, G. Britovsek, J. Cairney, C. A. Eckert, W. J. Frederick, J. P. Hallett, D. J. Leak, C. L. Liotta, J. R. Mielenz, R. Murphy, R. Templer, T. Tschaplinski, *Science* 2006, 311, 484–489.
7. J. C. Serrano, R. Luque, A. Sepulveda-Escribano, *Chem. Soc. Rev.* 2011, 40, 5266–5281.
8. A. Demirbas, *Energy Conv. Management* 2001, 42, 1357–1378.
9. A. V. Bridgwater, *Appl. Catal. A* 1994, 116, 5–47.
10. R. Luque, Designer nanomaterials for the production of energy and high added value chemicals, *Ideas in chemistry and molecular sciences: advances in nanotechnology, materials and devices,* Ed. Bruno Pignataro, Wiley-VCH, Weinheim, Germany, 2010, Chapter 2, 23–64.

Section I

Nanomaterials for Energy
Storage and Conversion

2 Green Carbon Nanomaterials

From Biomass to Carbon

Maria-Magdalena Titirici

CONTENTS

2.1 INTRODUCTION

In the early part of the 20th century, many industrialized materials such as solvents, fuels, synthetic fibers, and chemical products were made from plant/crop-based resources (Figure 2.1) [1,2]. Unfortunately, this is no longer the case and most of today's industrial materials including fuels, polymers, chemicals, carbons, pharmaceuticals, packing, construction, and many others are being manufactured from fossil-based resources. Humankind is still living in a world where petroleum resources have the absolute power. However, crude oil resources are rapidly diminishing. It is predicted that this will lead to serious conflicts in the world related to distribution and control. What is even of more concern, is that essentially such fossil fuel–derived products eventually end up as CO_2 in the Earth's atmosphere. Several important findings of climate research have been confirmed in recent decades and are finally accepted as facts by the scientific community. These include indeed a rapid increase in the carbon dioxide concentrations in the atmosphere during the last

(a)   (b)

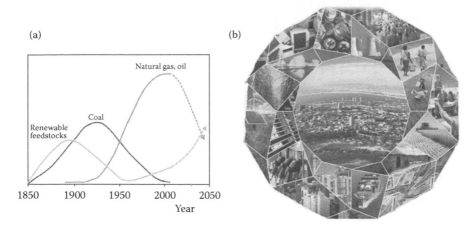

FIGURE 2.1 **(See color insert.)** (a) Raw materials basis of chemical industry in historical perspective. (Taken with permission from F. W. Lichtenthaler, S. Peters, *Comptes Rendus Chimie* 2004, 7, 65–90.) (b) A view on sustainable materials for a sustainable future. (Taken with permission from M. L. Green, L. Espinal, E. Traversa, E. J. Amis. *MRS Bulletin* 2012, 37, 303–309.)

150 years, from 228 ppm to the 2007 level of 383 ppm [3]. This increase is our own fault and is due to burning of fossil fuels.

Thus, what will the world look like in 2050? It is believed that if we continue relying on fossil fuels, we may face an ecological collapse of unprecedented scale due to degradation of natural capital and loss in ecosystem services. However, we have the capability to reverse this dark and worrisome perspective of an ecological fiasco and shape a future where we could live in harmony with nature. For this to happen, scientists have the most important responsibility and joint efforts from multidisciplinary scientific fields are of utmost significance to achieve this goal.

One of the most important issues is the production of renewable energy to cure our addiction to oil. Solar and wind energy are expected to play the most important role in the future. Available solar and wind energy depends strongly on geography and local climate and varies strongly with season, time of day, and weather. This creates additional subsidiary challenges of cost efficient energy storage and transportations. This is why high performing materials in smart grids, batteries, fuel cells, solar cells, gas storage, or efficient catalysts to convert renewable resources in transportation fuels need to be developed.

The paradigm shift from petroleum hydrocarbons to biobased feedstock provides remarkable opportunities for the chemical processing industry and production of sustainable materials able to perform the above mentioned functions strongly linked with a sustainable future [4].

Nature offers an abundance of opportunities for shaping structural and functional materials in its wide variety of raw materials including carbohydrates, nucleotides, and proteins. In this respect, Koopman et al. emphasized the importance of developing new starting materials from biomass from an industrial point of view [5]. Biomass is the most abundant renewable resource on Earth. An approximate estimation of terrestrial biomass growth amounts to 118 billion tons/year, dried [6]. About 14 billion tons of biomass/year are produced in agricultural cycles, and out of this, about 12 billion tons/year are essentially discharged as waste. Obviously, there is enough biomass available at almost no cost to be used in many different ways. I will point out here three directions for biomass transformation into useful products while focusing on the last one.

1. The greatest potential of biomass utilization is the generation of biofuels as a sustainable alternative for transportation with no CO_2 emissions. This can be achieved either by fermentation [7], gasification [8], or catalytic liquefaction [9], and it represents the topic of this book.
2. One aspect of green chemistry refers to the use of biomass to provide alternative starting materials for the production of chemicals, vitamins, pharmaceuticals, colorants, polymers, and surfactants [10]. Industrial white biotechnology highlights the use of microorganisms to provide the chemicals. It also includes the use of enzyme catalysis to yield pure products and consume less energy [11]. Examples using these techniques include composite materials such as polymeric foams and biodegradable elastomers generated from soybean oil and keratin fibers [12]. Plastics such as polylactic acid [13] along with biomass-based polyesters [14], polyamides [15], and polyurethanes [16,17] have also been developed. The list of such biomass derived products, commercially available or under development, is obviously much larger, but a comprehensive review is out of the scope of this thesis [18].
3. Work on conversion of biomass and municipal waste into carbon is still rare, but a significantly growing research topic. This is not surprising given the enormous potential of carbon to solve many of the challenges associated with sustainable technologies described earlier and outlined in Figure 2.1b.

Carbon (meaning "coal" in Latin) is one of the most widespread and versatile elements in nature and is responsible for our existence today. Humans have been using carbon since the beginning of our civilization. Carbon exists in nature in different allotrope forms from diamond to graphite and amorphous carbon. With the development of modern technology and the need for better performing materials, a larger number of new carbon materials with well-defined nanostructures have been synthesized by various physical and chemical processes, such as fullerenes, carbon nanotubes (CNTs), graphitic onions, carbon coils, carbon fibers, and others. Carbon materials have been awarded prizes three times in the last 15 years: fullerenes, the 1996 Nobel Prize in chemistry; CNTs, the 2008 Kavli Prize in nanoscience; graphene, the 2010 Nobel Prize in physics. To date, it is probably fair to say that researches on carbon materials are encountering the most rapid development period, which I would like to call "The Back to Black" period.

Despite its wide spreading and natural occurrence on Earth, carbon has been mainly synthesized from fossil-based precursors with sophisticated and energy-consuming methodologies, having as a consequence the generation of toxic gases and chemicals. Pressures of an evolving sustainable society are encouraging and developing awareness among the materials science community of a need to introduce and develop novel media technology in the most benign, resource-efficient manner possible. Especially the preparation of porous carbon materials from renewable resources is a quickly recognized area, not only in terms of application/economic advantages but also with regard to a holistic sustainable approach to useful porous media synthesis. Carbon has been created from biomass from the very beginning, throughout the process of coal formation. Nature is mastering the production of carbon from biomass, and we only need to translate it into a synthetic process.

In this chapter, I will provide a short overview on the state of the art concerning the production of green carbon materials and, wherever possible, point out their utilization as heterogeneous catalysts for the production of biofuels. Biomass-derived carbon materials, especially hydrothermal carbons, are biomass-derived, providing thus a holistic approach that could, in principle, be easily integrated into future biorefinery schemes.

2.2   GREEN CARBON MATERIALS

By "green carbon," I mean materials that are synthesized from renewable and highly abundant precursors using as little energy consumption as possible (e.g., low temperatures) and avoiding the use and generation of toxic and polluting substances. In addition, they should perform important technological tasks. These prerequisites are not trivial to achieve. Below I will provide some examples from the literature where the synthesis of such materials was targeted.

2.2.1   CARBON NANOTUBES AND GRAPHITIC NANOSTRUCTURES

Many potential applications have been proposed for carbon nanotubes, including conductive and high-strength composites, energy storage and energy conversion devices, sensors, field emission displays and radiation sources, hydrogen storage media, and nanometer-sized semiconductor devices, probes, and interconnects [19]. Some of these applications are now realized in products. Others are demonstrated in early and advanced devices, and one, hydrogen storage, is clouded by controversy. Nanotube cost, polydispersity in nanotube type, and limitations in processing and assembly methods are important barriers for some applications.

The demand for this raw material in the nanotechnology revolution is rising explosively. As this trend continues and nanomaterials become simple commodities, mundane production issues, such as the limitation of available resources, the cost of production materials, and the amount and cost of energy used in nanomaterial synthesis, will become the key cost drivers and bottlenecks. Many efforts have been made to find simple technologies for the mass production of CNTs at low cost. A review on this topic has been recently published by Dang Sheng Su [20]. For mass production, the catalyst is considered the key factor for CNT growth. Transition metals, Fe, Co, Ni, V, Mo, La, Pt, and Y, are active for CNT synthesis [21]. Any effective production process that leads to a large reduction in costs will lead to a breakthrough of CNT applications. Investigations into new inexpensive feedstocks as well as more efficient catalyst/support combinations suitable for the mass production of CNTs are required.

In one example, Mount Etna lava was used as a catalyst and support for the synthesis of nanocarbon [22]. The main component is Si (SiO_2, 48 wt%), whereas the total amount of iron as Fe_2O_3 is as high as 11 wt%, distributed among silicate phases and Fe–Ti oxides (Figure 2.2). The presence of iron oxide particles in the porous structure of Etna lava make these materials promising for the growth and immobilization of CNFs. For the CVD growth (700°C), the crushed lava powder was introduced into a horizontal quartz reactor and reduced with hydrogen before CVD treatment. Ethylene was used as a carbon source. A mixture of CNFs and CNTs grown on lava rock was obtained (Figure 2.2), the nanofibers being dominant. TEM analysis revealed that the CNFs and CNTs obtained from lava exhibited a graphitic wall structure. The diameter distribution of the obtained CNTs and CNFs was broad, ranging from a few nanometers to several micrometers.

Although the estimated volume of emitted lava was about 10–11 million cubic meters, there are still issues associated with the global availability of such a catalyst. The advantage of using lava, which avoids the wet chemical preparation of an iron catalyst, is also challenged by issues such as collection, transportation, and purification that may consume additional energy.

In another example, the same group used a special type of red soil from Croatia as a catalyst support for the synthesis of nanocarbons. The composition of the soil

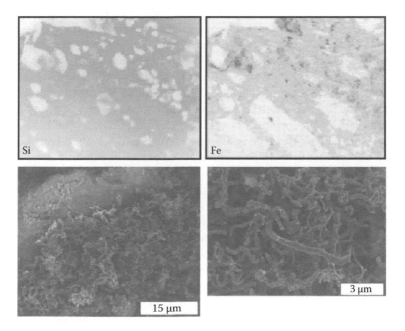

FIGURE 2.2 **(See color insert.)** SEM images showing the elemental distribution of Si and Fe in lava stone granulate (top) and CNFs grown on lava (bottom). (D. S. Su. *ChemSusChem.* 2009. 2. 1009–1020. Copyright Wiley-VCH Verlag GmbH & Co. KGaA. Reproduced with permission.)

was a mixture of Al, Fe, Si, Ca, and Mg oxides. Ethylene was used as carbon source for the CNT growth through a CVD process. CNFs grown on the red soil were found to exhibit a broad diameter distribution. The quality of the CNFs was comparable to that produced using lava rock as catalyst/support [20].

Endo and his group used garnet sand pulverized from natural garnet stones (Ube Sand Kogyo, 1.4 USD/kg) as a catalyst and cheap urban household gas ($1.1 m^{-3}) as a carbon source for the CVD process [23]. After CVD, the 200-mm-sized granulates of garnet powder (Figure 2.3a) were coarsened to about 400 mm (Figure 2.3b) and were covered with CNTs (Figure 2.3c and d). About 25%–30% of the weight from the sand–CNT composite corresponded to the CNTs. The produced CNTs had diameters typically in the range of 20–50 nm and exhibited well-ordered structures with large-diameter hollow cores (Figure 2.3e and f). The graphitization degree of the walls was much higher than that of the CNFs prepared with lava and soil catalysts, and in addition, the resulting CNTs could be very easily separated from the garnet sand, simply by an ultrasonic bath in a water suspension.

Other low-cost natural catalysts used in the production of CNTs were bentonite [24] natural minerals such as forsterite, disposide, quartz, magnesite, and brucite [25] or biomass-derived activated carbons previously modified with Fe by an impregnation method [26,27]. The later method resulted in hierarchically structured carbon, consisting of CNFs supported on activated carbon.

In another study, the intrinsic iron content of biomass-derived ACs (especially from palm kernel shell, coconut, and wheat straw) was directly used as a catalyst for CNF synthesis [28]. The step involving preparation of Fe particles on the activated carbon was circumvented, and the overall process was simplified.

Thus far, only examples of how low-cost and naturally abundant catalysts have been successfully integrated in the production of carbon nanotubes were presented. However, the precursors used were gases of fossil fuel origin. The natural materials originating from biomass, such as coal, natural gas, or biomass itself, can be also used as a carbon source for greener nanocarbon synthesis.

The feasibility of producing CNTs and fullerenes from Chinese coals has been investigated [29]. When used as a carbon source, camphor ($C_{10}H_{16}O$), a botanical carbon material, was reported to be a highly efficient CNT precursor requiring an exceptionally low amount of catalyst in a CVD process [30]. CNTs can be also obtained by heating grass in the presence of a suitable amount of oxygen [31]. Fabrication of CNTs with carbohydrates could be expected when all the other possibilities have been tested. It is interesting that the well-known formation mechanism of CNTs (i.e., generating active carbon atomic species followed by assembling them into CNTs) cannot be applied for biomass. Tubular cellulose in grass is directly converted into CNTs during the heat treatment.

In respect with developing different methods rather than CVD for the CNTs production, hydrothermal treatment represents a "greener" solution to nanotubes production [32], provided that the precursors also belong to the same category. Calderon Moreno et al. used the hydrothermal process to reorganize amorphous carbon at moderate temperature (600°C) and pressures of 100 MPa into nanographitic structures such as nanotubes and nanofibers. They provided evidence that carbon atoms rearrange to form curved graphitic layers during hydrothermal treatment. The growth of graphitic multiwall structures in hydrothermal conditions takes place

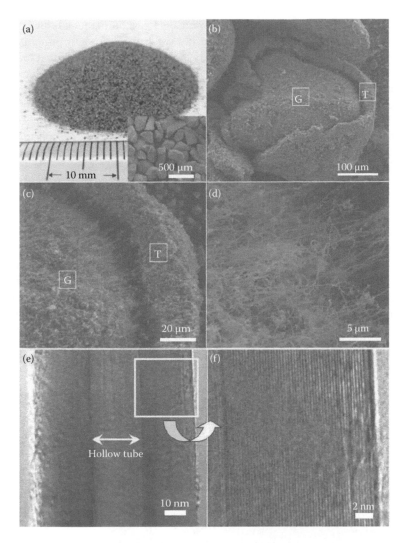

FIGURE 2.3 (a) Photograph of the garnet sand used to produce nanotubes; the inset is an SEM image showing the average diameter of the sand particles (average size ca. 200 mm). (b–d) SEM images of the CNTs grown on the surface of the garnet sand particles (in parts b and c, G and T indicate the garnet particle and CNT, respectively; part d corresponds to the CNTs only). (e, f) TEM images showing the central hollow core of a typical as-grown CNT (e) and the highly linear and crystalline lattice of the wall (f). (M. Endo, K. Takeuchi, Y. A. Kim, K. C. Park, T. Ichiki, T. Hayashi, T. Fukuyo, S. Linou, D. S. Su, M. Terrones, M. S. Dresselhaus. *ChemSusChem* 2008, 1, 820–822. Copyright Wiley-VCH Verlag GmbH & Co. KGaA. Reproduced with permission.)

by different mechanisms than in the gas phase. Hydrothermal conditions provide a catalytic effect caused by the reactivity of hot water that allows the graphitic sheets to grow, move, curl, and reorganize bonds at much lower temperatures than in the vapor phase in inert atmospheres. Such reorganization is induced by the physical tendency to reach a more stable structure with lower energy by reducing the number of dangling bonds in the graphitic sheets. The mechanism by which amorphous carbon rearranges into curled graphitic cells in the hot hydrothermal fluid is complex and involves the debonding of graphitic clusters from the bulk carbon material in hydrothermal conditions. Closed graphitic lattices can be favored at increasing temperatures or more chemically reactive environments (Figure 2.4).

An interesting and low-cost approach to high-quality multiwall carbon nanotubes is reported by Pol et al. who described a solvent-free process that converts polymer wastes such as low- and high-density polyethylene into multiwalled carbon nanotubes via thermal dissociation in the presence of chemical catalysts (cobalt acetate) in a closed system under autogenic pressure [34]. The readily available used/waste high-density polyethylene is introduced for the fabrication of the MWCNTs. The digital image of such feedstock is shown in Figure 2.5a. The grocery bags are extruded from a machine that works in the following manner: for the length of the bag, polyethylene molecules (inset Figure 2.5a) are arranged in the long chain direction, allowing maximum lengthwise stretch and possessing greater strength. As shown in Figure 2.5b, the MWCNTs grew outward forming bunches of 2- to 3-mm size. Each bunch is comprised of hundreds of MWCNTs growing outward.

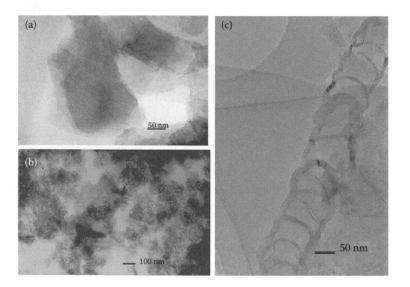

FIGURE 2.4 (a) HRTEM micrograph of the amorphous carbon particles used as starting material; (b) HRTEM micrograph of the bulk microstructure after hydrothermal treatment, showing the interconnected nanocells formed by curled graphitic walls; (c) a chain of connected cells illustrating how the graphitic carbon form a single interconnected structure with multiple individual nanocells. (Reprinted from *Carbon* 39, J. M. Calderon Moreno, T. Fujino, M. Yoshimura, 618–621, Copyright 2001, with permission from Elsevier.)

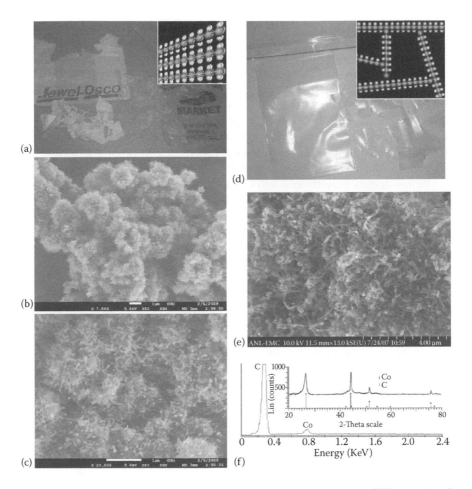

FIGURE 2.5 (a) Digital image of HDPE (inset shows the arrangement of PE groups) polymer wastes; (b) field emission scanning electron microscope (FE-SEM) image; (c) HRSEM image of as-prepared MWCNTs using a mixture of high-density polyethylene and CoAc; (d) digital image of low-density polyethylene (inset shows the arrangement of PE groups) polymer wastes; (e) FE-SEM of MWCNTs prepared from low-density polyethylene; and (f) energy dispersive spectroscopy (EDS) measurements of as-prepared MWCNTs fabricated from low-density polyethylene with CoAc catalyst (inset shows powder X-ray diffraction [XRD] pattern). (V. G. Pol, P. Thiyagarajan, *Journal of Environmental Monitoring* 2010, 12, 455–459. Reproduced by permission of The Royal Society of Chemistry.)

Under the above-mentioned experimental conditions, polyolefins will reduce to carbon, further producing MWCNTs around the cobalt nanocatalyst obtained from the dissociation of CoAc. The diameters of the MWCNTs are 80 nm; a length of more than a micron (Figure 2.5c) is observed within 2 h of the initial reaction time; thus, the growth of MWCNTs is a function of reaction time. A higher percentage of low-density polyethylene was used for making soft, transparent grocery sacks (Figure 2.5d), shrink/stretch films, pond liners, construction materials, and agriculture film.

In low-density polyethylene, the molecules of polyethylene are randomly arranged (inset Figure 2.5d). The as-formed MWCNTs obtained from the thermolysis of waste low-density polyethylene in the presence of CoAc catalyst in a closed system are shown in Figure 2.5e. The MWCNTs were randomly grown during 2 h of reaction time, not analogous to high-density polyethylene. The dissociation of low-density polyethylene with CoAc catalyst also created ~1000 psi pressure. In both cases, the grown MWCNTs were tipped with nanosized metallic Co particles. Transmission electron micrographs further confirmed the hollow tubular structures of MWCNTs. The energy dispersive spectroscopy (EDS) (Figure 2.5f) and X-ray diffraction (XRD) pattern (inset Figure 2.5f) of MWCNTs prepared from the mixture of low-density polyethylene and CoAc confirmed that the MWCNTs are comprised of graphitic carbon and trapped cobalt. It needs to be mentioned that in the absence of the catalysts, micrometer-sized hard spheres were obtained instead [35].

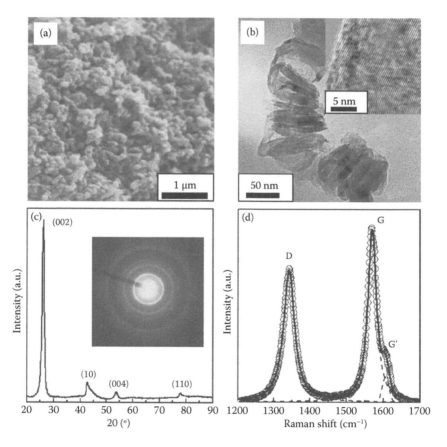

FIGURE 2.6 Structural characteristics of the graphitic carbon nanostructures obtained from the cellulose-derived hydrochar sample. (a) SEM microphotograph, (b) TEM image (Inset: HRTEM image), (c) XRD pattern (inset: selected area electron diffraction pattern), and (d) first-order Raman spectrum. (Reprinted from *Chemical Physics Letters* 490, M. Sevilla, A. B. Fuertes, 63–68, Copyright 2010, with permission from Elsevier.)

Given that polyethylene-based used plastics need hundreds of years to degrade in atmospheric conditions and innovative solutions are required for polymer waste, this technology represents a very environmentally friendly and low-cost method to produce carbon nanotubes.

Graphitic carbon nanostructures have been synthesized from cellulose by Sevilla via a simple methodology that essentially consists of (i) hydrothermal treatment of cellulose at 250°C and (ii) impregnation of the carbonaceous product with a nickel salt followed by thermal treatment at 900°C [36]. The formation of graphitic carbon nanostructures seems to occur by a dissolution–precipitation mechanism in which amorphous carbon is dissolved in the catalyst nanoparticles and then reprecipitated as graphitic carbon around the catalyst particles. The subsequent removal of the nickel nanoparticles and amorphous carbon by oxidative treatment leads to graphitic nanostructures with a coil morphology. This material exhibits a high degree of crystallinity and large and accessible surface area (Figure 2.6).

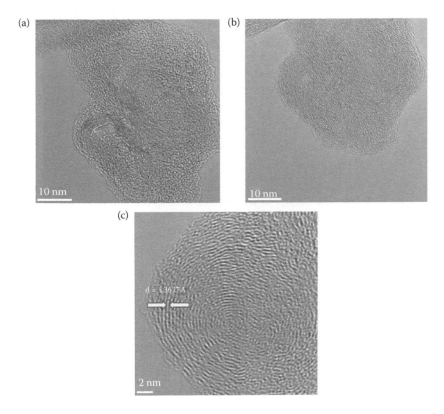

FIGURE 2.7   HRTEM images of the carbon material derived from collagen waste by treating at 1000°C for 8 h. (a, b) Polyhedral and spherical onion-like nanostructures showing the presence of graphitic layers with significant defects. (c) Spherical carbon nanoonion structure showing highly defective shells separated by 0.3363 nm. (M. Ashokkumar, N. T. Narayanan, A. L. Mohana Reddy, B. K. Gupta, B. Chandrasekaran, S. Talapatra, P. M. Ajayan, P. Thanikaivelan, *Green Chemistry* 2012, 14, 1689–1695. Reproduced by permission of The Royal Society of Chemistry.)

Ashokkumar et al. recently produced onion-like nitrogen dopes graphitic structures by simple high carbonization of collagen, a waste derivative from leather industries (Figure 2.7). The leather industry generates voluminous amounts of protein wastes at a level of 600 kg/ton of skins/hides processed, as leather processing is primarily associated with purification of a multicomponent skin to obtain a single protein, collagen. This synthetic route from biowaste raw material provides a cost-effective alternative to existing chemical vapor deposition methods for the synthesis of functional nanocarbon materials and presents a sustainable approach to tailor nanocarbons for various applications [37].

To summarize this subsection, some progress has been achieved in the synthesis of CNTs and graphitic nanostructures using either natural catalysts and/or natural precursors. Several studies have shown that natural materials can be used for the synthesis of nanomaterials, aimed at developing low-cost, environmentally benign, and resource-saving processes for large-scale production. Catalyst-free CNTs have been also successfully synthesized from amorphous carbon under hydrothermal conditions. The examples provided show promising potential and interesting perspective on nanocarbon syntheses using these inexpensive resources. Unfortunately, when using such low-cost technologies, uniform diameters and homogeneous structures are difficult to achieve. Although the investigations were performed to look for a cost-effective method for mass production of CNTs, studies regarding the sustainability of using such natural organic materials are still required.

2.2.2 GRAPHENE, GRAPHENE OXIDE, AND HIGHLY REDUCED GRAPHENE OXIDE

Since winning the Nobel Prize in physics in 2010, graphene is the new star of carbon science. Obviously, graphene is not a new material, and it is known to form graphite by parallel stacking, as well as fullerenes and carbon nanotubes by rolling into 2D nanostructures. The delay in its discovery as an individual material can be partially attributed to the single-atom-thick nature of the graphene sheet, which was initially thought to be thermodynamically unstable [38]. However, graphene is not only stable but also exhibits impressive electronic and mechanical properties (charge-carrier mobility = 250,000 cm^2 V^{-1} s^{-1} at room temperature [39], thermal conductivity = 5000 W m^{-1} K^{-1} [40], and mechanical stiffness = 1 TPa [41].

Chemical exfoliation strategies such as sequential oxidation–reduction of graphite often result in a class of graphene-like materials best described as highly reduced graphene oxide (HRG) [42,43], with graphene domains, defects, and residual oxygen-containing groups on the surface of the sheets. Indeed, none of the currently available methods for graphene production yields bulk quantities of defect-free sheets.

In general, methods for producing graphene and HRG can be classified into five main classes: (1) mechanical exfoliation of a single sheet of graphene from bulk graphite using Scotch tape [44,45], (2) epitaxial growth of graphene films [46], (3) chemical vapor deposition (CVD) of graphene monolayers [47], (4) longitudinal "unzipping" of CNTs [40,48], and (5) reduction of graphene derivatives, such as graphene oxide and graphene fluoride [49,50], which in turn can be obtained from the chemical exfoliation of graphite.

Among all these methods, chemical reduction of exfoliated graphite oxide (GO), a soft chemical synthesis route using graphite as the initial precursor, is the most efficient approach toward the bulk production of graphene-based sheets at low cost. Stankovich et al. [49,51] and Wang et al. [52] were among the first to carry out the chemical reduction of exfoliated graphene oxide sheets with hydrazine hydrate and hydroquinone as the reducing agents, respectively.

Since these first reports, significant effort has been made to find greener technologies to reduce exfoliated graphene oxide sheets to defect free graphene. Xia et al. reported an electrochemical method as an effective tool to modify electronic states via adjusting the external power source to change the Fermi energy level of electrode materials surface. This represents a facile and fast approach to the synthesis of high-quality graphene nanosheets in large scale by electrochemical reduction of the exfoliated GO at a graphite electrode, while the reaction rate can be accelerated by increasing the reduction temperature [53].

Other sustainable methods for the reduction of GO involved photochemical [54,55], sugars [56], iron [57], Zn powder [58], vitamin C [59], microwave [60], baker's yeast [61], phenols from tea [62], bacteria [63,64], gelatin [65], supercritical alcohols [66], and others.

Despite all these milder ways toward GO reduction and although it could become an industrially important method to produce graphene, until now the quality of this liquid exfoliated graphene is still lower than mechanically exfoliated graphene due to the destruction of the basal plane structure during the oxidation and incomplete removal of the functional groups. In addition, the oxidation of graphite is a tedious method involving very aggressive substances such as $KMnO_4$, $NaNO_3$, and H_2SO_4.

Recently, many research groups have published several CVD methods for growing large-sized graphene on wafers. For the growth of epitaxial graphene on single-crystal silicon carbide (SiC) [46], the cost of this graphene is high due to the price of the 4H-SiC substrate. Also, metals such as copper [67], nickel [47,68], iron [69], cobalt [70], and platinum [71] have been used as catalytic substrates to grow mono-, bi-, or multilayer graphene. The CVD method is limited to fossil-based gaseous carbon sources such as methane or acetylene.

The group of Tour has come up with a solution to the use of gas precursors and showed that large area, high-quality graphene with controllable thickness can be grown from different solid carbon sources—such as polymer films or small molecules—deposited on a metal catalyst substrate at temperatures as low as 800°C. Both pristine graphene and doped graphene were grown with this one-step process using the same experimental setup [72]. The same group expanded this concept to any solid precursor such as waste food and insects (e.g., cookies, chocolate, grass, plastics, roaches, and dog feces) to grow graphene directly on the backside of a Cu foil at 1050°C under H_2/Ar flow (Figure 2.8) [73]. The nonvolatile pyrolysed species were easily removed by etching away the front side of the Cu. Analysis by Raman spectroscopy, X-ray photoelectron spectroscopy, ultraviolet visible spectroscopy, and transmission electron microscopy indicates that the monolayer graphene derived from these carbon sources is of high quality. Through this method, low-valued foods and negative-valued solid wastes are successfully transformed into high valued graphene, which brings new solutions for recycling of carbon from impure sources.

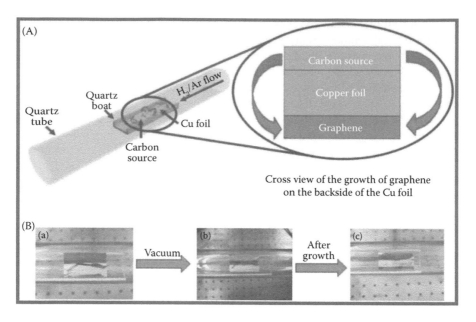

FIGURE 2.8 **(See color insert.)** (A) Diagram of the experimental apparatus for the growth of graphene from food, insects, or waste in a tube furnace. On the left, the Cu foil with the carbon source contained in a quartz boat is placed at the hot zone of a tube furnace. The growth is performed at 1050°C under low pressure with a H₂/Ar gas flow. On the right is a cross view that represents the formation of pristine graphene on the backside of the Cu substrate. (B) Growth of graphene from a cockroach leg. (a) One roach leg on top of the Cu foil. (b) Roach leg under vacuum. (c) Residue from the roach leg after annealing at 1050°C for 15 min. The pristine graphene grew on the bottom side of the Cu film (not shown). (Reprinted with permission from G. Ruan, Z. Sun, Z. Peng, J. M. Tour, *ACS Nano* 2011, 5, 7601–7607. Copyright 2011 American Chemical Society.)

Hermenegildo García and coworkers showed that chitosan, an N-containing biopolymer can form high-quality films on glass, quartz, metals, and other hydrophilic surfaces. Pyrolysis of chitosan films under argon at 800°C and under inert atmosphere gives rise to high-quality single layer N-doped graphene films (over 99% transmittance) as evidenced by XPS, Raman spectroscopy, and TEM imaging [74].

Ruiz-Hitzky et al. demonstrated the possibility of preparing graphene-like materials from natural resources such as sucrose (table sugar) and gelatin assembled to silica-based porous solids without a requirement for reducing agents. The resulting materials show interesting characteristics, such as simultaneous conducting behavior afforded by the sp² carbon sheets, together with chemical reactivity and structural features, provided by the silicate backbone, which are of interest for diverse high-performance applications. The formation mechanism of supported graphene is still unclear, with further studies being needed to optimize its preparation following these green processes [75].

Much progress has been done to date in the synthesis of sustainable graphene-derived materials. Given that the filed is relatively new, it is expected that new synthetic breakthroughs are soon to come for the large-scale, low-cost synthesis of defect-free graphene. I believe that graphene will continue to play an important role in materials science when associated with applications related to its exceptional physical properties. However, for many of the applications described lately in the literature such as adsorption, catalysis, or energy storage, graphene in its pure form is not necessary and other carbon materials perform as well. In addition, the word "graphene" is too easily used in many of the recent publications for structures that are in fact just disordered graphite and that have been known in the literature for many years.

2.2.3 ACTIVATED CARBONS

Thus far, we discussed crystalline forms of carbons such as carbon nanotubes and graphenes. Activated carbons belong to the amorphous carbon category. Activated carbons are by far the oldest and most numerous category of materials prepared from renewable resources. A comprehensive description is behind the scope of this chapter. Many reviews exist in the literature on this topic [76]. Activated carbons are prepared either by chemical or physical activation from biomass or waste precursors at temperatures between 600°C and 900°C. They are microporous and used mainly for adsorption processes (i.e., water purification) and recently in supercapacitors [77] and gas storage [78]. One main disadvantage of activated carbons is the impossibility to predict their resulting porosity and to control their pore properties.

2.2.4 STARBONS

The "Starbon" technology was developed in the group of Professor James Clark at the University of York, and it is based on the transformation of nanostructured forms of polysaccharide biomass into more stable porous carbonaceous forms for high value applications [79]. This approach opens new routes to the production of porous materials and represents a green alternative to traditional materials based on templating methods. The principle of this methodology relies on the generation of porous polysaccharide precursors that can then be carbonized to preserve the porous structure.

This synthetic strategy was initially focused on the use of mesoporous forms of the polysaccharide starch (from where the name is derived) but evolved into a generic tunable polysaccharide-based route. The technology involves (1) native polymer expansion via polysaccharide aqueous gel preparation, (2) production of solid mesoporous polysaccharide, via solvent exchange/drying, and (3) thermal carbonization.

The resulting carbon-based materials are highly porous and mechanically stable in the temperature preparation range from 150°C to 1000°C. At a temperature >700°C, the carbonization process leads to the synthesis of robust mesoporous carbons with a wide range of technologically important applications, including heterogeneous catalysis, water purification, separation media, as well as potential future applications in energy generation and storage applications.

The three main production stages are shown in Figure 2.9. Starch (typically from high amylose cornstarch) is transformed into a gel by heating in water. The resulting viscous solution is cooled to 5°C for typically 1 to 2 days to yield a porous gel.

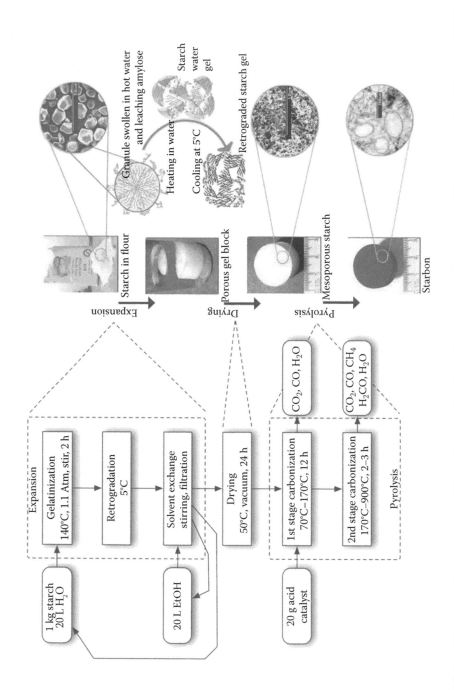

FIGURE 2.9 Diagrammatic representation of the main processing steps in the production of starch-derived Starbon materials. (Taken from X. W. Lou, J. S. Chen, P. Chen, L. A. Archer, *Chemistry of Materials* 2009, *21*, 2868–2874.)

Water in the gel is then exchanged with the lower surface tension solvent ethanol. The resulting material is then filtered and may be oven dried to yield a predominantly mesoporous starch with a surface area of typically 180–200 m^2 g^{-1} [80,81]. In the final stage, the mesoporous starch can be doped with a catalytic amount of an organic acid (e.g., p-toluenesulfonic acid).

The surface area of the as-prepared materials increased with increasing the carbonization temperature from 293 m^2/g at 300°C up to 600 m^2/g to 800°C. A positive aspect of this technology, similar to hydrothermal carbons, is the fact that the surface polarity and the porosity can be modulated with temperature.

Inspired from systematic studies on the starch system, the same authors investigated the use of other linear polysaccharides in the preparation of second-generation Starbon materials. It was anticipated that the utilization of differing polysaccharide structures and functionality may allow access to materials with differing textural properties and nanomorphological properties compared with the original starch-derived materials. The other two polysaccharides investigated were alginic acid and pectin.

Alginic acid is a complex, seaweed-derived acidic polysaccharide with a linear polyuronide block copolymer structure. Nonporous native alginic acid may be transformed into a highly mesoporous aerogel (S$_{BET}$ ca. 320 m^2 g^{-1}; V$_{meso}$ ca. 2.50 cm^3 g^{-1}; pore size ca. 25 nm), presenting an acidic accessible surface using the same methodology employed for the preparation of porous starches [82]. N$_2$ sorption analysis of alginic acid–derived Starbons demonstrated the highly mesoporous nature of these materials particularly at low carbonization temperatures. Isotherms presented type IV reversible hysteresis, whereas mesoporous volumes contracted with increasing carbonization temperatures up to 500°C, where porous properties were stabilized and maintained to 1000°C (Figure 2.10e). TEM images (Figure 2.10a–d) demonstrate the organization of a rod-like morphology into mesoscale-sized domains, generating the large mesopore volumes observed from N$_2$ sorption studies. Materials could also be prepared up to 1000°C with no decrease in the quality of the textural properties or alteration in the structural morphology. This approach to the generation of second-generation Starbon materials used no additive catalyst and simply relied on the decomposition of the acidic polysaccharide itself to initiate the carbonization process.

Pectin, an inexpensive, readily available, and multifunctional polyuronide occurs as a major cell wall component in land plants. Common commercial sources include fruit skins, a major commercial waste product. Gelation of native citrus pectin in water and subsequent recrystallization yielded a semitransparent gel, which was converted to the porous aerogel via solvent exchange/drying outlined above [83]. Supercritical CO$_2$ extraction of ethanol was found to yield a low density ($\rho = 0.20$ g cm^{-1}), highly porous powder aerogel (Figure 2.11, PI-powder). Addition of hydrochloric acid yielded a viscous solution, which could be poured into any desired shaped vessel and cured at room temperature to yield a very dimensionally strong gel, which upon water removal (via solvent exchange/supercritical CO$_2$ drying) yielded extremely low-density materials ($\rho = 0.07$ g cm^{-3}, Figure 2.11 PMI-monolith).

Direct heating of the pectin aerogels under an inert atmosphere allowed direct access to the carbonaceous materials. Promisingly, the resulting Starbons-type materials prepared from the two different pectin aerogel precursors, presented

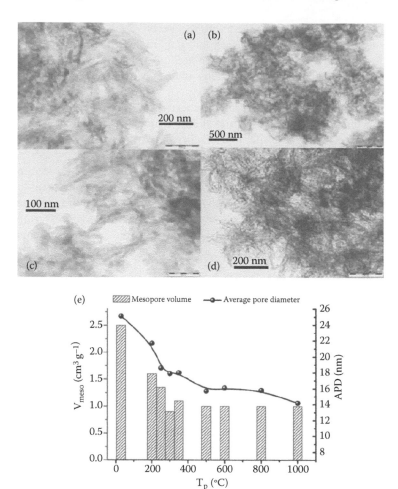

FIGURE 2.10 TEM images of alginic acid (AS1)–derived Starbon materials at T_p = (a) 300°C, (b and c) 500°C, and (d) 1000°C. (e) Impact of increasing carbonization temperature on the mesoporous properties of alginic acid–derived Starbon materials. (R. J. White, C. Antonio, V. L. Budarin, E. Bergstroem, J. Thomas-Oates, J. H. Clark, *Advanced Functional Materials* 2010, 20, 1834–1841. Copyright Wiley-VCH Verlag GmbH & Co. KGaA. Reproduced with permission.)

significantly different textural and mesoscale morphologies compared with materials prepared from either alginic acid or acid-doped starch.

Pectin-derived Starbon materials demonstrate the flexibility of this material synthesis approach in terms of textural and morphological properties and also further exemplifies the impact of polysaccharide structure and the metastable gel state necessary to generate the mesoporosity in these materials. The utilization of pectin provides two extra routes (excitingly here from the same polysaccharide) for the production of second generation Starbon materials with not only differing porous properties, but with remarkably variable mesoscale morphology, accessible via a

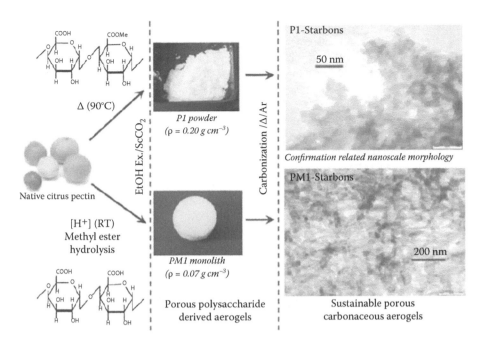

FIGURE 2.11 **(See color insert.)** Representation of routes to porous polysaccharide-derived materials from pectin, and the corresponding TEM images of pectin-derived Starbon-type materials. (R. J. White, V. L. Budarin, J. H. Clark, *Chemistry—A European Journal* 2010, 16, 1326–1335. Copyright Wiley-VCH Verlag GmbH & Co. KGaA. Reproduced with permission.)

simple change in the gelation route and the resulting chemical modification of polysaccharide structure (Figure 2.11).

Thus, using various polysaccharide precursors from plat material, Clark, White, and coworkers successfully prepared well-defined porous carbon materials with tunable porosity, morphology, and surface groups. This of course opens the doors to many different applications.

Starbon materials have been intensively applied as heterogeneous catalysts. The ability to adjust the surface properties and hydrophobicity/hydrophilicity balance of mesoporous Starbons gives the possibility to achieve highly active, selective, and reusable water-tolerant solid acids. Thus, using Starbons as solid catalysis modified with sulfonic groups, the researchers from York successfully esterified various substrates (e.g., succinic (SA), fumaric, levulinic, and itaconic acids) in aqueous ethanol, providing high conversions and selectivities to their respective esters [84]. The rates of esterification of diacids (succinic, fumaric, and itaconic) were found to be between 5 and 10 times higher for our Starbon acids compared with those of commercial solid acids (e.g., zeolites, sulfated zirconias, acidic clays and resins) or microporous commercial sulfonated carbons (DARCOs and NORITs).

Starbon acids were also found to have a temperature-dependent optimum of catalytic activity (which could be controlled by the preparation temperature of the parent Starbons and consequently by modification of its surface properties) as well as a substrate dependent catalytic activity maximum. Starbon acid activities peaked at ca. 400°C, 450°C, and 550°C for succinic, fumaric, and itaconic acids, respectively, with sharply reduced activities below or above this preparation temperature.

The same sulfonated carbons proved also to be efficient catalysts for preparation of aromatic amides via N-acylation of amines under microwave irradiation [85]. Quantitative conversions of starting material were typically achieved in 5–15 min with very high selectivities to the target product, applicable to a wide range of compounds (including aromatic and aliphatic amines), substituents, and acids. Starbon acids provided starkly improved activities compared with other acid catalysts including zeolites, Al-MCM-41, and acidic clays.

The resulting Starbon materials have been also successfully hybridized with various nanoparticles and applied in heterogeneous catalysis for various applications. A comprehensive review on this topic is found in [86].

Another very interesting and important application is the application of the alginic acid derived materials prepared at 1000°C as stationary phase in liquid chromatography for the separation of a mixture of carbohydrates [82]. The separation potential was demonstrated for the representative highly polar neutral sugars glucose (mono-), sucrose (di-), raffinose (tri-), stachyose (tetra-), and verbascose (pentasaccharide). This allowed the generation of designer PGC-type stationary phases, whereby the surface polarity could be moderated by selecting the carbonization temperature employed to control the degree or extent of the graphitic structure development.

Starbon technology provides a useful and sustainable route to highly mesoporous carbonaceous materials. Flexibility in terms of preparation temperature provides carbonaceous materials with tunable surface chemistry properties, arguably a material feature not accessible via conventional hard or soft template routes. By the selection of gelation conditions, polysaccharide type, and carbonization temperature, a wide range of carbonaceous materials may be synthesized using inexpensive and readily available renewable sugar-based precursors. Drawbacks of this technology are the lack of well-defined pore size, most of the materials exhibiting broad pore size distributions. In addition, the pore properties are unpredictable as in the case of activated carbons. This should be in principle overcome by the use of either hard or soft templates (preferably also derived from biomass), which should lead to hierarchically porous materials.

Another disadvantage of this technology is that it is limited to the use of polysaccharides in their pure form, which requires additional isolation and purification from the derived biomass parent material.

2.2.5   USE OF IONIC LIQUIDS IN THE SYNTHESIS OF CARBON MATERIALS

Recently, it has been shown that carbon materials can also be obtained by the direct carbonization of some particular ionic liquids (ILs). Thus, Paraknowitsch et al. designed a set of ILs entirely composed of C, N, and H atoms using a combination of nitrogen-containing cations (i.e., 1-ethyl-3-methylimidazolium (EMIm) or 3-methyl-1-butylpyridine (3-MPB)) and anions (i.e., dicyanamide (DCA)) [87]. Within the same context, Lee et al. designed ILs composed of different cations that contained imidazolium groups [not only EMIm but also 1-butyl-3 methylimidazolium (BMIm)

FIGURE 2.12 (See color insert.) Reaction scheme of the trimerization of a nitrile-containing anion, leading to the formation of an extended framework. (J. S. Lee, X. Wang, H. Luo, S. Dai, *Advanced Materials* 2010, 22, 1004–1007. Copyright Wiley-VCH Verlag GmbH & Co. KGaA. Reproduced with permission.)

and 1,3-bis(cyanomethyl)imidazolium (BCNIm)] and anions that contain nitrile groups (e.g., $[C(CN)_3]^-$) (Figure 2.12) [88].

The nitrogen-rich character of these ILs allowed, by direct combustion, to obtain nitrogen-doped carbons with remarkable nitrogen contents of up to 18 at% [88]. Interestingly, the authors demonstrated that the carbonization yield depended on the nitrile character of the anions so that the resulting carbon network can be cross-linked via both cations and anions.

This approach based on the use of ILs as carbonaceous precursors was also applied to the preparation of porous carbons with high surface area. Kuhn et al. first reported on an ionothermal polymerization method using a molten salt ($ZnCl_2$) and simple aromatic nitriles (e.g., 4,4-dicyanobiphenyl and 4-cyanobiphenyl), which provided carbonaceous polymer networks with well-defined bimodal micro- and (nonperiodic) mesoporosity [89]. Later on, Lee et al. used ILs composed of nitrile-functionalized imidazolium-based cations (e.g., [BCNIm]$^+$ or 1-cyanomethyl-3-methylimidazolium [MCNIm]$^+$) and non-nitrile–functionalized anions (e.g., [Tf2N]$^-$ and [beti]$^-$), with surface areas of up to 780 m^2 g^{-1} [90]. Obviously, anions release was detrimental in terms of both carbonization yields and nitrogen contents.

Textural properties without compromising nitrogen contents can be also obtained with the aid of traditional structural directing agents. The use of porous silica nanoparticles promoted more so than a 10-fold increase in the surface area of the resulting carbons as compared with the ones obtained in the absence of templates [87] (ca. 1500 versus 70 m^2 g^{-1}, respectively).

More intriguing is the use of ILs as solvents for the formation of a silica gel from either hydrolysis or condensation of regular orthosilicate precursors (e.g., TEOS) or by coagulation of LUDOX-silica nanoparticles. Upon carbonization of the ILs and subsequent silica dissolution hierarchical porous (with pores at both the nano- and microscale), carbon monoliths could be obtained (Figure 2.12) [87,91,92]. The application of this approach to ILs composed of DCA (as anion) and a long-alkyl-chain pyridinium derivative (as cation) has recently resulted in the formation of composite (silica- and nitrogen-doped carbon) microparticles with a well-defined mesoporous structure (Figure 2.13) [93].

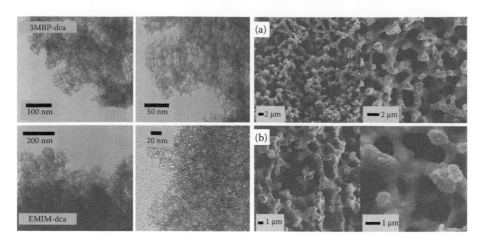

FIGURE 2.13 Left: TEM images of Ludox-templated IL-derived materials. (Top) 3MBP-DCA, (bottom) EMIM-DCA. Right: SEM images of (a) silica monolith template and (b) N-doped carbon monolith. (J. P. Paraknowitsch, A. Thomas, M. Antonietti, *Journal of Materials Chemistry* 2010, 20, 6746–6758. Reproduced by permission of The Royal Society of Chemistry.)

However, the "green" character of burning ionic liquids is questionable. Even if such materials showed good properties for various applications especially related to energy storage, the high cost of the precursor does not really pay off.

More valuable methodologies are thus in which the ion liquids can be used as recyclable solvents. The lack of vapor pressure that characterizes ILs is what provides their "green" character and makes them interesting alternatives for replacing highly volatile organic solvents in synthetic processes. Besides, their excellent solubilization properties especially for biomass-derived components should make them very suitable for the production of various materials from biomass. All these, in combination with their very high thermal stability, should make them ideal "solvents" for high-temperature carbonization reactions in ionic liquids.

Cooper et al. [94] was the first who reported a new type of solvothermal synthesis in which ILs were used as both the solvent- and the structure-directing agent in the synthesis of zeolites. This methodology has been termed ionothermal synthesis, and since then, it has become one of the most widely used synthetic strategies among the zeolite community. It was also extended to the synthesis of metal–organic frameworks (MOFs), covalent organic frameworks (COFs), polymer organic frameworks (POFs), porous silicas, nanoparticles, polymers, and others. For more details, see refs. 95–98.

In the context of this chapter, ionic liquids have been also used for "ionothermal carbonization." Titirici and Taubert reported that metal-containing ILs can simultaneously play the role of structural directing agent, catalyst for carbonization, and solvent [99]. Interestingly, the ionic liquids can be fully recovered at the end of each carbonization process without any effect on their chemical structure. A variety of carbohydrates were used as a carbon source (i.e., D-glucose, D-fructose, D-xylose, and starch), whereas 1-butyl-3-methylimidazolium tetrachloroferrate(III), [Bmim] [FeCl$_4$] as a reusable solvent, and catalyst. The carbon materials derived from these different carbohydrates were similar in terms of particle size (Figure 2.14) and chemical composition, possessing relatively high surface areas from 44 to 155 m^2 g^{-1} after ionothermal processing, which could be significantly increased to >350 m^2 g^{-1} by further thermal treatment (e.g., post-carbonization at 750°C). CO$_2$ and N$_2$ sorption analysis, combined with Hg intrusion porosimetry, revealed a promising hierarchical pore structuring to these carbon materials. The ionic liquid [Bmim][FeCl$_4$] had a triple role: it acted as both a soft template to generate the characterized pore structure, a solvent, and a catalyst resulting in enhanced ionothermal carbon yields. Importantly from a process point of view, the ionic liquid could be successfully recovered and reused.

The group of Dai et al. used a protic ionic liquid [DMFH][Tf$_2$N] (*N,N*-dimethyl-*N*-formylammonium bis(trifluoromethylsulfonyl)imide) for the synthesis of ionothermal carbons from glucose and fructose at low temperature and ambient pressure.

The observed results were similar with those of Titirici, wherein the ionic liquid induces porosity in the resulting carbons while the carbonization yield is significantly increased [100].

Although only these two publications exist currently in the literature, this methodology could represent an interesting direction for the synthesis of sustainable carbon materials. The carbonization can take place at atmospheric pressure in a standard flask while real biomass precursors can be employed. Especially cellulose is known to be solubilized by protic ionic liquids. In addition, as it was already reported that

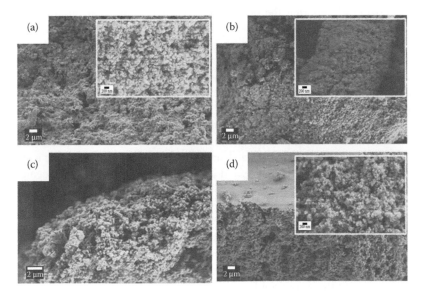

FIGURE 2.14 SEM micrographs of as-synthesized ionothermal carbons: (a) C-glucose, (b) C-fructose, (c) C-xylose, and (d) C-starch (the scale bar in the inset is 200 nm). (Taken with permission from Z.-L. Xie, R. J. White, J. Weber, A. Taubert, M. M. Titirici, *Journal of Materials Chemistry* 2011, *21*, 7434–7442.)

the ionic liquids can efficiently catalyze the production of hydroxymethylfurfural from biomass [101], which can then significantly increase the HTC yield. By designing tailor-made ionic liquids, it should be possible to control the porosity as well as introduce various functions into the resulting iono-carbons. Metal-containing ionic liquids should offer the possibility to produce various interesting nanocomposites under the appropriate synthetic conditions. Furthermore, the use of ionic liquids will be far more justified in such a procedure compared with their irreversible conversion in various materials (i.e., carbons, polymers).

A related class of ionic liquids named deep-eutectic solvents (DESs) are obtained by complexion of quaternary ammonium salts with hydrogen bond donors [102]. DESs share many characteristics of conventional ILs (e.g., nonreactive with water, nonvolatile, biodegradable) while offering certain advantages. For instance, the preparation of eutectic mixtures in a pure state is accomplished more easily than for ILs. There is no need for post-synthesis purification as the purity of the resulting DES will only depend on the purity of its individual components. The low cost of those eutectic mixtures based on readily available components [103,104] makes them particularly desirable (more so than conventional ILs) for large-scale synthetic applications. However, a close inspection of recent literature revealed that their use in materials synthesis is sporadic compared with ILs. This situation is currently slightly changing and different authors consider DESs as the next generation of ILs. They can act as true solvent-template-reactant systems, where the DES is at the same time the precursor, the template, and the reactant medium for the fabrication of a desired material with a defined morphology or chemical composition. A very comprehensive review on the topic was recently published by the group of Francisco del Monte [105].

Carriazo et al. reported the preparation of a DES based on the mixture of resorcinol and choline chloride. Polycondensation with formaldehyde resulted in the formation of monolithic carbons with a bimodal porosity comprised of both micropores and large mesopores of ca. 10 nm (Figure 2.15) [106]. The morphology of the resulting carbons consisted of a bicontinuous porous network built from highly cross-linked clusters that aggregated and assembled into a stiff, interconnected structure. This morphology is typical for carbons obtained via spinodal decomposition [107]. Carriazo et al. hypothesized that one of the components forming the DES (e.g., resorcinol) is acting as a precursor of the polymer phase whereas the second one (e.g., choline chloride) acts as a structural directing agent following a synthetic mechanism based on DES rupture and controlled delivery of the segregated SDA into the reaction mixture.

The wide range of DESs that can be prepared provides a remarkable versatility to this synthetic approach. For instance, Carriazo et al. also reported that the use of ternary DESs composed of resorcinol, urea, and choline chloride results in carbons with surface areas of nearly 100 m^2 g^{-1} higher than those of carbons obtained from binary DESs composed of resorcinol and choline chloride. Urea was here partially incorporated into the resorcinol-formaldehyde (RF) network (upon its participation in polycondensation reactions), the release of which (during carbonization) resulted in the above-mentioned increase in the surface area.

The same group used DESs composed of resorcinol, 3-hydroxypyridine, and choline chloride [108]. In this case, DESs played multiple roles in the synthetic process:

FIGURE 2.15 Top panel: SEM micrographs of RFRC-DES (left, bar is 5 mm) and RFRUC-DES (right, bar is 2 mm) gels. Insets show a picture of the monolithic RF gels. Bottom panel: SEM micrographs of CRC-DES (left, bar is 1 mm) and CRUC-DES (right, bar is 1 mm) monoliths. Insets show a picture of the respective monolithic carbons and TEM micrographs of CRC-DES (left, bar is 50 nm) and CRUC-DES (right, bar is 150 nm). R = resorcinol; C = chlorine chloride; U = urea; F = formaldehyde. (Reprinted with permission from D. Carriazo, M. A. C. Gutiérrez, M. L. Ferrer, F. del Monte, *Chemistry of Materials* 2010, *22*, 6146–6152.)

liquid medium ensuring reagents homogenization; structural-directing agent responsible for the achievement of the hierarchically porous structure; source of carbon and nitrogen (Figure 2.16). The formation of a polymer-rich phase upon resorcinol and 3-hydroxypyridine polycondensation promotes DES rupture and choline chloride segregation into a spinodal-like decomposition process. Interestingly, the resulting carbons exhibited a combination of surface areas and nitrogen contents (from ca. 550 to 650 m^2 g^{-1} and from ca. 13 to 5 at% for carbonization temperatures ranging from 600°C to 800°C, respectively) that, unless traditional SDAs are also used, had never been attained by synthetic processes carried out in ILs.

It is finally worth noting the "green" character of this process because of the absence of residues and/or byproducts eventually released after the synthetic process; i.e., one of the components forming the DES (e.g., resorcinol, mixtures of resorcinol and hydroxypyridine or urea, and furfuryl alcohol) becomes the material itself

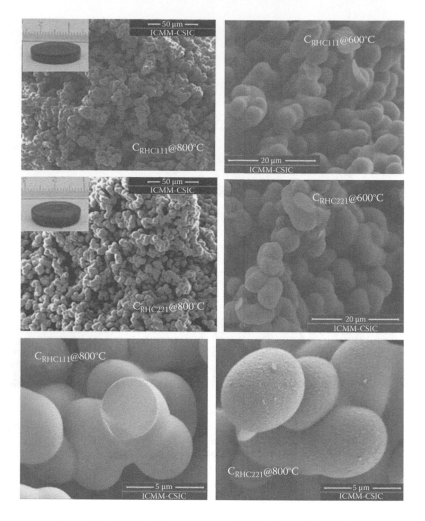

FIGURE 2.16   SEM micrographs of CRHC-DES and CRHC-DES obtained after thermal treatments at 600°C and 800°C. Insets show pictures of the CRHC-DES and CRHC-DES monoliths obtained after thermal treatment at 800°C. C, chlorine chloride; R, resorcinol; H,3-hydroxypyridine. (K. Katuri, M. L. Ferrer, M. C. Gutierrez, R. Jimenez, F. del Monte, D. Leech, *Energy & Environmental Science* 2011, 4, 4201–4210. Reproduced by permission of The Royal Society of Chemistry.)

with high yields of conversion (within the 60%–80% range), whereas the second one (e.g., single choline chloride in resorcinol-based synthesis) is fully recovered and can be reused in subsequent reactions. However, using DES based on carbohydrates (see ref. [109]) should improve even further the green character of this methodology for the future production of carbon materials. Such DESs were found to serve some basic function in living cells and organisms. They include sugars, some amino acids, choline, and some organic acids such as malic acid, citric acid, lactic acid, and succinic acid. Taking the plant metabolomics data Verpoorte and coworkers have

TABLE 2.1
List of Natural ILs and DES

Combination	Molar Ratio
Citric acid/choline chloride	1:2, 1:3
Malic acid/choline chloride	1:1, 1:2, 1:3
Maleic acid/choline chloride	1:1, 1:2, 1:3
Aconitic acid/choline chloride	1:1
Glc/chlorine chloride/water	1:1:1
Fru/choline chloride/water	1:1:1
Suc/choline chloride/water	1:1:1
Citric acid/Pro	1:1, 1:2, 1:3
Malic acid/Glc	1:1
Malic acid/Fru	1:1
Malic acid/Suc	1:1
Citric acid/Glc	2:1
Citric acid/trehalose	2:1
Citric acid/Suc	1:1
Maleic acid/Glc	4:1
Maleic acid/Suc	1:1
Glc/Fru	1:1
Fru/Suc	1:1
Glc/Suc	1:1
Suc/Glc/Fru	1:1:1

collected over recent years, they could clearly see similarities with the synthetic ILs. The above-mentioned major cellular constituents are perfect candidates for making ILs and DES. The authors made various combinations of these candidates (Table 2.1 [109]). Such "natural deep eutectic solvents" (NADES) could be a potential interesting source for new and existing sustainable carbon materials, which will be surely exploited in the near future.

2.2.6 HYDROTHERMAL CARBONIZATION (HTC)

The last green carbonization technique I would like to present here is a technology that converts biomass or biomass derivatives into carbonaceous materials in water at mild temperatures (130°C–250°C) under autogenic pressures and subcritical conditions. Although this methodology was developed almost 100 years ago [110], its full potential as a synthetic route for carbon materials having important applications in several crucial fields of the 21st century, such as catalysis, energy storage, CO_2 sequestration, water purification, agriculture, has been revealed only in the past few years with main contributions from the author of this current book chapter [111,112]. This section will offer an overview of the recently developed HTC technology including formation mechanism and final chemical structure, porous materials, heteroatom doping, hybrid carbon–inorganic materials, and various applications, among which is the utilization of HTC materials as solid catalysts for biofuels production.

2.2.6.1 Formation Mechanism

The formation of HTC-based materials is a very complex process due to the multitude of simultaneous reactions occurring in the autoclave during the hydrothermal treatment. Glucose was selected as a good model carbon source and ^{13}C-enriched samples were prepared [113]. ^{13}C NMR was used to investigate the chemical composition and local structure of the hydrothermally treated materials. Several standard MAS-NMR techniques (singe pulse (SP), cross-polarization (CP), and inversion recovery cross-polarization (IRCP)), which are mainly employed in the study of carbon structures, have been used to identify the amount and type of sp² and sp³ carbon sites. In addition, bond filtering techniques based on J-coupling (insensitive nuclei enhanced by polarization transfer (INEPT)) provided a clear-cut distinction between protonated and quaternary carbons. Finally, the connectivity between carbon species was identified using a 2D [13]C–[13]C correlation experiment based on double-quantum excitation of through-space dipolar-coupled nuclei. It was found that about 60% of the carbon atoms belong to a cross-linked furan-based structure. Furan moieties are directly linked either via the R-carbon or via sp²- or sp³-type carbon groups, where cross-linking can occur. Additional cross-linking sites are located at the -carbons of the furan ring (Figure 2.17).

HTC was proposed to form upon the dehydration of hexoses (glucose and fructose) to HMF and of pentoses (xylose) to furfural [114]. Once these compounds are formed, they then undergo a very complex chemical cascade involving a simultaneous combination of ring-opening reactions to create diketones that can further undergo aldol-type condensations with the furan ring, while Diels–Alder reactions may lead to more aromatic features, with concurrent polycondensation reactions also occurring. In summary, a set of dehydrations, polymerization (condensation and addition), and aromatization reactions will lead to the final chemical structure of

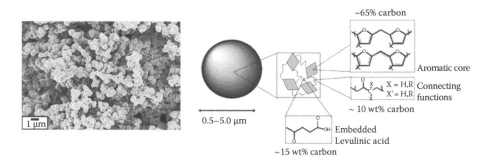

FIGURE 2.17   SEM micrograph of HTC obtained from glucose at 180°C after 12 h and a schematic representation of its chemical structures base on an NMR study. (Reprinted with permission from N. Baccile, G. Laurent, C. Coelho, F. Babonneau, L. Zhao, M.-M. Titirici, *Journal of Physical Chemistry C* 2011, 115, 8976–8982. Copyright 2011 American Chemical Society.)

HTC, which is composed of condensed furan rings bridged by aliphatic regions with terminal hydroxyl and carbonyl functional groups (Figure 2.17). Upon the "polymerization" of HMF or furfural, nucleation takes place followed by growth of the particles upon further incorporation of HMF-derived monomers, leading to spherically shaped particles.

Recently, Titirici, Falco, and Baccile embarked on a project to examine the HTC of microcrystalline cellulose as well as of real lignocellulosic biomass (i.e., rye straw) and compare the chemical structure of cellulose and rye straw HTC materials to the ones derived from glucose [115]. ^{13}C solid-state NMR was used once more for the characterization and comparison of HTC materials derived from glucose, cellulose, and rye straw. In the case of glucose, changing the processing temperature allows control of both the particle diameter and particle-size distribution. The final carbon chemical structure can be switched from a carbonaceous polyfuran rich in oxygen-containing functional groups to a carbon network of extensive aromatic domains. On the other hand, in the case of cellulose and generally of lignocellulosic biomass, a fundamental difference in the HTC mechanism is observed. The polyfuranic intermediate, which is characteristic for glucose-derived HTC materials at either low processing temperatures or short reaction times, cannot be observed in cellulose derived HTC carbons. In contrast, the biomass-derived HTC materials show a well-developed aromatic nature, formed already during the early stages of HTC reaction. Thus, under mild hydrothermal operating conditions (180°C–250°C) cellulose reacts according to a reaction scheme that can be more associated with the classical pyrolysis process, even if the exact chemical paths are not clear yet. This study also showed how lignin is only mildly affected by HTC process and how its presence influences the final material structuring. Therefore, converting cellulose and lignocellulosic biomasses into HTC with the same chemical structure and morphology such as glucose remains a very challenging task.

2.2.6.2 Porous HTC Materials

One of the main disadvantages of the HTC materials is that they present limited porosity and surface area. For enhanced performance, such textural nanoscale porosity is highly desirable. There are many well established technologies to produce porous carbons among which the most common ones are chemical activation [77d] and templating methodologies [116].

2.2.6.2.1 Chemical Activation

The first effect of chemical activation on HTC carbons is a complete loss of initial spherical morphology and the formation of bulk-like carbon structures with rough surfaces due to the presence of micropores. Falco and Lozano-Castello have investigated the development of porosity in HTC materials derived from glucose, cellulose, and rye straw using KOH as an activating agent at 750°C and a KOH ratio of 3 [117]. Overall, this analysis highlights that the HTC temperature extensively affects the porosity of the derived ACs. HTC carbons synthesized at higher temperatures (e.g., 280°C) generate ACs with a lower porosity development and narrower pore size distributions (PSDs) while the ones produced at 180°C–240°C, upon KOH activation, develop a much larger porosity characterized by a greater mesopores fraction.

These trends can be explained by taking into account the dependence of the chemical nature of HTC carbons upon the synthesis temperature: higher temperatures generate HTC carbons with a higher degree of aromatization resulting in enhanced chemical stability and structural order. As observed for the hydroxide activation of several coals, both features are detrimental to the reactivity of the carbon substrate leading to a reduced porosity development.

Additionally, the KOH ratio affects the porous properties of HTCs. Low KOH ratio values (2:1) produce a lower extent of activation and favor the formation of microporous carbons with a narrow PSD. Using higher amounts of KOH leads to higher porosity development, as indicated by larger surface area and total pore volume values. However, their increase is partially the result of pore widening, since the pore-volume distribution is clearly shifted more toward the supermicropore and mesopore ranges. These observations are in agreement with the trends generally observed during KOH activation of petroleum-derived carbonaceous substrates [118], confirming the similar reactivity of HTC carbons during chemical activation and their resultant high suitability as AC precursors.

2.2.6.2.2 Nanocasting, Natural Templates

In the literature, there are a multitude of techniques for introducing porosity into carbon materials [119]. In hard templating, a preformed porous solid is infiltrated with the carbon precursor and the carbonization occurs inside the pores, followed by original scaffold removal, yielding a carbon porous replica. This technique was intensively used for the replication of ordered silica materials (e.g., SBA-15 [120]) into the corresponding ordered porous materials CMK-3 and CMK-5 [121]. More recently, a soft-templating approach in carbon was developed by Zhao et al. [122] and Dai et al. [119], who demonstrated a direct route to ordered carbons via organic–organic block copolymer self-assembly (e.g., Pluronic: F127) and suitable aromatic carbon precursor (e.g., phloroglucinol, resorcinol).

One of the main limitations regarding the chemistry of these reported carbons is the relatively chemically condensed pore walls/surfaces, which inhibits facile post-chemical modification. Therefore, it would be advantageous from a sustainable and economic footing if renewable precursors could be employed in a simple, scalable (potentially carbon neutral) process (i.e., HTC) in the direct synthesis of such ordered porous carbonaceous materials. This has been demonstrated by combining the HTC of glucose with hard templates for the production of mesoporous spheres (Figure 2.18a) [123], ordered carbon materials (HTC-CMK-3, -5) (Figure 2.18b) [124], tubular carbon (Figure 2.18c) [125], and hollow carbon spheres [126] (Figure 2.18d).

However, the hard templating using silica is tedious because of problematic infiltration procedures as well as template removal, which normally involve harsh and non-environmentally friendly conditions. Soft templating was also successfully implemented in HTC. The use of Pluronic amphiphilic block copolymers as structure-directing agents and fructose as a carbon source enabled the production of HTC materials with well-defined ordered micro- or mesopores (Figure 2.18e, f) [127].

For example, a porous nitrogen-doped carbon was originally prepared using the crustacean shells of shrimps and lobsters as natural templates Here, the $CaCO_3$ nanoparticles served as removable hard templates, whereas chitin as a

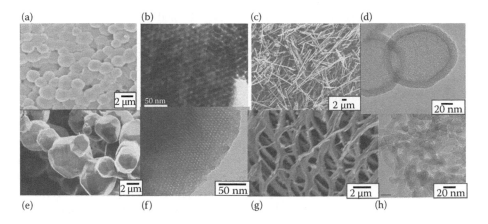

FIGURE 2.18 Porous HTCs: (a) Mesoporous HTC microspheres obtained by replication of silica using glucose. (M.-M. Titirici, A. Thomas, M. Antonietti, *Advanced Functional Materials* 2007, 17, 1010–1018. Copyright Wiley-VCH Verlag GmbH & Co. KGaA. Reproduced with permission.) (b) Ordered mesoporous HTC obtained by replication of SBA–15 using furfural. (M.-M. Titirici, A. Thomas, M. Antonietti, *Journal of Materials Chemistry* 2007, 17, 3412–3418. Reproduced by permission of The Royal Society of Chemistry.) (c) HTC tubular carbons obtained using a macroporous alumina membrane as template and furfural as a carbon source. (Reprinted with permission from S. Kubo, I. Tan, R. J. White, M. Antonietti, M.-M. Titirici, *Chemistry of Materials* 2010, 22, 6590. Copyright 2010 American Chemical Society.) (d) HTC hollow spheres obtained using latex nanoparticles as sacrificial templates and glucose as a carbon source. (Reprinted with permission from R. J. White, K. Tauer, M. Antonietti, M.-M. Titirici, *Journal of the American Chemical Society* 2011, 132, 17360–17363. Copyright 2011 American Chemical Society.) (e, f) Micromesoporous ordered HTC particles obtained in the presence of amphiphilic block copolymers as soft templates and fructose as a carbon source. (Reprinted with permission from S. Kubo, R. J. White, N. Yoshizawa, M. Antonietti, M.-M. Titirici, *Chemistry of Materials* 23, 2011, 4882–4885. Copyright 2011 American Chemical Society.) (g) Nitrogen-doped porous carbon obtained using shrimp/lobster food wastes after hydrothermal carbonization and removal of $CaCO_3$; (h) porous nitrogen-doped carbon aerogel obtained using the natural interactions between carbohydrates and gelating proteins (i.e., ovalbumin). (R. J. White, N. Yoshizawa, M. Antonietti, M.-M. Titirici, *Green Chemistry* 2011, 13, 2428–2434. Reproduced by permission of The Royal Society of Chemistry.)

(nitrogen-doped) carbon source (Figure 2.18g) [128]. Another unique procedure to produce porous nitrogen-doped carbons was based on the natural interactions between carbohydrates (glucose) and gelating proteins. Depending on the applied conditions, either nitrogen-doped carbon nanoparticles [129] or monolithic hierarchically porous nitrogen-doped carbons were obtained in the absence of sacrificial templates (Figure 2.18h) [130].

2.2.6.3 Heteroatom-Doped HTCs

The introduction of dopants into carbon materials is a useful method of altering their physical and chemical properties. Successful work thus far includes the superior performance of doped carbons in applications such as electrode materials for catalysis

[131] or energy storage [132], stationary phases [133], and chemoselective adsorption [134].

The majority of research in this field has focused on nitrogen doping, which is known to induce favorable changes in the carbon material, e.g., increased conductivity [135] and high activity in the oxygen reduction reaction (ORR) in fuel cells [136]. The synthesis of nitrogen-doped carbons has been achieved via a variety of pathways, such as posttreatment of carbon with ammonia [137], amines or urea [138], and also more direct approaches using acetonitrile [139], pyrrole [140], polyacetonitrile [132], or polyaniline [141] as starting products.

Concerning sustainability, most of the aforementioned synthesis methods for heteroatom-doped carbon materials show drawbacks in the, often harsh, reaction conditions used. To avoid these aspects, a hydrothermal carbonization route was chosen to produce materials with similar properties in a sustainable fashion.

Thus, nitrogen-doped carbon microspheres were obtained by hydrothermally treating naturally occurring and nitrogen containing compounds, such as glycine [113], N-acetyl glucosamine [142], or chitosan [143]. Albumin, a glycoprotein, was also used as a structure-directing additive in the HTC of glucose to produce nitrogen-doped carbon aerogels [130].

A very important feature of using nitrogen precursors or additives in HTC is the possibility of covalently incorporating the nitrogen via Maillard chemistry. The Maillard reaction is not one specific reaction but rather refers to a group of reactions occurring between reducing sugars and amino containing molecules. In the process, hundreds of different (flavor) compounds can be created, which can then break down to form even more compounds, etc. It is therefore impossible to provide a definite mechanistic pathway [144]. As previously stated, HTC is formed via HMF. HMF can interact with amino acids/proteins and with Maillard reaction intermediates in countless possible combinations of reactants and reaction cascades, which of course increase with higher reactivity of the additive (i.e., in how many different ways it can react). We will not describe here in detail such possibilities but only say that either aliphatic nitrogen (i.e., amines) or nitrogen already incorporated into a heterocycle (i.e., pyrrole) can be formed under the employed hydrothermal conditions. Because of the low HTC temperatures employed (130°C–280°C), Maillard reactions can take place, and thus the nitrogen heteroatom is covalently bonded to the carbon network. Therefore, upon further calcination under an inert atmosphere (to increase the conductivity, i.e., for electrochemical applications) the same amount of dopant will be maintained in the resulting carbons. This is not the case for the direct pyrolysis of N-containing precursors or reagents where most of the nitrogen is lost as volatile gas. Therefore, HTC represents a powerful modality to introduce large amounts of nitrogen as well as to mediate the type of nitrogen at low temperatures. We also need to mention that upon further pyrolysis all the nitrogen groups are in an aromatic environment, as pyridinic, pyridonic, or quaternary nitrogen groups. Such nitrogen-doped HTCs already showed improved electric and thermal conductivities as compared with the nondoped materials [145]. They have also been successfully used as selective adsorbents for CO_2 [146], electrodes for supercapcitors [142], or electrocatalysts for the oxygen reduction reaction (ORR) in fuel cells [147].

Complementing nitrogen as a dopant, sulfur is receiving increasing attention in current carbon materials research. Overall, literature reports suggest that nitrogen is the dopant of preference concerning the tuneability of electronic properties of the carbon material, whereas sulfur, due to its large size, has been used more for applications where its easily polarizable lone pairs (and thus chemical reactivity) are of importance.

Sulfur-doped carbon materials have, for example, shown beneficial effects on the selective adsorption of waste metals [134] and the desulfurylation of crude oil [148]. The synthesis of these sulfur-doped materials generally involves the pyrolysis of sulfur-containing polymer-based carbons [134,148] and also arc vaporization in the presence of sulfur containing compounds such as thiophenes [136].

Concerning the combined incorporation of sulfur and nitrogen within the same material, only a few reports exist in the literature. Sulfur-assisted growth of carbon nanotubes by chemical vapor deposition of acetonitrile was shown to increase the nitrogen doping levels as well as the magnetic properties of the nanotubes [149]. Choi et al. synthesized heteroatom doped carbon materials by the pyrolysis of amino acid/metal chloride composites. They obtained sulfur doping levels of 2.74 wt% using cysteine and were able to show that materials containing both nitrogen and sulfur increased the material's ORR activity in acidic media, relative to undoped or purely nitrogen-doped carbons [150]. These results lead us to the conclusion that dual heteroatom-doped carbon materials may be a highly desirable class of materials.

Therefore, solely sulfur as well as both sulfur and nitrogen have been introduced into HTC materials. For the sulfur-doped materials, 2-thiophene carboxaldehyde (TCA) has been used whereas the dual sulfur and nitrogen-doped materials have been using cysteine (C) or S-(2-thienyl)-L-cysteine (TC) as additive during the HTC of glucose.

Contrary to N, S cannot undergo Maillard during HTC. However, also in this case HTC plays the role of the mediator in the future heteroatom chemical environment. Sulfur can be incorporated either via nucleophilic or electrophilic substitutions into the final HTC structure. Addition of cysteine gave rise to a material with more pending functional groups such as thiols, amides, and sulfonates. Addition of thienyl-cysteine, on the other hand, resulted in more structurally bound sulfur, such as in thiophenes. After pyrolysis at 900°C, the resulting materials showed almost three times higher specific conductivity than a corresponding undoped sample made from pure glucose, as well as a highly increased interlayer distance of, presumably, buckled carbon sheets [151]. The sulfur binding motif is relevant with respect to chemical, mechanical, and electrical properties arising from different heteroatom binding states. Pending surface thiol groups could, for example, be useful for the adsorption of metal nanoparticles in a catalyst support material, but are expected to be dominant for the low temperature cysteine series, whereas structurally bound sulfur and nitrogen are known to alter electronic properties such as conductivity.

When porosity was introduced into these materials, they exhibited impressive electrocatalytic activity for the ORR reaction in fuel cells [147]. In acidic conditions all doped aerogels showed very good stability compared with a platinum-based catalyst, as well as an activity that is still much better than ordinary carbon supports, but not competitive to the noble metal systems.

2.2.6.4 HTC–Inorganic Hybrids

Controlled synthesis of carbonaceous nanocomposites has become a hot research area and achieved many important results because of their combined and improved properties with highly potential values in many fields. Two main methodologies can be distinguished for the production of hydrothermal carbon nanocomposites: post-modification and in situ synthesis.

Many carbon–inorganic materials have been prepared using the hydrothermal technology, and due to space constraints, I cannot mention all these examples. I will limit instead to a few examples. Regarding coating of preformed inorganic nanoparticles, some silicon/carbon (Figure 2.19a) as well as LiFePO4/carbon (Figure 2.19b) materials have been reported. Both of these composites were successfully applied as anode [152], respectively, cathode [153] in Li-ion batteries.

FIGURE 2.19   (a) Si nanoparticle coated hydrothermally with a thin layer of carbon from glucose. (Y.-S. Hu, R. Demir-Cakan, M. M. Titirici, J. O. Muller, R. Schlogl, M. Antonietti, J. Maier, *Angewandte Chemie International Edition* 2008, 47, 1645–1649. Copyright Wiley-VCH Verlag GmbH & Co. KGaA. Reproduced with permission.) (b) LiFePO4 coated with carbon. (R. D.-C. Jelena Popovic, J. Tornow, M. Morcrette, D. S. Su, M. A. Robert Schlögl, M.-M. Titirici, *Small* 2011, 8, 1127–1135. Copyright Wiley-VCH Verlag GmbH & Co. KGaA. Reproduced with permission.) (c) Nitrogen-doped carbon obtained from lobster waste and loaded with Pt nanoparticles; (d) TEM micrograph of Ag/HTC nanocables obtained from starch and AgNO_3 at 160°C for 12 h. (X. J. C. S. H. Yu, L. L. Li, K. Li, B. Yu, M. Antonietti, H. Coelfen, *Advance Materials* 2004, 16, 1636. Copyright Wiley-VCH Verlag GmbH & Co. KGaA. Reproduced with permission.) (e) Pd/HTC core/shell particles. (P. Makowski, R. D. Cakan, M. Antonietti, F. Goettmann, M.-M. Titirici, *Chemical Communications* 2008, 999–1001. Reproduced by permission of The Royal Society of Chemistry.) (f) TiO_2/HTC composite. (L. Zhao, X. Chen, X. Wang, Y. Zhang, W. Wei, Y. Sun, M. Antonietti, M.-M. Titirici, *Advance Materials* 2010, 22, 3317–3321. Copyright Wiley-VCH Verlag GmbH & Co. KGaA. Reproduced with permission.) (g) Low-magnification SEM picture of the iron oxide hollow spheres obtained after calcinations, indicating their homogeneity. (M.-M. Titirici, M. Antonietti, A. Thomas, *Chemistry of Materials* 2006, 18, 3808–3812. Reproduced by permission of The Royal Society of Chemistry.) (h) SnO_2 mesoporous spheres obtained after removal of HTC from the resulting composite. (R. Demir-Cakan, Y. S. Hu, M. Antonietti, J. Maier, M.-M. Titirici, *Chemistry of Materials* 2008, 20, 1227–1229. Reproduced by permission of The Royal Society of Chemistry.)

The important advantage of HTC materials is, as previously mentioned, their rich surface functional groups, such as hydroxyl, aldehyde, and carboxyl groups. These surface functional groups have shown remarkable reactivity toward various inorganic precursors. For example, based on the redox reaction, they could in situ reduce and stabilize noble-metal ions forming very fine noble-metal nanostructures, such as Ag, Au, Pt, and Pd [154–156]. Titirici, White, and Palkovitz reported the loading of Pt nanoparticles onto nitrogen-doped carbon materials (Figure 2.19c) [157]. The resulting composites were used as efficient catalysts for the oxidation of methane to methanol with superior performance to the standard dichlorobipyrimidyl platinum (II) complex reported by Periana and coworkers [158].

When noble–metal ions are first added into the HTC process, novel carbonaceous-encapsulated core–shell nanocomposites could be successfully produced in a one-step HTC process. The size of the resulting composites could be carefully controlled by changing the reagent ratios and reaction conditions. The advantages of this one-step HTC process include the following:

(i) The carbonization process and the reduction and growth of noble metal ions proceed simultaneously
(ii) The noble–metal ions can catalyze the carbonization process, and
(iii) It is the one-step process that can potentially be upscaled.

One example is the synthesis of silver-carbonaceous nanocables (Figure 2.19d) [159]. These nanocables, with lengths up to 10 mm and diameters of ~200–500 nm, contain pentagonal-shaped silver nanowire cores and tend to branch or fuse with each other.

Pd@C core–shell nanoparticles have also been successfully synthesized by the one-step HTC process [160] and have shown selective catalytic capability for the batch partial hydrogenation of hydroxyl aromatic derivatives (Figure 2.19e). In a similar fashion, using titanium isopropoxide as an inorganic precursor and glucose as an HTC precursor, mesoporous TiO_2/HTC composites were successfully synthesized (Figure 2.19f). Such composites exhibited strong photocatalytic activity under visible light, which was not observed for the pure TiO_2 materials [161].

Titirici et al. has also reported a generalized method to produce metal oxide hollow spheres using various inorganic salts during the hydrothermal carbonization of glucose [162]. After the synthesis, the carbon is burned off while the metal ions agglomerate around into hollow carbon nanoparticles (Figure 2.19g, i.e., Fe_2O_3 hollow spheres). In a similar fashion, HTC can also be used to incorporate preformed SnO_2 nanoparticles onto HTC spheres. Similarly, after the removal of carbon, mesoporous SnO_2 spheres can be obtained (Figure 2.19h) [163]. Such hollow and porous metal oxides can have a wide range of applications from sensors to electrodes in batteries and others.

2.2.6.5 Applications of HTC Materials

The most appealing feature of HTC is the fact that it represents an easy, green, and kilograms-scalable process allowing the production of various carbon and hybrid nanostructures with practical applications on a price base that is comparatively lower than corresponding petrochemical processes. Even though relatively in their infancy, HTC materials have already found numerous applications including soil enrichment, catalysis, water purification, energy storage, and CO_2 sequestration. In the following paragraphs, examples where HTC-based materials have proved to be not only sustainable, but to possess extraordinary properties that in some cases surpass those of current "golden standards" will be summarized.

2.2.6.5.1   Agricultural Applications

Much attention is currently focused on obtaining charcoal from slow pyrolysis processes with the final aim of adding it to soils as a carbon sink and as a means of improving soil productivity or filtration of percolating soil water. In this context, the charred material is denoted "biochar" [164]. By analogy, the charred product obtained via the HTC process can be called "hydrochar" [165]. The advantage of converting biomass into hydrochar via the HTC process is that it can transform wet input material into carbonaceous solids at relatively high yields without the need for an energy-intensive drying step before or during the process. This opens up the feedstock choice to a variety of nontraditional sources: wet animal manures, human waste, sewage sludge, municipal solid waste, as well as aquaculture, and algal residues. These feedstock represent large, continuously generated, renewable streams that require some degree of management, treatment, and/or processing to ensure protection to the environment.

To predict the behavior of different hydrochars in soil, it is very important to understand the final chemical structure of the employed biomass-derived carbon. In this regard, several groups have compared the chemical and physical properties of the materials obtained using the same precursor by pyrolysis or HTC: swine manure [166], corn stover [167], or carbohydrates and cellulose in general [168]. Those works show that pyrolysis provides arene-rich chars, whereas HTC gives rise to furan-rich chars, which contain abundant oxygen groups as well. Although there are many reports on the effect of the addition of biochar to soils, only Steinbeiss et al. [169] and Rillig et al. [170] have investigated the impact of hydrochar. Their results evidence that hydrochar should be carefully tested and optimized before applications in the field are undertaken.

2.2.6.5.2   Adsorption

Liquid phase. Adsorption is by far the most frequent application of activated carbons. Contrary to activated carbons, hydrothermal carbons produced at T = 180°C do not have any microporosity unless further thermally treated or chemically activated. This is, however, counterbalanced by the high number of oxygenated groups located at their surface, which can promote adsorption. Furthermore, these functionalities can be easily tuned by further surface modifications. Thus, as shown by Demir-Cakan et al., the addition of small amounts of vinyl organic monomers into the HTC of D-glucose leads to the production of carbon materials rich in carboxyl functionalities [171]. The synthesized materials were investigated in adsorption experiments for the removal of heavy metals from aqueous solutions. The adsorption capacity was as high as 351.4 mg/g for Pb (II) and 88.8 mg/g for Cd (II), which is well beyond ordinary sorption capacities, proving the efficiency of the materials to bind and buffer ions, or more specifically to remove heavy metal pollutants. Afterward, Chen et al. followed a different approach for the introduction of carboxylic groups into the HTC products, consisting of their posttreatment in air [172]. In this way, the adsorption capacity of the sample treated in air at 300°C was 326.1 ± 3.0 mg/g for Pb^{2+} and 150.7 ± 2.7 mg/g for Cd^{2+}, values 3 and 30 times higher than that of HTC material.

Liu et al. compared the performance of a hydrothermal carbon (or hydrochar) and a biochar obtained from pinewood in the removal of Cu^{2+} [173]. The main difference between both materials was the oxygen content (34.8% for the hydrochar and 3.8% for the biochar), the porosity development being low and similar in both cases. When tested in the adsorption of Cu^{2+}, the hydrochar had an estimated maximum adsorption capacity of 4.46 mg/g vs. 2.75 mg/g for the biochar. The adsorption of Cu^{2+} took place through an ion-exchange mechanism with the H^+ of the oxygen-containing groups, which was manifested in a decrease of the pH of the solution.

Gas adsorption. CO_2 capture and H_2 storage are two of the current hot topics in energy and environmental science. Their physisorption in porous solids have attracted keen interest and is one of the most promising alternatives. As previously mentioned, HTC materials lack porosity development, which hinders their application in such porosity/surface area-sensitive applications. Sevilla et al. have circumvented that inconvenience by applying a chemical activation process with KOH (Figure 2.1), proving that such a process is a powerful tool for the creation of porosity in HTC materials [174,175]. They applied the procedure to HTC materials derived from glucose, starch, furfural, cellulose, and eucalyptus sawdust, achieving large apparent surface areas, up to ~3000 m^2/g, and pore volumes in the 0.6- to 1.4-cm^3/g range. Those materials were further characterized by narrow micropore size distributions in the supermicropore range (0.7–2 nm). Tuning of the PSD was achieved through the modification of the activation temperature (600°C–850°C) and the amount of KOH used (KOH/HTC weight ratio = 2 or 4) (Figure 2.20). When

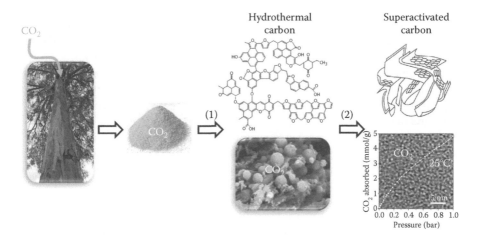

FIGURE 2.20 **(See color insert.)** Schematic illustration of the overall synthesis procedure for HTC-based activated carbons and their application in CO_2 capture: (1) hydrothermal carbonization at 230°C–250°C (2 h), and (2) chemical activation with KOH.

tested as hydrogen stores in cryogenic conditions (−196°C), those materials exhibited capacities of ~2.5 wt% at 1 bar and in the 5.3- to 5.6-wt% range at 20 bar [175]. These hydrogen uptakes are in most cases superior to those obtained for other activated carbons with large surface area under similar conditions. With regards to CO_2 capture, it was analyzed at a pressure of 1 bar and three adsorption temperatures (0°C, 25°C, and 50°C) [174]. The HTC-based activated carbons exhibited CO_2 capture capacities up to 6.6 mmol CO_2/g at 0°C, 4.8 mmol/g at 25°C, and 3.6 mmol/g at 50°C. The outstanding adsorption uptakes are ascribed to the fact that a large fraction of the porosity corresponds to narrow micropores, which have strong adsorption potentials that enhance their filling by the CO_2 molecules. Additionally, the CO_2 adsorption was very fast, whereas N_2 adsorption was slower and much lower. In this way, the $[CO_2/N_2]$ selectivity measured under equilibrium conditions was 5.4. Furthermore, those sorbents could be easily, quickly, and fully regenerated by flowing a He stream at 25°C.

Another interesting approach in relation to the use of HTC materials for CO_2 capture is based on their functionalization with amine groups, which exhibit a high affinity to CO_2. In this respect, Titirici et al. reported the CO_2 capture by means of an amine-rich HTC product [146]. This material was prepared by a two-step process: (a) hydrothermal carbonization of glucose in the presence of small amounts of acrylic acid and (b) functionalization of carboxylic-rich HTC products with triethylamine. This aminated HTC material shows high CO_2 capture capacities (up to 4.3 mmol CO_2/g at −20°C). More important, these materials exhibited a very high $[CO_2/N_2]$ selectivity, up to 110 at 70°C.

2.2.6.5.3   Energy Storage

Energy storage is necessary to (i) make viable the use of renewable sources, such as wind or solar, as their supply is not continuous, (ii) increase the efficiency of the use of energy (power leveling), (iii) protect against power loss, and (iv) use of mobile equipment.

Supercapacitors and Li-ion batteries are the two "stars" of energy storage devices. Whereas supercapacitors are of great interest for high-power electrochemical energy storage devices, Li-ion batteries stand out for long-term energy supply.

Supercapacitors. This is another surface area-dependent application as in the case of gas storage. Therefore, works related to the use of HTC materials such as supercapacitors electrodes are based on a chemical activation process following the HTC treatment.

Wei et al. analyzed the supercapacitor performance in organic electrolyte (1 M $TEABF_4$ in acetonitrile) of HTC cellulose (C), starch (S), and sawdust (W) activated at 700°C and 800°C with a KOH/sample weight ratio of 4 [176]. The performance of those materials was spectacular, recording the highest capacitance ever reported for porous carbons in a symmetric two-electrode configuration using such electrolyte, i.e., 236 F/g (100 F/cm³) at 1 mV/s (Figure 2.21a). It exceeded the specific capacitance of commercial activated carbons optimized for EDLC applications, such as YP-17D, by 100%. Furthermore, the samples were capable of retaining 64% to 85% of the capacitance when the current density was increased from 0.6 to 20 A/g (Figure 2.2b).

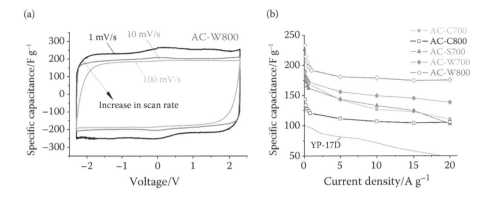

FIGURE 2.21 Electrochemical characterization of HTC-derived activated carbons in 1 M TEABF₄ solution in acetonitrile at room temperature: (a) cyclic voltammograms (CV) of the activated carbon obtained from sawdust at 800°C (AC-W800) and (b) capacitance retention with current density in comparison with that of commercially available YP-17D activated carbon. (L. Wei, M. Sevilla, A. B. Fuertes, R. Mokaya, G. Yushin, *Advanced Energy Materials* 2011, 1, 356–361. Copyright Wiley-VCH Verlag GmbH & Co. KGaA. Reproduced with permission.)

Titirici et al. have analyzed the capacitance behavior of N-containing HTC carbons activated with KOH (weight ratio of KOH to hydrothermal carbons = 1–4 and T = 600°C) [142]. Those materials exhibited excellent electrochemical performance in KOH 6 M and H₂SO₄ 1 M, achieving specific capacitances up to 220 and 300 F/g at a current density of 0.1 A/g in the basic and acidic electrolyte respectively. This superior capacitance is due to the combination of EDLC capacitance and pseudocapacitance arising from redox reactions of the nitrogen functionalities. Thus, humps were detected at around 0.5 V vs. SCE. Additionally, good capacitance retention at high current density (4 A/g) was observed, which proves good conductivity and quick charge propagation in both electrolytes.

Li-ion batteries. Carbon is the most used negative electrode material in Li-ion batteries. In this regard, HTC materials have also been used as electrodes in Li-ion batteries. The first study performed by Huang et al. demonstrated that the reversible lithium insertion/extraction capacity of this kind of material is much higher than the theoretical capacity of graphitized carbonaceous materials although the performance diminishes with increasing the C rate [177]. Following these reports, Tang et al. have investigated the performance of hollow HTC nanospheres as anodes in Li⁺ [178] as well as Na⁺ [179] ion batteries. The reversible capacity at a 1C rate could reach up to 370 mA h/g. Even at the very high rate of 50C (18.6 A/g), a capacity of ~100 mA h/g is still maintained. This value is much higher than traditional graphite electrodes (almost negligible at such a high rate).

Besides the use of pure HTC materials as electrodes in Li-ion batteries, great importance has been given to the utilization of HTC/inorganic composites as electrode materials. In this regard, Hu, Titirici et al. successfully in situ coated commercially available Si nanoparticles with a thin layer (10 nm) of HTC material via the conversion of D-glucose [152]. The resulting composite was further carbonized to increase the conductivity of the carbon layer resulting in a Si/SiOx/C composite with a markedly improved cyclability compared with pure Si. The reversible capacity was as high as 1100 mA h/g at a current density of 150 mA/g. Similarly, Demir-Cakan, Titirici et al. synthesized mesoporous SnO_2 microspheres using HTC of furfural in the presence of SnO_2 nanoparticle sols [163]. The stable reversible capacities were ~370 and 200 mA h/g at the current densities of 1 and 2 A/g.

The development of improved cathode materials is recognized as even more challenging than anode materials. In this respect, olivine $LiFePO_4$ is considered one of the most promising cathode materials for Li-ion batteries. Urchin-like hierarchical mesocrystals of pristine $LiFePO_4$, as well as carbon coated $LiFePO_4$ composites, have been synthesized by Popovic, Titirici et al. using a simple one-step solvothermal method [180]. The lithium storage performance of the pure $LiFePO_4$ was compared with that of its carbon-coated counterparts, proving to be superior.

2.2.6.5.4 Energy Production: Fuel Cells

Fuel cells are electrochemical devices that convert chemical energy from a fuel into electric energy continuously, as chemicals constantly flow into the cell. In both cathode and anode, catalysts are necessary for the outcome of the electrochemical reactions at low temperature. Those catalysts are composed in most of the cases of metal nanoparticles (Pt, Pt/Ru, Pd) dispersed over a support, which normally consists of carbon materials. The first to explore the use of a HTC-based carbon as electrocatalyst support in fuel cells was Yang et al. [181]. They deposited 10 wt% Pt nanoparticles over carbon spherules obtained by HTC of sucrose at 190°C and posttreatment

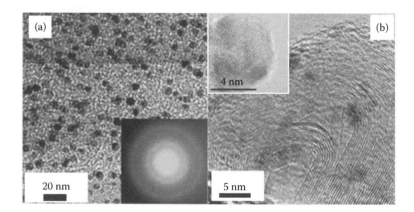

FIGURE 2.22 (a) HRTEM image of Pt nanoparticles deposited over HTC hard carbon spherules (inset: selected area electron diffraction pattern of Pt nanoparticles), (b) HRTEM image of PtRu nanoparticles deposited over carbon nanocoils obtained from HTC sucrose (inset: detail of a PtRu nanoparticle showing the cubic structure). (Reprinted from *Journal of Power Sources* 171, M. Sevilla, G. Lota, A. B. Fuertes, 546–551. Copyright 2007, with permission from Elsevier.)

at 1000°C (Figure 2.22). Those catalysts had a better performance than Pt/Vulcan in methanol electrooxidation due to the higher utilization of Pt particles as a result of higher accessibility. Other authors, like Kim et al. [182], Sevilla et al. [183], and Joo et al. [184] applied a catalytic graphitization step to increase the crystallinity of HTC carbon, as electrical conductivity is one of the key properties of a carbon electrocatalyst support. Most of those electrocatalysts outperformed Pt/Vulcan or PtRu/Vulcan electrocatalysts.

2.2.6.5.5 Heterogeneous Catalysis

Given the depletion of fossil fuels and the alarming facts associated with global warming there is an increasing interest in discovering new and efficient catalysts for the chemical conversion of biomass into biofules [186]. Homogeneous acid catalysts, such as sulfuric acid, are commonly employed. However, these catalysts have several drawbacks, such as corrosion and toxicity problems, costly and inefficient procedures for separating them from the products, and the need to neutralize the waste streams. These problems could be solved by developing heterogeneous solid-acid catalysts, which could then be more easily and efficiently separated from the products, enabling their reuse. However, most solid-acid catalysts reported so far are expensive or involve complex synthetic procedures, which impede their commercialization. These include acid zeolites, mesostructured silica functionalized with sulfonic groups, tungstated zirconia, sulfated zirconia, sulfonated polymers (Amberlyst-15), and Nafion-based composites [187]. In this respect, HTC presents again multiple advantages. It is low-cost, based on biomass precursors itself, stable, its porosity can be tuned and it is already acidic. In addition, its surface functionality can help its further modification with stronger acidic groups such as SO_3H groups.

Thus, Sevilla Fuertes et al. used the sulfonation of carbonaceous microspheres obtained by the hydrothermal carbonization of glucose (Figure 2.23A). This synthetic strategy circumvents gas-phase pyrolysis, thereby avoiding the emission of harmful gases, and yields a solid acid comprised of spherical particles of uniform, micrometer-regime size (Figure 2.23B). The activity of this sulfonated carbon catalyst toward the esterification of oleic acid with ethanol, a typical reaction in the synthesis of biodiesel, was investigated. Figure 2.23c shows the formation of ethyl oleate during reaction at 55°C. For comparison, results for equivalent amounts of sulfuric acid, p-toluenesulfonic acid, and Amberlyst-15 are also shown. In the absence of catalyst (blank experiment) the ethyl oleate yield was only 3.5% after 24 h. As expected, the homogeneous catalysts (sulfuric acid and p-toluenesulfonic acid) showed the highest activities, but they lack the advantages of solid acids pointed out before. The HTC with sulfonic groups sample exhibits a higher activity than Amberlyst-15 despite the fact that the latter has a higher density of sulfonic groups. This may be because the sulfonated carbon microspheres are highly hydrophilic. This facilitates the adsorption of a large amount of hydrophylic molecules such as ethanol and favors the access of the reactants to the $-SO_3H$ sites.

In a similar approach, Xiao et al. produced novel biacidic carbon via one-step hydrothermal carbonization of glucose, citric acid, and hydroxyethylsulfonic acid at 180°C for only 4 h. The novel carbon had an acidity of 1.7 mmol/g with the carbonyl to sulfonic acid groups molar ratio of 1:3. The catalytic activities of the carbon were investigated,

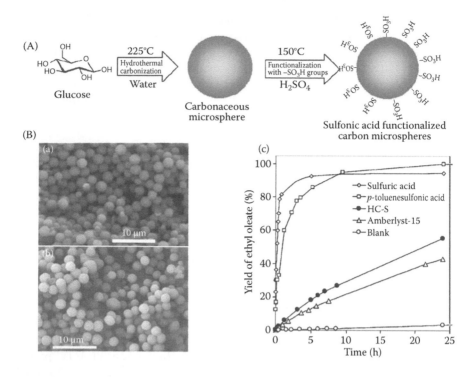

FIGURE 2.23 (A) Schematic illustration of the synthetic procedure for obtaining carbon microspheres functionalized with –SO$_3$H groups; (B) SEM images of (a) the hydrothermally carbonized glucose and (b) the sulfonated sample; Time courses of the esterification of oleic acid with ethanol. The yield of ethyl oleate is based on oleic acid. Reaction temperature: 55°C. (J. A. Macia-Agullo, M. Sevilla, M. A. Diez, A. B. Fuertes, *ChemSusChem* 2010, 3, 1352–1354. Copyright Wiley-VCH Verlag GmbH & Co. KGaA. Reproduced with permission.)

through esterification and oxathioketalization. The results showed that the carbon owned the comparable activities to sulfuric acid, which indicated that the carbon holds great potential for the green processes related to biomass conversion [189].

I believe that the HTC materials produced from biomass could play in the future an important role in the conversion of lignocellulosic biomass into useful products and biofuels. First, about 30%–40% of the products resulting from HTC of biomass are left in the liquid phase [190]. Thus, if part of the HTC solid materials making the other 60%–70% fraction could be used to competently convert the liquid phase into useful products, the efficiency of this process will be greatly improved. Furthermore, the ratio of liquid to solid fraction as well as their composition could be controlled throughout appropriate synthesis conditions (i.e., temperature, pressure, catalyst, additives). In this respect, it has been previously shown that various "Starbon" materials, a process with many similarities to HTC regarding the flexibility of the resulting carbons [79], can act as efficient catalysts with tuned selectivity depending on the preparation method and the reaction conditions [80,191].

Other examples of the use of HTC as catalysts supports in heterogeneous catalysis include the hydrogenation of hydroxyl aromatic derivatives [160], dry methane reforming and partial oxidation of methane [157], and Fisher–Tropsch reaction [192].

2.2.6.5.6 Photocatalysis and Others

In the past decade, much research effort has focused on extending the photoresponse of the TiO_2 to the visible region. Among these attempts, TiO_2 doping, either with main group elements or transition metals, has been the most important approach for improving the photocatalytic performance of the catalyst. In this regard, HTC has been used to produce carbon doped TiO_2 with visible light activity [161]. It was shown by Titirici's group that the surface of nanometre sized carbon materials can also show collective polarization modes, and therefore, these optical absorption transitions are feasible to sensitize TiO_2, which then acts as a novel "dyad"-type structure, with an improved TiO_2 hole reactivity, while the electron is taken up by the carbon component. This results in an improved photocatalytic activity over the complete spectral range.

Other applications of HTC materials that have already been explored include magnetic applications (magnetic separation [193], contrast agents in MRI [194]), sensors [195], and bioapplications (bioimaging [196] and drug delivery [197]).

The underlying motivation for doing research on HTC materials is the ever-growing need to reduce the dependence on fossil fuels and move toward more sustainable technologies. The aim of this subchapter was to give an overview of a "palette" of functional nanomaterials suitable for a wide range of essential and sustainable applications. We strongly hope that throughout this chapter we could convince the reader that HTC, a simple and sustainable technique for carbon material formation, is a viable and powerful addition to the toolbox of carbon-based technologies.

2.3 SUMMARY

In this chapter, we have presented several classes of sustainable carbon materials prepared either from renewable precursors, natural catalysts, or economic design, ranging from crystalline nanostructures such as carbon nanotubes and graphenes to amorphous activated carbons, Starbons, and hydrothermal carbonization. Especially for Starbons and hydrothermal carbons we have tried to point out also the numerous important applications that emerged from these technologies.

Great progress has been done in the synthesis of greener carbon materials, and it is highly likely that green carbon materials will ultimately overcome current favored fossil-based materials and thus contribute to more sustainable technologies.

REFERENCES

1. J. P. H. van Wyk, *Trends in Biotechnology* 2001, *19*, 172–177.
2. F. W. Lichtenthaler, S. Peters, *Comptes Rendus Chimie* 2004, *7*, 65–90.
3. S. Rahmstort, J. Morgan, A. Levermann, K. Sach, in *Global Sustainability—A Nobel Cause* (Ed.: H. J. Schellnhuber, M. Molina, N. Stern, V. Huber and S. Kadner), Cambridge University Press, Cambridge, UK, 2010, p. 68.

4. M. L. Green, L. Espinal, E. Traversa, E. J. Amis, *MRS Bulletin* 2012, *37*, 303–309.
5. R. J. Koopmans, *Soft Matter* 2006, *2*, 537–543.
6. O. Bobleter, *Progress in Polymer Science* 1994, *19*, 797–841.
7. (a) K. A. Gray, L. S. Zhao, M. Emptage, *Current Opinion in Chemical Biology* 2006, *10*, 141–146; (b) A. T. W. M. Hendriks, G. Zeeman, *Bioresource Technology* 2009, *100*, 10–18; (c) J. Lee, *Journal of Biotechnology* 1997, *56*, 1–24; (d) Y. Lin, S. Tanaka, *Applied Microbiology and Biotechnology* 2006, *69*, 627–642; (e) L. R. Lynd, *Annual Review of Energy and the Environment* 1996, *21*, 403–465; (f) L. R. Lynd, W. H. van Zyl, J. E. McBride, M. Laser, *Current Opinion in Biotechnology* 2005, *16*, 577–583; (g) L. R. Lynd, P. J. Weimer, W. H. van Zyl, I. S. Pretorius, *Microbiology and Molecular Biology Reviews* 2002, *66*, 506–577; (h) M. Ni, D. Y. C. Leung, M. K. H. Leung, *International Journal of Hydrogen Energy* 2007, *32*, 3238–3247; (i) O. J. Sanchez, C. A. Cardona, *Bioresource Technology* 2008, *99*, 5270–5295; (j) Y. Sun, J. Y. Cheng, *Bioresource Technology* 2002, *83*, 1–11; (k) C. E. Wyman, *Annual Review of Energy and the Environment* 1999, *24*, 189–226; (l) J. Zaldivar, J. Nielsen, L. Olsson, *Applied Microbiology and Biotechnology* 2001, *56*, 17–34.
8. (a) A. V. Bridgwater, *Fuel* 1995, *74*, 631–653; (b) A. Demirbas, *Progress in Energy and Combustion Science* 2007, *33*, 1–18; (c) L. Devi, K. J. Ptasinski, F. Janssen, *Biomass & Bioenergy* 2003, *24*, 125–140; (d) Y. Matsumura, T. Minowa, B. Potic, S. R. A. Kersten, W. Prins, W. P. M. van Swaaij, B. van de Beld, D. C. Elliott, G. G. Neuenschwander, A. Kruse, M. J. Antal, *Biomass & Bioenergy* 2005, *29*, 269–292; (e) P. McKendry, *Bioresource Technology* 2002, *83*, 37–46; (f) P. McKendry, *Bioresource Technology* 2002, *83*, 55–63; (g) A. A. Peterson, F. Vogel, R. P. Lachance, M. Froeling, M. J. Antal, Jr., J. W. Tester, *Energy & Environmental Science* 2008, *1*, 32–65; (h) D. Sutton, B. Kelleher, J. R. H. Ross, *Fuel Processing Technology* 2001, *73*, 155–173.
9. (a) D. M. Alonso, J. Q. Bond, J. A. Dumesic, *Green Chemistry* 2010, *12*, 1493–1513; (b) Y.-C. Lin, G. W. Huber, *Energy & Environmental Science* 2009, *2*, 68–80; (c) P. S. Nigam, A. Singh, *Progress in Energy and Combustion Science* 2011, *37*, 52–68; (d) R. C. Saxena, D. Seal, S. Kumar, H. B. Goyal, *Renewable & Sustainable Energy Reviews* 2008, *12*, 1909–1927; (e) A. Sivasamy, K. Y. Cheah, P. Fornasiero, F. Kemausuor, S. Zinoviev, S. Miertus, *Chemsuschem* 2009, *2*, 278–300; (f) X. Tong, Y. Ma, Y. Li, *Applied Catalysis A: General* 2010, *385*, 1–13; (g) S. Van de Vyver, J. Geboers, P. A. Jacobs, B. F. Sels, *Chemcatchemistry* 2011, *3*, 82–94; (h) M. M. Yung, W. S. Jablonski, K. A. Magrini-Bair, *Energy & Fuels* 2009, *23*, 1874–1887.
10. P. T. Anastas, M. M. Kirchhoff, *Accounts of Chemical Research* 2002, *35*, 686–694.
11. (a) M. Adamczak, U. T. Bornscheuer, W. Bednarski, *European Journal of Lipid Science and Technology* 2009, *111*, 800–813; (b) H. W. Blanch, B. A. Simmons, D. Klein-Marcuschamer, *Biotechnology Journal* 2011, *6*, 1086–1102; (c) S. P. S. Chundawat, G. T. Beckham, M. E. Himmel, B. E. Dale, in *Annual Review of Chemical and Biomolecular Engineering, Vol 2*, (Ed.: J. M. Prausnitz), 2011, pp. 121–145; (d) I. A. Hoell, G. Vaaje-Kolstad, V. G. H. Eijsink, in *Biotechnology and Genetic Engineering Reviews, Vol 27*, (Ed.: S. E. Harding), 2010, pp. 331–366.
12. C. K. Hong, R. P. Wool, *Journal of Applied Polymer Science* 2005, *95*, 1524–1538.
13. K. M. Nampoothiri, N. R. Nair, R. P. John, *Bioresource Technology* 2010, *101*, 8493–8501.
14. S. Gharbi, J. P. Andreolety, A. Gandini, *European Polymer Journal* 2000, *36*, 463–472.
15. A. Mitiakoudis, A. Gandini, *Macromolecules* 1991, *24*, 830–835.
16. S. Boufi, M. N. Belgacem, J. Quillerou, A. Gandini, *Macromolecules* 1993, *26*, 6706–6717.
17. C. Pavier, A. Gandini, *European Polymer Journal* 2000, *36*, 1653–1658.

18. (a) A. Brandelli, D. J. Daroit, A. Riffel, *Applied Microbiology and Biotechnology* 2010, *85*, 1735–1750; (b) C. Briens, J. Piskorz, F. Berruti, *International Journal of Chemical Reactor Engineering* 2008, *6*; (c) R. Fischer, N. Emans, *Transgenic Research* 2000, *9*, 279–299; (d) R. Harun, M. Singh, G. M. Forde, M. K. Danquah, *Renewable & Sustainable Energy Reviews* 2010, *14*, 1037–1047; (e) A. S. Mamman, J.-M. Lee, Y.-C. Kim, I. T. Hwang, N.-J. Park, Y. K. Hwang, J.-S. Chang, J.-S. Hwang, *Biofuels Bioprod. Biorefining* 2008, *2*, 438–454; (f) A. Pandey, C. R. Soccol, *Brazilian Archives of Biology and Technology* 1998, *41*, 379–389; (g) O. Pulz, W. Gross, *Applied Microbiology and Biotechnology* 2004, *65*, 635–648; (h) R. R. Singhania, A. K. Patel, C. R. Soccol, A. Pandey, *Biochemical Engineering Journal* 2009, *44*, 13–18; (i) D. Sipkema, R. Osinga, W. Schatton, D. Mendola, J. Tramper, R. H. Wijffels, *Biotechnology and Bioengineering* 2005, *90*, 201–222; (j) Y. Tokiwa, B. P. Caiabia, *Canadian Journal of Chemistry—Revue Canadienne De Chimie* 2008, *86*, 548–555.

19. R. H. Baughman, A. A. Zakhidov, W. A. de Heer, *Science* 2002, *297*, 787–792.

20. D. S. Su, *ChemSusChem* 2009, *2*, 1009–1020.

21. A.-C. Dupuis, *Prog. Mater. Sci.* 2005, *50*, 929–961.

22. D. S. Su, X.-W. Chen, *Angewandte Chemie* 2007, *119*, 1855–1856.

23. M. Endo, K. Takeuchi, Y. A. Kim, K. C. Park, T. Ichiki, T. Hayashi, T. Fukuyo, S. Linou, D. S. Su, M. Terrones, M. S. Dresselhaus, *ChemSusChem* 2008, *1*, 820–822.

24. A. Rinaldi, J. Zhang, J. Mizera, F. Girgsdies, N. Wang, S. B. A. Hamid, R. Schlogl, D. S. Su, *Chemical Communications* 2008, 6528–6530.

25. S. Kawasaki, M. Shinoda, T. Shimada, F. Okino, H. Touhara, *Carbon* 2006, *44*, 2139–2141.

26. D. S. Su, X. Chen, G. Weinberg, A. Klein-Hofmann, O. Timpe, S. B. A. Hamid, R. Schlögl, *Angewandte Chemie International Edition* 2005, *44*, 5488–5492.

27. X.-W. Chen, D. S. Su, S. B. A. Hamid, R. Schlögl, *Carbon* 2007, *45*, 895–898.

28. X. W. Lou, J. S. Chen, P. Chen, L. A. Archer, *Chemistry of Materials* 2009, *21*, 2868–2874.

29. (a) A. Dosodia, C. Lal, B. P. Singh, R. B. Mathur, D. K. Sharma, *Fullerenes Nanotubes and Carbon Nanostructures* 2009, *17*, 567–582; (b) Q. Jieshan, L. Yongfeng, L. Yunpeng, W. Tonghua, Z. Zongbin, Z. Ying, L. Feng, C. Huiming, *Carbon* 2003, *41*, 2170–2173; (c) R. B. Mathur, C. Lal, D. K. Sharma, *Energy Sources Part A Recovery Utilization and Environmental Effects* 2007, *29*, 21–27; (d) J. S. Qiu, Y. F. Li, Y. P. Wang, W. Li, *Fuel Processing Technology* 2004, *85*, 1663–1670; (e) J. S. Qiu, F. Zhang, Y. Zhou, H. M. Han, D. S. Hu, S. C. Tsang, P. J. F. Harris, *Fuel* 2002, *81*, 1509–1514; (f) J. S. Qiu, Y. Zhou, L. N. Wang, S. C. Tsang, *Carbon* 1998, *36*, 465–467; (g) J. S. Qiu, Y. Zhou, Z. G. Yang, D. K. Wang, S. C. Guo, S. C. Tsang, P. J. F. Harris, *Fuel* 2000, *79*, 1303–1308; (h) J. S. Qiu, Y. Zhou, Z. G. Yang, L. N. Wang, F. Zhang, S. C. Tsang, P. J. S. Harris, *Molecular Materials* 2000, *13*, 377–384; (i) M. Z. Wang, F. Li, *New Carbon Materials* 2005, *20*, 71–82; (j) M. A. Wilson, H. K. Patney, J. Kalman, *Fuel* 2002, *81*, 5–14; (k) X.-F. Zhao, J.-S. Qiu, Y.-X. Sun, C. Hao, T.-J. Sun, L.-W. Cui, *New Carbon Materials* 2009, *24*, 109–113; (l) W. Zhi, W. Bin, G. Qianming, S. Huaihe, L. Ji, *Materials Letters* 2008, *62*, 3585–3587.

30. M. Kumar, T. Okazaki, M. Hiramatsu, Y. Ando, *Carbon* 2007, *45*, 1899–1904.

31. Z. H. Kang, E. B. Wang, B. D. Mao, Z. M. Su, L. Chen, L. Xu, *Nanotechnology* 2005, *16*, 1192.

32. Y. Gogotsi, J. A. Libera, M. Yoshimura, *Journal of Materials Research* 2000, *15*, 2591–2594.

33. J. M. Calderon Moreno, T. Fujino, M. Yoshimura, *Carbon* 2001, *39*, 618–621.

34. V. G. Pol, P. Thiyagarajan, *Journal of Environmental Monitoring* 2010, *12*, 455–459.

35. S. V. Pol, V. G. Pol, D. Sherman, A. Gedanken, *Green Chemistry* 2009, *11*, 448–451.

36. M. Sevilla, A. B. Fuertes, *Chemical Physics Letters* 2010, *490*, 63–68.
37. M. Ashokkumar, N. T. Narayanan, A. L. Mohana Reddy, B. K. Gupta, B. Chandrasekaran, S. Talapatra, P. M. Ajayan, P. Thanikaivelan, *Green Chemistry* 2012, *14*, 1689–1695.
38. N. D. Mermin, *Physical Review* 1968, *176*, 250–254.
39. M. Orlita, C. Faugeras, P. Plochocka, P. Neugebauer, G. Martinez, D. K. Maude, A. L. Barra, M. Sprinkle, C. Berger, W. A. de Heer, M. Potemski, *Physical Review Letters* 2008, *101*, 267601.
40. A. A. Balandin, S. Ghosh, W. Bao, I. Calizo, D. Teweldebrhan, F. Miao, C. N. Lau, *Nano Letters* 2008, *8*, 902–907.
41. C. Lee, X. Wei, J. W. Kysar, J. Hone, *Science* 2008, *321*, 385–388.
42. S. Park, J. An, I. Jung, R. D. Piner, S. J. An, X. Li, A. Velamakanni, R. S. Ruoff, *Nano Letters* 2009, *9*, 1593–1597.
43. S. Park, J. An, R. D. Piner, I. Jung, D. Yang, A. Velamakanni, S. T. Nguyen, R. S. Ruoff, *Chemistry of Materials* 2008, *20*, 6592–6594.
44. K. S. Novoselov, D. Jiang, F. Schedin, T. J. Booth, V. V. Khotkevich, S. V. Morozov, A. K. Geim, *Proceedings of the National Academy of Sciences of the United States of America* 2005, *102*, 10451–10453.
45. K. S. Novoselov, A. K. Geim, S. V. Morozov, D. Jiang, Y. Zhang, S. V. Dubonos, I. V. Grigorieva, A. A. Firsov, *Science* 2004, *306*, 666–669.
46. C. Berger, Z. Song, X. Li, X. Wu, N. Brown, C. Naud, D. Mayou, T. Li, J. Hass, A. N. Marchenkov, E. H. Conrad, P. N. First, W. A. de Heer, *Science* 2006, *312*, 1191–1196.
47. K. S. Kim, Y. Zhao, H. Jang, S. Y. Lee, J. M. Kim, K. S. Kim, J.-H. Ahn, P. Kim, J.-Y. Choi, B. H. Hong, *Nature* 2009, *457*, 706–710.
48. L. Jiao, L. Zhang, X. Wang, G. Diankov, H. Dai, *Nature* 2009, *458*, 877–880.
49. S. Stankovich, D. A. Dikin, R. D. Piner, K. A. Kohlhaas, A. Kleinhammes, Y. Jia, Y. Wu, S. T. Nguyen, R. S. Ruoff, *Carbon* 2007, *45*, 1558–1565.
50. L. J. Cote, F. Kim, J. Huang, *Journal of the American Chemical Society* 2008, *131*, 1043–1049.
51. S. Stankovich, R. D. Piner, X. Chen, N. Wu, S. T. Nguyen, R. S. Ruoff, *Journal of Materials Chemistry* 2006, *16*, 155–158.
52. G. Wang, J. Yang, J. Park, X. Gou, B. Wang, H. Liu, J. Yao, *The Journal of Physical Chemistry C* 2008, *112*, 8192–8195.
53. H. L. Guo, X. F. Wang, Q. Y. Qian, F. B. Wang, X. H. Xia, *Acs Nano* 2009, *3*, 2653–2659.
54. Y. H. Ding, P. Zhang, Q. Zhuo, H. M. Ren, Z. M. Yang, Y. Jiang, *Nanotechnology* 2011, *22*.
55. T.-P. Fellinger, R. J. White, M.-M. Titirici, M. Antonietti, *Advanced Functional Materials* 2012, *22*, 3254–3260.
56. C. Z. Zhu, S. J. Guo, Y. X. Fang, S. J. Dong, *Acs Nano* 2010, *4*, 2429–2437.
57. Z.-J. Fan, W. Kai, J. Yan, T. Wei, L.-J. Zhi, J. Feng, Y.-M. Ren, L.-P. Song, F. Wei, *Acs Nano* 2011, *5*, 191–198.
58. Y. Liu, Y. Li, M. Zhong, Y. Yang, Y. Wen, M. Wang, *Journal of Materials Chemistry* 2011, *21*, 15449–15455.
59. Z. Sui, X. Zhang, Y. Lei, Y. Luo, *Carbon* 2011, *49*, 4314–4321.
60. W. Kai, F. Tao, Q. Min, D. Hui, C. Yiwei, S. Zhuo, *Applied Surface Science* 2011, *257*, 5808–5812.
61. P. Khanra, T. Kuila, N. H. Kim, S. H. Bae, D.-S. Yu, J. H. Lee, *Chemical Engineering Journal* 2012, *183*, 526–533.
62. Y. Wang, Z. Shi, J. Yin, *Acs Applied Materials & Interfaces* 2011, *3*, 1127–1133.
63. G. Wang, F. Qian, C. W. Saltikov, Y. Jiao, Y. Li, *Nano Research* 2011, *4*, 563–570.
64. E. C. Salas, Z. Sun, A. Lüttge, J. M. Tour, *Acs Nano* 2010, *4*, 4852–4856.
65. K. Liu, J.-J. Zhang, F.-F. Cheng, T.-T. Zheng, C. Wang, J.-J. Zhu, *Journal of Materials Chemistry* 2011, *21*, 12034–12040.

66. E. B. Nursanto, A. Nugroho, S.-A. Hong, S. J. Kim, K. Y. Chung, J. Kim, *Green Chemistry* 2011, *13*, 2714–2718.
67. X. Li, W. Cai, J. An, S. Kim, J. Nah, D. Yang, R. Piner, A. Velamakanni, I. Jung, E. Tutuc, S. K. Banerjee, L. Colombo, R. S. Ruoff, *Science* 2009, *324*, 1312–1314.
68. A. Reina, X. Jia, J. Ho, D. Nezich, H. Son, V. Bulovic, M. S. Dresselhaus, J. Kong, *Nano Letters* 2008, *9*, 30–35.
69. D. Kondo, S. Sato, K. Yagi, N. Harada, M. Sato, M. Nihei, N. Yokoyama, *Applied Physics Express* 2010, *3*, 025102.
70. H. Ago, Y. Ito, N. Mizuta, K. Yoshida, B. Hu, C. M. Orofeo, M. Tsuji, K.-I. Ikeda, S. Mizuno, *Acs Nano* 2010, *4*, 7407–7414.
71. B. J. Kang, J. H. Mun, C. Y. Hwang, B. J. Cho, *Journal of Applied Physics* 2009, *106*, 104309–104306.
72. Z. Sun, Z. Yan, J. Yao, E. Beitler, Y. Zhu, J. M. Tour, *Nature* 2010, *468*, 549–552.
73. G. Ruan, Z. Sun, Z. Peng, J. M. Tour, *Acs Nano* 2011, *5*, 7601–7607.
74. A. Primo, P. Atienzar, E. Sanchez, J. M. Delgado, H. Garcia, *Chemical Communications* 2012, *48*, 9254–9256.
75. E. Ruiz-Hitzky, M. Darder, F. M. Fernandes, E. Zatile, F. J. Palomares, P. Aranda, *Advanced Materials* 2011, *23*, 5250–5255.
76. (a) T. Ahmad, M. Rafatullah, A. Ghazali, O. Sulaiman, R. Hashim, *Journal of Environmental Science and Health Part C—Environmental Carcinogenesis & Ecotoxicology Reviews* 2011, *29*, 177–222; (b) Y. Chen, Y. Zhu, Z. Wang, Y. Li, L. Wang, L. Ding, X. Gao, Y. Ma, Y. Guo, *Advances in Colloid and Interface Science* 2011, *163*, 39–52; (c) A. Demirbas, *Journal of Hazardous Materials* 2009, *167*, 1–9; (d) O. Ioannidou, A. Zabaniotou, *Renewable & Sustainable Energy Reviews* 2007, *11*, 1966–2005; (e) S.-H. Lin, R.-S. Juang, *Journal of Environmental Management* 2009, *90*, 1336–1349; (f) M. Mayhew, T. Stephenson, *Environmental Technology* 1997, *18*, 883–892; (g) A. R. Mohamed, M. Mohammadi, G. N. Darzi, *Renewable & Sustainable Energy Reviews* 2010, *14*, 1591–1599; (h) G. Rodriguez, A. Lama, R. Rodriguez, A. Jimenez, R. Guillen, J. Fernandez-Bolanos, *Bioresource Technology* 2008, *99*, 5261–5269; (i) G. Skodras, I. Diamantopouiou, A. Zabaniotou, G. Stavropoulos, G. P. Sakellaropoulos, *Fuel Processing Technology* 2007, *88*, 749–758; (j) Y. Uraki, S. Kubo, *Mokuzai Gakkaishi* 2006, *52*, 337–343; (k) H. Yu, G. H. Covey, A. J. O'Connor, *International Journal of Environment and Pollution* 2008, *34*, 427–450.
77. (a) E. Frackowiak, F. Beguin, *Carbon* 2001, *39*, 937–950; (b) A. G. Pandolfo, A. F. Hollenkamp, *Journal of Power Sources* 2006, *157*, 11–27; (c) P. Simon, Y. Gogotsi, *Nat. Mater.* 2008, *7*, 845–854; (d) L. L. Zhang, X. S. Zhao, *Chemical Society Reviews* 2009, *38*, 2520–2531.
78. (a) S. Choi, J. H. Drese, C. W. Jones, *Chemsuschem* 2009, *2*, 796–854; (b) D. Lozano-Castello, J. Alcaniz-Monge, M. A. de la Casa-Lillo, D. Cazorla-Amoros, A. Linares-Solano, *Fuel* 2002, *81*, 1777–1803.
79. R. J. White, V. Budarin, R. Luque, J. H. Clark, D. J. Macquarrie, *Chemical Society Reviews* 2009, *38*, 3401–3418.
80. V. L. Budarin, J. H. Clark, R. Luque, D. J. Macquarrie, R. J. White, *Green Chemistry* 2008, *10*, 382–387.
81. V. Budarin, J. H. Clark, J. J. E. Hardy, R. Luque, K. Milkowski, S. J. Tavener, A. J. Wilson, *Angewandte Chemie International Edition* 2006, *45*, 3782–3786.
82. R. J. White, C. Antonio, V. L. Budarin, E. Bergstroem, J. Thomas-Oates, J. H. Clark, *Advanced Functional Materials* 2010, *20*, 1834–1841.
83. R. J. White, V. L. Budarin, J. H. Clark, *Chemistry—A European Journal* 2010, *16*, 1326–1335.
84. V. Budarin, R. Luque, D. J. Macquarrie, J. H. Clark, *Chemistry—A European Journal* 2007, *13*, 6914–6919.

85. R. Luque, V. Budarin, J. H. Clark, D. J. Macquarrie, *Green Chemistry* 2009, *11*, 459–461.
86. R. J. White, R. Luque, V. L. Budarin, J. H. Clark, D. J. Macquarrie, *Chemical Society Reviews* 2009, *38*, 481–494.
87. J. P. Paraknowitsch, J. Zhang, D. Su, A. Thomas, M. Antonietti, *Advanced Materials* 2010, *22*, 87.
88. J. S. Lee, X. Wang, H. Luo, S. Dai, *Advanced Materials* 2010, *22*, 1004–1007.
89. P. Kuhn, A. Forget, J. Hartmann, A. Thomas, M. Antonietti, *Advanced Materials* 2009, *21*, 897–901.
90. J. S. Lee, X. Wang, H. Luo, G. A. Baker, S. Dai, *Journal of the American Chemical Society* 2009, *131*, 4596–4597.
91. X. Wang, S. Dai, *Angewandte Chemie International Edition* 2010, *49*, 6664–6668.
92. J. P. Paraknowitsch, A. Thomas, M. Antonietti, *Journal of Materials Chemistry* 2010, *20*, 6746–6758.
93. J. P. Paraknowitsch, Y. Zhang, A. Thomas, *Journal of Materials Chemistry* 2011, *21*, 15537–15543.
94. E. R. Cooper, C. D. Andrews, P. S. Wheatley, P. B. Webb, P. Wormald, R. E. Morris, *Nature* 2004, *430*, 1012–1016.
95. E. R. Parnham, R. E. Morris, *Accounts of Chemical Research* 2007, *40*, 1005–1013.
96. A. Taubert, Z. Li, *Journal of the Chemical Society, Dalton Transactions* 2007, 723–727.
97. R. E. Morris, *Chemical Communications* 2009, 2990–2998.
98. Z. Ma, J. Yu, S. Dai, *Advanced Materials* 2010, *22*, 261–285.
99. Z.-L. Xie, R. J. White, J. Weber, A. Taubert, M. M. Titirici, *Journal of Materials Chemistry* 2011, *21*, 7434–7442.
100. J. S. Lee, R. T. Mayes, H. Luo, S. Dai, *Carbon* 2010, *48*, 3364–3368.
101. (a) J. B. Binder, R. T. Raines, *Journal of the American Chemical Society* 2009, *131*, 1979–1985; (b) S. Hu, Z. Zhang, Y. Zhou, B. Han, H. Fan, W. Li, J. Song, Y. Xie, *Green Chemistry* 2008, *10*, 1280–1283; (c) C. Lansalot-Matras, C. Moreau, *Catalysis Communications* 2003, *4*, 517–520; (d) C. Moreau, A. Finiels, L. Vanoye, *Journal of Molecular Catalysis A—Chemical* 2006, *253*, 165–169; (e) H. Zhao, J. E. Holladay, H. Brown, Z. C. Zhang, *Science* 2007, *316*, 1597–1600.
102. A. P. Abbott, G. Capper, D. L. Davies, R. K. Rasheed, V. Tambyrajah, *Chemical Communications* 2003, 70–71.
103. A. P. Abbott, D. Boothby, G. Capper, D. L. Davies, R. K. Rasheed, *Journal of the American Chemical Society* 2004, *126*, 9142–9147.
104. C. A. Nkuku, R. J. LeSuer, *The Journal of Physical Chemistry B* 2007, *111*, 13271–13277.
105. D. Carriazo, M. C. Serrano, M. C. Gutierrez, M. L. Ferrer, F. del Monte, *Chemical Society Reviews* 2012, *41*, 4996–5014.
106. D. Carriazo, M. A. C. Gutiérrez, M. L. Ferrer, F. del Monte, *Chemistry of Materials* 2010, *22*, 6146–6152.
107. M. Takenaka, T. Izumitani, T. Hashimoto, *Journal of Chemical Physics* 1993, *98*, 3528–3539.
108. K. Katuri, M. L. Ferrer, M. C. Gutierrez, R. Jimenez, F. del Monte, D. Leech, *Energy & Environmental Science* 2011, *4*, 4201–4210.
109. Y. H. Choi, J. van Spronsen, Y. Dai, M. Verberne, F. Hollmann, I. W.C.E. Arends, G.-J. Witkamp, R. Verpoorte, *Plant Physiol* 2011, *156*, 1701–1705.
110. F. Bergius, *Naturwissenschaften* 1928, *16*, 1–10.
111. M.-M. Titirici, R. J. White, C. Falco, M. Sevilla, *Energy & Environmental Science* 2012, *5*, 6796–6822.
112. M.-M. Titirici, M. Antonietti, *Chemical Society Reviews* 2010, *39*, 103–116.
113. N. Baccile, G. Laurent, C. Coelho, F. Babonneau, L. Zhao, M.-M. Titirici, *Journal of Physical Chemistry C* 2011, *115*, 8976–8982.

114. M. M. Titirici, M. Antonietti, N. Baccile, *Green Chemistry* 2008, *10*, 1204–1212.
115. C. Falco, N. Baccile, M. M. Titirici, *Green Chemistry* 2011, *13 (11)*, 3273–3281.
116. A. H. Lu, F. Schuth, *Advanced Materials* 2006, *18*, 1793–1805.
117. C. Falco, J. P. Marco-Lozar, D. Salinas-Torres, E. Morallón, D. Cazorla-Amorós, M. M. Titirici, D. Lozano Castello, Tailoring the porosity of chemically activated hydrothermal carbons: Influence of the Precursor and hydrothermal carbonization temperature, *Carbon*, 2013, accepted, in press.
118. A. Linares-Solano, D. Lozano-Castello, M. A. Lillo-Rodenas, D. Cazorla-Amoros, in *Chemistry and Physics of Carbon, Vol. 30*, CRC Press, Taylor & Francis Group, Boca Raton, FL, 2008, pp. 1–62.
119. C. Liang, S. Dai, *Journal of the American Chemical Society* 2006, *128*, 5316–5317.
120. D. Y. Zhao, J. L. Feng, Q. S. Huo, N. Melosh, G. H. Fredrickson, B. F. Chmelka, G. D. Stucky, *Science* 1998, *279*, 548–552.
121. S. H. Joo, R. Ryoo, M. Kruk, M. Jaroniec, *Journal of Physical Chemistry B* 2002, *106*, 4640–4646.
122. F. Zhang, Y. Meng, D. Gu, Y. Yan, C. Yu, B. Tu, D. Zhao, *J. Am. Chem. Soc.* 2005, *127*, 13508–13509.
123. M.-M. Titirici, A. Thomas, M. Antonietti, *Advanced Functional Materials* 2007, *17*, 1010–1018.
124. M.-M. Titirici, A. Thomas, M. Antonietti, *Journal of Materials Chemistry* 2007, *17*, 3412–3418.
125. S. Kubo, I. Tan, R. J. White, M. Antonietti, M.-M. Titirici, *Chemistry of Materials* 2010, *22*, 6590.
126. R. J. White, K. Tauer, M. Antonietti, M.-M. Titirici, *Journal of the American Chemical Society* 2011, *132*, 17360–17363.
127. S. Kubo, R. J. White, N. Yoshizawa, M. Antonietti, M.-M. Titirici, *Chemistry of Materials*, 2011, *23*, 4882–4885.
128. R. J. White, M. Antonietti, M.-M. Titirici, *Journal of Materials Chemistry* 2009, *19*, 8645–8650.
129. N. Baccile, M. Antonietti, M.-M. Titirici, *ChemSusChem* 2010, *3*, 246–253.
130. R. J. White, N. Yoshizawa, M. Antonietti, M.-M. Titirici, *Green Chem.* 2011, *13*, 2428–2434.
131. K. P. Gong, F. Du, Z. H. Xia, M. Durstock, L. M. Dai, *Science* 2009, *323*, 760.
132. T. Iijima, K. Suzuki, Y. Matsuda, *Synth. Met* 1995, *73*, 9.
133. C. West, C. Elfakir, M. Lafosse, *J. Chromatogr., A* 2010, *1217*, 3210.
134. C. Petit, G. W. Peterson, J. Mahle, T. J. Bandosz, *Carbon* 2010, *48*.
135. S. H. Lim, H. I. Elim, X. Y. Gao, A. T. S. Wee, W. Ji, J. Y. Lee, J. Lin, R. Pietrzak, H. Wachowska, P. Nowicki, *Energy Fuels* 2006, *20*, 1275.
136. K. A. Kurak, A. B. Anderson, *Journal of Physical Chemistry* C 2009, *113*, 6730.
137. F. Jaouen, M. Lefevre, J. P. Dodelet, M. Cai, *Journal of Physical Chemistry B* 2006, *110*, 5553.
138. R. Pietrzak, H. Wachowska, P. Nowicki, *Energy Fuels* 2006, 1275.
139. P. H. Matter, L. Zhang, U. S. Ozkan, *Journal of Catalysis* 2006, *239*, 83–96.
140. S. Glenis, A. J. Nelson, M. M. Labes, *Journal of Applied Physics* 1996, *80*, 5404.
141. L. Li, E. Liu, Y. Yang, H. Shen, Z. Huang, X. Xiang, *Mater. Lett* 2010, *64*, 2115.
142. L. Zhao, L.-Z. Fan, M.-Q. Zhou, H. Guan, S. Qiao, M. Antonietti, M.-M. Titirici, *Advanced Materials* 2010, *22*, 5202–5206.
143. L. Zhao, N. Baccile, S. Gross, Y. Zhang, W. Wei, Y. Sun, M. Antonietti, M.-M. Titirici, *Carbon* 2010, *48*, 3778–3787.
144. H. Steinhart, *Angewandte Chemie International Edition* 2005, *44*, 7503.
145. L. Zhao, R. Crombez, F. P. Caballero, M. Antonietti, J. Texter, M.-M. Titirici, *Polymer* 2010, *51*, 4540–4546.

146. L. Zhao, Z. Bacsik, N. Hedin, W. Wei, Y. Sun, M. Antonietti, M.-M. Titirici, *Chemsuschem* 2010, *3*, 840–845.
147. S.-A. Wohlgemuth, R. J. White, M.-G. Willinger, M.-M. Titirici, M. Antonietti, *Green Chemistry* 2012, *14*, 1515–1523.
148. M. K. M. Seredych, T. J. Bandosz, *ChemSus Chem* 2011, *4*, 139–147.
149. T. X. Cui, R. Lv, Z. H. Huang, F. Y. Kang, K. L. Wang, D. H. Wu, *Nanoscale Res. Lett* 2011, *6*, 77.
150. C. H. Choi, S. H. Park, S. I. Woo, *Green Chemistry* 2011, *13*, 406.
151. S.-A. Wohlgemuth, F. Vilela, M.-M. Titirici, M. Antonietti, *Green Chemistry* 2012, *14*, 741–749.
152. Y.-S. Hu, R. Demir-Cakan, M. M. Titirici, J. O. Muller, R. Schlogl, M. Antonietti, J. Maier, *Angewandte Chemie International Edition* 2008, *47*, 1645–1649.
153. R. D.-C. Jelena Popovic, J. Tornow, M. Morcrette, D. S. Su, M. A. Robert Schlögl, M.-M. Titirici, *Small* 2011, *8*, 1127–1135.
154. X. M. Sun, Y. D. Li, *Advanced Materials* 2005, *17*, 2626–2630.
155. Q. Hai-Sheng, M. Antonietti, Y. Shu-Hong, *Advanced Functional Materials* 2007, *17*, 637–643.
156. H.-W. Liang, W.-J. Zhang, Y.-N. Ma, X. Cao, Q.-F. Guan, W.-P. Xu, S.-H. Yu, *Acs Nano* 2011, *5*, 8148–8161.
157. M. Soorholtz, R. J. White, M. M. Titirici, M. Antonietti, R. Palkovits, F. Schüth, *Chemical Communications* 2012, *49*, 240–242.
158. D. J. T. R. A. Periana, S. Gamble, H. Taube, T. Satoh, H. Fujii, *Science* 1998, *280*, 560–564.
159. X. J. C. S. H. Yu, L. L. Li, K. Li, B. Yu, M. Antonietti, H. Coelfen, *Advanced Materials* 2004, *16*, 1636.
160. P. Makowski, R. D. Cakan, M. Antonietti, F. Goettmann, M.-M. Titirici, *Chemical Communications* 2008, 999–1001.
161. L. Zhao, X. Chen, X. Wang, Y. Zhang, W. Wei, Y. Sun, M. Antonietti, M.-M. Titirici, *Advanced Materials* 2010, *22*, 3317–3321.
162. M.-M. Titirici, M. Antonietti, A. Thomas, *Chemistry of Materials* 2006, *18*, 3808–3812.
163. R. Demir-Cakan, Y.-S. Hu, M. Antonietti, J. Maier, M.-M. Titirici, *Chemistry of Materials* 2008, *20*, 1227–1229.
164. J. Lehmann, M. C. Rillig, J. Thies, C. A. Masiello, W. C. Hockaday, D. Crowley, *Soil Biology & Biochemistry* 2011, *43*, 1812–1836.
165. C. Kammann, S. Ratering, C. Eckhard, C. Mueller, *Journal of Environmental Quality* 2012, *41*, 1052–1066.
166. X. Cao, K. S. Ro, M. Chappell, Y. Li, J. Mao, *Energy & Fuels* 2010, *25*, 388–397.
167. A. B. Fuertes, M. C. Arbestain, M. Sevilla, J. A. Macia-Agullo, S. Fiol, R. Lopez, R. J. Smernik, W. P. Aitkenhead, F. Arce, F. Macias, *Aust J Soil Res*, 2010, *48*, 618–626.
168. C. Falco, F. Perez Caballero, F. Babonneau, C. Gervais, G. Laurent, M.-M. Titirici, N. Baccile, *Langmuir* 2011.
169. S. Steinbeiss, G. Gleixner, M. Antonietti, *Soil Biology & Biochemistry* 2009, *41*, 1301–1310.
170. M. C. Rillig, M. Wagner, M. Salem, P. M. Antunes, C. George, H.-G. Ramke, M.-M. Titirici, M. Antonietti, *Applied Soil Ecology* 2010, *45*, 238–242.
171. R. Demir-Cakan, N. Baccile, M. Antonietti, M.-M. Titirici, *Chemistry of Materials* 2009, *21*, 484–490.
172. Z. Chen, L. Ma, S. Li, J. Geng, Q. Song, J. Liu, C. Wang, H. Wang, J. Li, Z. Qin, S. Li, *Applied Surface Science* 2011, *257*, 8686–8691.
173. Z. Liu, F.-S. Zhang, J. Wu, *Fuel* 2010, *89*, 510–514.
174. M. Sevilla, A. B. Fuertes, *Energy & Environmental Science* 2011, *4*, 1765–1771.

175. M. Sevilla, A. B. Fuertes, R. Mokaya, *Energy & Environmental Science* 2011, *4*, 1400–1410.
176. L. Wei, M. Sevilla, A. B. Fuertes, R. Mokaya, G. Yushin, *Advanced Energy Materials* 2011, *1*, 356–361.
177. Q. Wang, H. Li, L. Chen, X. Huang, *Carbon* 2001, *39*, 2211–2214.
178. K. Tang, R. J. White, X. Mu, M.-M. Titirici, P. A. van Aken, J. Maier, *ChemSusChem* 2012, *5*, 400–403.
179. T. Kun, F. Lijun, R. J. White, Y. Linghui, M. M. Titirici, M. Antonietti, J. Maier, *Advanced Energy Materials* 2012, *2*, 873–877.
180. J. Popovic, R. Demir-Cakan, J. Tornow, M. Morcrette, D. S. Su, R. Schlögl, M. Antonietti, M.-M. Titirici, *Small* 2011, *7*, 1127–1135.
181. R. Yang, X. Qiu, H. Zhang, J. Li, W. Zhu, Z. Wang, X. Huang, L. Chen, *Carbon* 2005, *43*, 11–16.
182. P. Kim, J. Joo, W. Kim, J. Kim, I. Song, J. Yi, *Catalysis Letters* 2006, *112*, 213–218.
183. M. Sevilla, C. Sanchís, T. Valdés-Solís, E. Morallón, A. B. Fuertes, *Electrochimica Acta* 2009, *54*, 2234–2238.
184. J. B. Joo, Y. J. Kim, W. Kim, P. Kim, J. Yi, *Catalysis Communications* 2008, *10*, 267–271.
185. M. Sevilla, G. Lota, A. B. Fuertes, *Journal of Power Sources* 2007, *171*, 546–551.
186. J. N. Chheda, G. W. Huber, J. A. Dumesic, *Angewandte Chemie International Edition* 2007, *46*, 7164–7183.
187. J. I. J. A. Melero, G. Morales, *Green Chemistry* 2009, *11*, 1285–1308.
188. J. A. Macia-Agullo, M. Sevilla, M. A. Diez, A. B. Fuertes, *ChemSusChem* 2010, *3*, 1352–1354.
189. X. Huiquan, G. Yingxue, L. Xuezheng, Q. Chenze, *Journal of Solid State Chemistry* 2010, *183*, 1721–1725.
190. N. D. Berge, K. S. Ro, J. Mao, J. R. V. Flora, M. A. Chappell, S. Bae, *Environmental Science & Technology* 2011, *45*, 5696–5703.
191. (a) V. L. Budarin, J. H. Clark, R. Luque, D. J. Macquarrie, *Chemical Communications* 2007, 634–636; (b) R. Luque, L. Herrero-Davila, J. M. Campelo, J. H. Clark, J. M. Hidalgo, D. Luna, J. M. Marinas, A. A. Romero, *Energy & Environmental Science* 2008, *1*, 542–564; (c) M. J. Gronnow, R. Luque, D. J. Macquarrie, J. H. Clark, *Green Chemistry* 2005, *7*, 552–557.
192. G. Yu, B. Sun, Y. Pei, S. Xie, S. Yan, M. Qiao, K. Fan, X. Zhang, B. Zong, *Journal of the American Chemical Society* 2010, *132*, 935–937.
193. Z. Zhang, H. Duan, S. Li, Y. Lin, *Langmuir* 2010, *26*, 6676–6680.
194. G. Tian, Z. J. Gu, X. X. Liu, L. J. Zhou, W. Y. Yin, L. Yan, S. Jin, W. L. Ren, G. M. Xing, S. J. Li, Y. L. Zhao, *Journal of Physical Chemistry C* 2011, *115*, 23790–23796.
195. X.-L. Li, T.-J. Lou, X.-M. Sun, Y.-D. Li, *Inorganic Chemistry* 2004, *43*, 5442–5449.
196. S. R. Guo, J. Y. Gong, P. Jiang, M. Wu, Y. Lu, S. H. Yu, *Advanced Functional Materials* 2008, *18*, 872–879.
197. B. R. Selvi, D. Jagadeesan, B. S. Suma, G. Nagashankar, M. Arif, K. Balasubramanyam, M. Eswaramoorthy, T. K. Kundu, *Nano Letters* 2008, *8*, 3182–3188.

3 Carbon Materials and Their Energy Conversion and Storage Applications

Ji Liang, Ruifeng Zhou, Denisa Hulicova-Jurcakova, and Shi Zhang Qiao

CONTENTS

Carbon, one of the oldest elements recognized and utilized by human civilizations, is also one of the most abundant and versatile materials. It has been used since the prehistoric era as a charcoal for writing and has been intensively studied for hundreds of years. Nevertheless, carbon is still under intensive study because of its unique properties and extensive applications.

Most sources of carbon are abundant, therefore cheap. It is relatively easy to obtain activated carbon or carbon black from natural resources such as biomass and fossils. Diamond and graphite can also be gathered from nature or prepared by synthetic routes. In the last 30 years, the discovery of nanosized fullerenes, mesoporous carbons, carbon nanotubes, and graphenes has brought the carbon research to a new climax.

Thanks to the unique versatile features of carbon, it has been applied indispensably in almost every aspect of our daily life, industry, and research. In this chapter, we will focus on the various types of carbon materials (including conventional carbon blacks, activated carbons, and graphite as well as novel carbon materials such as mesoporous carbons and carbon nanomaterials) and discuss their energy-related applications including fuel cells, solar cells, photocatalysis, and lithium ion batteries as well as supercapacitors.

3.1 CARBONS APPLIED IN FUEL CELLS

Unlike conventional thermal power plants, which convert chemical energy to heat, mechanical energy, and finally electricity, fuel cells are devices that directly convert the chemical energy stored in fuels into the electricity by the redox reactions at both the anode and cathode, where the fuel and oxygen are oxidized and reduced, respectively. From this aspect, fuel cells are very promising for the next generation of portable power sources, due to the wide availability and high energy density of the fuels, as well as the high-energy conversion efficiency. Fuel cells can be classified into two categories, the low-temperature and high-temperature fuel cells. The low temperature ones include proton exchange membrane fuel cells (PEMFCs), alkaline fuel cell and phosphoric acid fuel cells (PAFCs), etc. High-temperature fuel cells such as solid oxide fuel cells (SOFCs) and molten carbonate fuel cells (MCFCs), which operate at temperatures greater than 600°C, utilize hydrogen, natural gas, hydrocarbon, or coals as fuels.[1] Among these fuel cells, the low temperature ones are promising for the future portable power source solution due to their mild working environment, high safety, and moderate mobility. On the other hand, high-temperature fuel cells have the potential to serve as high output power sources in power plants and industry.

In case of low-temperature fuel cells, fuel and oxygen can hardly be oxidized or reduced at low temperatures (from room temperature to about 100°C) without the presence of catalyst. The most commonly used catalysts are noble metals such as platinum (Pt), ruthenium (Ru), palladium (Pd) or their alloys. To better utilize these expensive and low reserved metals, they are usually synthesized as nanoparticles and loaded on various sorts of supports to avoid the possible aggregation or merging during the fuel cell operation. Carbon materials (carbon black or graphitic carbons) have been most widely used as the catalyst support since the 1970s. Recently, novel carbons including carbon nanomaterials and mesoporous carbons are drawing special attention due to their unique properties. In this section, a diverse range of carbons will be introduced as supports for catalyst metal particles for applications in fuel cells.

3.1.1 Conventional Carbon Blacks and Graphitic Carbons

Catalyst support for a fuel cell must satisfy several criteria, including a high specific surface area for sufficient catalyst loading and dispersion, appropriate pore size for smooth mass transfer of reactants and products, high electrical conductivity, as well as enough stability against electrochemical corrosion in the extremely harsh fuel cell environments. Among the three main types of catalyst supports including metallic materials, polymeric materials, and nonmetallic inorganic materials, carbon or carbon-based materials are the most favorable and promising candidates, as they meet most of the above mentioned criteria at an acceptably low cost.[2] Indeed since the first development of fuel cells in the early 1970s until the late 1990s, carbon blacks have been almost exclusively used as supports for the metallic nanoparticle catalysts.

Carbon black can be obtained by the thermal decomposing of hydrocarbon precursors (such as natural gas) at elevated temperatures in inert atmosphere or partial combustion in inadequate air supply. Up to date, several techniques have been developed to produce these carbon materials including furnace process, thermal process, and channel process.[3] Different types of carbon blacks have been developed so far for fuel cells as summarized in Table 3.1.

To this day, carbon blacks have been the most widely applied and most successfully commercialized catalyst supports for fuel cells due to their low price and relatively good mechanical/chemical properties. However, main drawbacks of carbon black should not be neglected such as its vulnerability to the electrochemical corrosion in a fuel cell environment, which is very harsh to facile the redox reactions. This carbon corrosion has been further worsened by the highly active metallic nanoparticles, which not only catalyze the oxygen reduction reactions (ORR) but also simultaneously promote the carbon oxidation.[12,13] In addition, the acidic/basic or even the neutral aqueous/vaporous conditions of a fuel cell can cause the carbon corrosion (oxidation) occurring at lower potentials than the theoretical 0.2 V versus reversible hydrogen electrode.[14] Furthermore, the elevated temperature/operating potential, which is necessary to obtain higher cell efficiency, can also enhance the electrochemical corrosion.[15–17]

TABLE 3.1
Physical Property of Several Commercially Available Carbon Blacks for Fuel Cells

Carbon Blacks	Supplier	Synthesis Route	Surface Area (m^2/g)	Particle Size (nm)
Vulcan XC-72	Cabot	Furnace black	250–260	20–50[4]
Black Pearl 2000	Cabot	Furnace black	1475–1500	15[5,6]
Denka Black	Denka	Acetylene black	58–65	40[5,6]
Shawinigan Black	Chevron	Acetylene black	70–90	40–50[6,7]
Conductex 975 Ultra	Columbian	Furnace black	250	24[5]
3250/3750/3950	Mitsubishi	—	240/800/1500	28/28/16[8]
Ketjen EC-300 J	Akzo Nobel	Furnace black	800[6]	30[9]
Ketjen EC-600 JD	Akzo Nobel	Furnace black	1270	30[10a]

[a] Ketjen EC-600 JD from Cabot Corp. has an average particle size of 34 nm.[11]

To overcome the corrosion issues of carbon blacks, highly crystallized, or in other words graphitized carbons have been sought and synthesized as catalyst supports due to their higher resistance against electrochemical corrosion.[12] The graphitization can be introduced and improved in amorphous carbons by heating in a high vacuum or an inert atmosphere at temperatures above 2000°C. Commercial carbon blacks have been treated in this way to achieve a higher graphitization degree.[18,19] The significant decrease of I_D/I_G ratio in the Raman spectra of Vulcan XC-72 after the heat treatment shown in Figure 3.1a has clearly shown the increment of its graphitization degree.

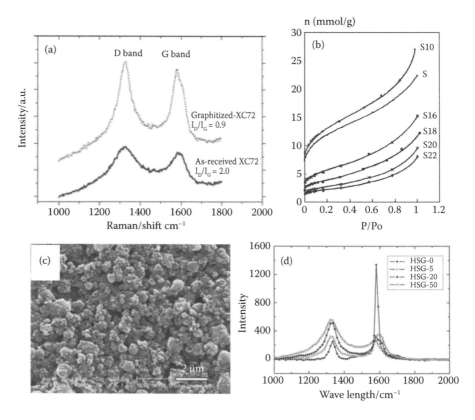

FIGURE 3.1 Feature changes of carbon black after different treatments. (a) Decrease of I_D/I_G of Vulcan XC-72 after treatment at 2200°C for 1 h in Ar. (From C.-C. Hung, P.-Y. Lim, J.-R. Chen and H. C. Shih, *Journal of Power Sources*, 2011, 196, 140–146.) (b) Changes of N_2 adsorption behaviors of carbon blacks treated at different temperatures (S10, 16, 18, 20, and 22 refers to 1000°C, 1600°C, 1800°C, 2000°C, and 2200°C, respectively, in He for 1 h) compared with the original sample (S). (From F. Coloma, A. Sepulvedaescribano and F. Rodriguezreinoso, *Journal of Catalysis*, 1995, 154, 299–305.) (c) Morphology of 50-h ball-milled natural graphite (a), and (d) Raman spectra of graphite ball-milled for different periods of time (HSG-0–HSG-50 represents samples grinded for 0–50 h). (From H.-Q. Li, Y.-G. Wang, C.-X. Wang and Y.-Y. Xia, *Journal of Power Sources*, 2008, 185, 1557–1562.)

Although the heat treatment improves the crystallinity, the rearrangement of carbon atoms in this process would also cause the deterioration of the microstructure, such as collapse or closing of the pores, which results in the decrease of the total surface area up to 60% (Figure 3.1b). The issue of maintaining high surface area while improving the crystallinity might be achieved in two ways. One way to avoid the shrinkage of pores at high-temperature treatments is to enlarge the pore size of the carbon before the heat treatment. This way, although the pores shrink during the heating process, they do not disappear and the surface area can thus be partially retained.[21,22]

Another approach for high-crystallinity carbon is the direct ball-milling of graphite into fine particles to obtain a high surface area.[20] Figure 7.1c shows micromorphology of the graphite particles that have been ball-milled for 1 h. The BET surface area increased from 7 m^2/g of the raw graphite to 580 m^2/g of graphite particles treated for up to 50 h. However, this harsh treatment reduces the crystallinity of graphite as evidenced by increasing ratios of I_D/I_G for ball-milled graphite compared with the original counterpart (Figure 3.1d).

3.1.2 NOVEL CARBON MATERIALS FOR FUEL CELL APPLICATIONS

Despite the moderate performance of carbon blacks in fuel cells, they suffer from a poorly defined and uncontrollable micro structure. Taking the most commonly used XC-72 as an example, macropores or mesopores only contribute to half of the total pore volume. The rest is from micropores less than 1 nm wide (Figure 3.2),[23] which has been believed to contribute little to the fuel cell performance because of their inaccessibility to the reactants.[24] In this regard, novel porous carbon materials have been developed to obtain more suitable and better defined pore structures for applications in fuel cells. At the same time carbon nanomaterials have been studied for this application due to their unique properties caused by their nanostructures.

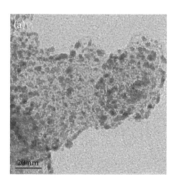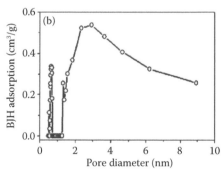

FIGURE 3.2   (a) Transmission electron micrograph of Vulcan XC-72 loaded with metallic Pt particles. (From M. Kim, J.-N. Park, H. Kim, S. Song and W.-H. Lee, *Journal of Power Sources*, 2006, 163, 93–97.) (b) Pore distribution of Vulcan XC-72 calculated by Barrett-Joyner-Halenda (BJH) method from nitrogen sorption isotherm. (From V. Raghuveer and A. Manthiram, *Electrochemical and Solid-State Letters*, 2004, 7, A336–A339.)

3.1.2.1 Mesoporous Carbons

Mesoporous carbon with the pore size between 2 to 50 nm is a promising candidate for the application in fuel cells as a catalyst support due to more suitable porosity than the microporous carbon black. Generally, mesoporous carbons can be divided into two groups according to their pore structures: (a) mesoporous carbons with highly ordered and uniform-sized pores and (b) disordered mesoporous carbons with irregular pores.

Mesoporous carbons can be prepared by either a hard template or soft template method. In the hard template fabrication, the synthesis procedure includes casting of a porous template with a carbon precursor such as sucrose, followed by high temperature treatment in an inert atmosphere or vacuum, and the final removal of the template. The first ordered mesoporous carbon (OMC) named CMK-1 was successfully obtained by nanocasting of mesoporous silica MCM-48 with sucrose.[28] Following this approach, carbons with larger pores, optimized for fuel cell applications, have been synthesized using the two-dimensional hexagonally ordered mesoporous silica SBA-15 as the template and sucrose as the carbon precursor (Figure 3.3a,b).[26,29]

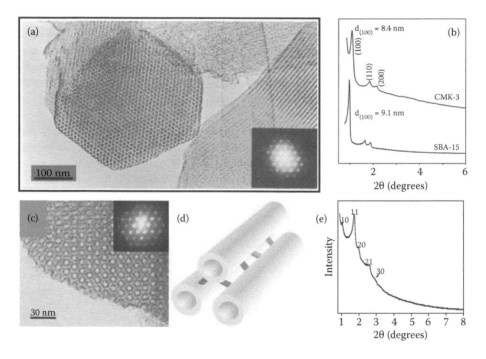

FIGURE 3.3 (a) TEM image of CMK-3 and corresponding electron diffraction pattern. (b) Small angle X-ray diffraction patterns of CMK-3 and parent template SBA-15, TEM image of CMK-3. (From S. Jun, S. H. Joo, R. Ryoo, M. Kruk, M. Jaroniec, Z. Liu, T. Ohsuna and O. Terasaki, *Journal of the American Chemical Society*, 2000, 122, 10712–10713.) (c–e) TEM image and electron diffraction pattern of the tubular CMK-5, the schematic model of CMK-5, and its small angle X-ray diffraction pattern confirming the hexagonal ordered arrangement of tubular carbon nanorods. (From S. H. Joo, S. J. Choi, I. Oh, J. Kwak, Z. Liu, O. Terasaki and R. Ryoo, *Nature*, 2001, 412, 169–172.)

Other forms of ordered mesoporous carbons have been successfully obtained from hard templates with both the two-dimensional[27,29] and the three-dimensional ordered pore structures.[30,31]

In the soft template synthesis route, organic–organic nanocomposite is employed to obtain OMCs with similar structure as those prepared by the hard template method, but without using the sacrificial silica as the hard template.[32] In this method, organic self-assembly between two or more sorts of carbon precursors in a liquid system has been employed to form potential carbon framework, and surfactant is commonly employed to introduce ordered pore structures. One example is the *Ia3d* structured OMC (denoted as C-FDU-14) prepared by this method.[33]

Mesoporous carbon with disordered pores can also be easily prepared using various carbon precursors and commercial silica colloidals with different diameters as hard templates, which makes it an easy way to obtain mesoporous carbon with a wide range of pore structures.[34–36] The carbon aerogel, another type of disordered mesoporous carbon, can be prepared by carbonization of organic aerogels, which are synthesized by sol–gel polycondensation of organic monomers.[37] These carbon aerogels vary from both microscopic structures and macroscopic forms.

Mesoporous carbon composed of hexagonally packed tubular carbon nanorods (CMK-5) was firstly selected as the catalyst support material for the Pt nanoparticles (Figure 3.3c, d and e).[27] Using this support, much finer Pt particles (<10 nm) could be obtained compared with the ones loaded on commercial carbon black (~30 nm). The mass activity of this Pt/CMK-5 composite catalyst for electrochemical reduction of oxygen was ~100 A/g, which was significantly higher than that obtained on Pt/carbon black (~5–10 A/g Pt). The advantage of the electrochemical activity on Pt/CMK-5 was attributed to the small and uniform Pt particles loaded on the mesoporous carbon as well as the unique structure of the support, which could withstand high loadings of the Pt catalyst.

CMK-3 has also been selected as a support for the metallic nanoparticles using post impregnation in CMK-3 or in-situ co-heating of the carbon and Pt precursors in the channels of SBA-15 during the preparation process, which acted as a nanoreactor.[38] In the later in-situ process, ultrafine Pt particles with diameters between 1 and 5 nm were obtained inside the nanochannels of SBA-15 even though the heating temperature was as high as 900°C during the carbonization of the sucrose carbon precursor. On the other hand, the former impregnation procedure led to the particle size as large as 50 nm at the same Pt loading. In the direct methanol fuel cell performance test of both the in situ prepared Pt/CMK-3 and electrochemical Pt/carbon black catalyst, a fuel cell using Pt/CMK-3 catalyst showed both higher open circuit potential and higher single cell output current density than the commercial Pt/carbon black catalyst. These characters indicated a higher methanol tolerance and reaction activity of the ultrafine Pt nanoclusters loaded on the high surface area mesoporous carbon. The extraordinary performance on the Pt/CMK-3 was attributed to more exposed (100) and (111) faces on the fine Pt particles, which are more tolerant to methanol and active for oxygen reduction reaction (ORR).

Apart from Pt loaded on CMK-3 or CMK-5, various other metal particles have been deposited on ordered mesoporous carbons as well as disordered mesoporous carbons, including Pt or Pt/X (X can be Pd, Ru, etc.), etc.[39,40] For the metal particles, the size is well controlled between 3 and 5 nm and their good dispersion is related to the well-developed porous structure of carbon support.

In most of the recent studies, metal nanoparticles loaded on mesoporous carbons generally show better catalytic performance for fuel oxidation and oxygen reduction in the fuel cell than the same metal catalyst loaded on commercial carbon blacks. The improvement in catalytic activity can be attributed to two aspects: (1) the smaller metal particle sizes loaded on high surface areas of mesoporous carbons; (2) regular and uniform pores on carbon networks. The former aspect can provide sufficient reaction active sites for the electrode reactions while the latter can make the mass transfer easier inside the channels. These features are favored for both metal dispersion and fuel/oxidant transfer.

On the other hand, a report has indicated a declined activity on catalysts loaded on mesoporous carbons.[39] Pt-Ru/CMK-3 was prepared by absorbing the metal particles dispersed in solvent. The metal particles were small (0.8–2.6 nm) and well dispersed on the carbon support but showed inferior performance compared with commercially available 20 wt% Pt-Ru/XC-72. The author ascribed this to the lower metal loading on the mesoporous carbon as well as the better Pt/Ru ratio on the commercial catalyst.

3.1.2.2 Carbon Nanomaterials

As aforementioned, one of the reasons that lead to the carbon corrosion in the fuel cells is low graphitic degree of commercial carbon blacks. This problem also exists in mesoporous carbons discussed above. To fabricate graphite with high surface area, ball milling of bulky graphite or high-temperature treatment of amorphous carbon has been performed. However, this procedure led to the product with decreased crystalline degree or uncontrollable microstructures. The appearance of the carbon nanomaterials (Figure 3.4) with both high crystallinity and high surface area has shed light on the potential application as a catalyst support.

Carbon nanotubes (CNTs), composed of one or more layers of rolled-up graphene, have attracted much attention since their discovery in 1991,[42] and they have been proven to be ideal candidates for numerous applications including fuel cells. According to the number of graphene walls, they can be classified into two groups: single-walled carbon nanotubes (SWCNTs), which are composed of a single graphene layer, and multi-walled carbon nanotubes (MWCNTs), which are composed of several coaxial graphene layers with interwall distance of approximately 0.34 nm.[43] The highly crystal nature of CNTs gives them both the excellent electrical and mechanical properties as well as high resistance to chemical corrosion.[44,45] However, in the catalyst loading process, the defect-free surface of CNTs is a disadvantage since the defects on carbon surface can act as the anchoring sites for the metal particles. To resolve this issue, it is necessary to modify the CNT surface and to produce suitable amount of active sites for the metal particle anchoring. In most cases, CNTs are treated with different oxidants or their mixture at elevated temperature to graft functional groups on their outer walls. In a typical process, CNTs are refluxed at 100°C or higher in concentrated sulfuric acid, nitric acid, or their mixtures,[3] which results in surface functionalization by –OH; –COOH and –CO groups with a ratio of approximately 4:2:1.[46] In some cases, more gentle treatment is adopted, such as reflux in diluted nitric acid, and this yields predominantly –OH

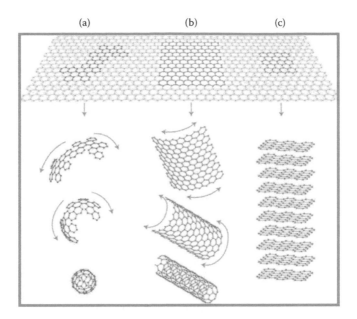

FIGURE 3.4   (See color insert.) Schematic formation of carbon nanomaterials: (a) fullerene (zero-dimensional), (b) carbon nanotubes (one-dimensional), and (c) graphene (two-dimensional). (From A. K. Geim and K. S. Novoselov, *Nature Materials*, 2007, 6, 183–191.)

surface functional groups on CNTs.[47] The surface functionalization of CNTs generally leads to smaller catalyst particle size (3–5 nm) and good dispersion when the loading amount is up to 30 wt%. This subsequently results in better electrode reaction catalytic performance compared with the catalyst supported on untreated CNTs.[48,49]

Apart from the two-step process in which CNTs surface functionalization is followed by catalyst loading, the surface of CNTs can also be modified at the same time when Pt is deposited.[50] Using glacial acetic acid as the deposition agent, surface functional groups were created on the surface of CNTs grown on carbon fibers while the reduction of Pt precursor and Pt metal particles deposition was happening concurrently. Through this additional CNTs surface modification, the Pt particles could be well controlled between 2 and 4 nm and were all monodispersed on the CNTs. Thus prepared Pt/CNT catalyst was tested for methanol oxidation as well as on a single cell, which showed 66% higher current density than the Pt/carbon black catalyst. These advantages in the electrochemical properties of this Pt/CNT catalyst could be attributed to the improved Pt dispersion on the catalyst support as well as the three-dimensional structure of the CNTs grown on the carbon fibers.

Electrochemical deposition of Pt particles on CNTs has been reported as well. Pt was electrochemically deposited on the surface of CNTs in a setup where CNTs were used as a working electrode with the Pt precursor as the electrolyte, under the potentiostatic conditions (−0.25 V versus SCE) as well as potentiodynamic conditions (stepped between 300 and 700 mV versus SCE).[51,52] The size of thus deposited Pt particles ranged from 1 to 3 nm on 4-aminobenzene modified MWNT and from 60 to 80 nm on the pristine MWCNTs.

Another deposition process involved the ultrasonic treatment of the mixture containing sulfuric acid/nitric acid/CNTs, to draft more surface functional groups compared with the CNTs functionalized in a mechanically stirred liquid phase.[53] The CNTs modified by this way were then deposited with Pt through a conventional Pt reduction process and Pt particles were very uniformly dispersed on the CNT surface. Excellent dispersion was attributed to the even distribution of the functional groups on CNT surface. The electrochemical test of this Pt/CNTs catalyst showed a hydrogen adsorption/desorption amount two times as high as that on commercial Pt/carbon black, indicating more electrochemical active sites on the Pt/CNTs catalyst and its potential to be applied in a PEMFC.

Sputtering technique can also be used to prepare Pt particles on CNTs.[54] A short time (30 s) and low current (10 mA) sputtering process can produce Pt particles loaded on CNTs with a much smaller size (2 nm) compared with the ones prepared through a chemical reduction method (2–5 nm).

Graphene is another carbon nanomaterial composed of one or several layers of hexagonally close packed carbon atoms. The flexible two-dimensional nature of graphene as well as its superior mechanical/electrical properties makes it a promising candidate as an electrochemical catalyst support. In most cases, graphene is prepared in the oxidized form (GO), which already presents surface functional groups for the metal nanoparticles anchoring. Using co-reduction of Pt precursor and GO with sodium tetrahydroborate as the reductant in sodium hydroxide aqueous solution, Pt was simultaneously deposited on the surface of reduced graphene.[55] As-synthesized Pt/graphene possessed an electrochemically active surface area of 44.6 m^2/g compared with 30.1 m^2/g of commercial Pt/carbon black. A half-cell using this catalyst also showed twice higher the reaction current toward methanol electrochemical oxidation than the commercial Pt/carbon black catalyst.

Apart from in situ reduction, graphene can also be reduced and functionalized before Pt deposition similar to CNTs. As an example, GO was reduced and expanded using a rapid thermal treatment.[56,57] This thermally treated graphene was then impregnated with chloroplatinic acid followed by heat treatment at 300°C in hydrogen to reduce the $PtCl_6^{2-}$ into metallic Pt nanoparticles. The electrochemically active surface area and catalytic performance toward ORR of thus-prepared catalysts were significantly improved in comparison with the Pt/carbon black. The performance enhancement obtained from the Pt/graphene catalyst was attributed to the small and stable Pt particles anchored on the thermally treated graphene surface.

Further surface decoration of graphene has also been conducted by using poly(diallyldimethylammonium chloride) (PDDA) as a functionalization agent.[58] PDDA is a positively charged polymer with a long-chain structure and can be permanently bonded on the graphitic surfaces through the π–π interaction.[59] The PDDA decorated graphene with a positively charged surface provided more stable anchoring sites for the negatively charged $PtCl_6^{2-}$ through the electrostatic forces. After the reduction of $PtCl_6^{2-}$ such catalyst showed similar ORR performance as the Pt/CNT or Pt/carbon black, but 2–3 times higher stability. The improved stability was attributed to (1) stronger forces between the metal nanoparticles and the graphene surface from the π–π interaction and (2) defect-less graphene surface, which is more chemical resistant compared with the MWCNT or carbon black surface.

Apart from Pt nanoparticles, other metals with a variety of shapes/structures loaded on carbon nanomaterials have been studied as the fuel cell catalyst in recent years, including Pt clusters,[60] Pd nanoparticles,[61] Pt-Ru alloy nanoparticles,[62] and Pt-Pd alloy nanoparticles.[63]

3.2 PHOTOELECTRICAL CELL

Harvesting energy from sunlight has been the dream of human beings for a long time. In fact, we have already been enjoying heat and food, whose energy origi-nates from the sun. Moreover, energy of hydropower and fossil fuel also comes from the sun. Nevertheless, direct conversion from solar energy into electric or chemical energy with high efficiency and low cost is still very demanding for the development of human beings nowadays, as it is unlimited, everywhere, and clean. Solar energy harvesting is especially important in mobile and portable electrical equipment that is not connected to power grids, e.g., vehicle, satellite, mobile phone, etc.

3.2.1 DYE-SENSITIZED SOLAR CELL

First dye-sensitized solar cell (DSSC) with acceptable efficiency was proposed by Grätzel in 1991.[64] Although the DSSC converts light into electricity with a lower efficiency than an Si based solar cell, it has a lot of advantages, the most important one its low cost. The mechanism of DSSC is illustrated in Figure 3.5.[65]

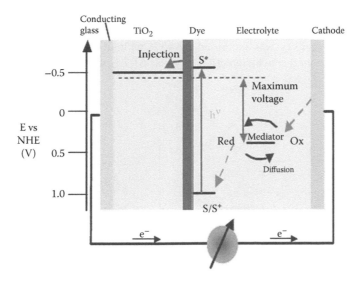

FIGURE 3.5 Principle of operation and energy level scheme of the DSSC. Potentials are referred to the normal hydrogen electrode (NHE). The open-circuit voltage of the solar cell corresponds to the difference between the redox potential of the mediator and the Fermi level of the nanocrystalline film indicated with a dashed line. (From M. Gratzel, *Journal of Photochemistry and Photobiology C: Photochemistry Reviews*, 2003, 4, 145–153.)

Briefly, ground-state electrons in a dye molecule are excited by absorbing energy from incident photons. The excited electrons then inject into TiO_2 (n-type semi-conductor) on an anode to make it negatively charged. On the other hand, the dye molecules grab electrons to neutralize themselves by oxidizing mediators in electro-lyte, which then diffuse to cathode to grab electrons to make it positively charged. Therefore, a voltage between the anode and the cathode is built.

Crucial components that decide the efficiency in a DSSC are photoanode (TiO_2 and dye) electrolyte and cathode (counter electrode). Carbon materials can play roles in DSSC in at least 3 places: additive in TiO_2 photoanode, catalyst in cathode, and transparent electrode.

TiO_2 is a broad band-gap semiconductor ($E_g = 3.2eV$) with a low conductivity. Low conductivity in an anode increases recombination of electron–hole pairs and therefore lowers the efficiency. Many efforts have been made to dope TiO_2 with carbon additives to improve its conductivity, as well as electron–hole separation. Kim[66] studied the effect of incorporating SWNTs in TiO_2 film electrodes on the properties of DSSCs. Compared with an unmodified solar cell, it is observed that the short-circuit photocurrent (J_{sc}) of the modified solar cells increases while the open-circuit photovoltage decreases slightly. The enhanced J_{sc} is correlated with increased electrical conductivity, light scattering, and concentration of free conduction band electrons. Brown[67] showed that CNT plays an important role in improving the charge separation, as the rate of back electron transfer between the oxidized sensitizer (Ru(III)) and the injected electrons becomes slower in the presence of the SWCNT scaffold (Figure 3.6). Other studies have shown a similar improving effect of CNT modified TiO_2 photoanode.[68–72]

Zhai[73] introduced graphene as 2D bridges into the nanocrystalline electrodes of DSSCs, which brought a faster electron transport and a lower recombination, together with a higher light scattering. On the basis of these advantages, the short-circuit current density was increased by 45% without sacrificing the open-circuit voltage, and the total conversion efficiency was 6.97%, which was increased by 39%,

(a)   (b)

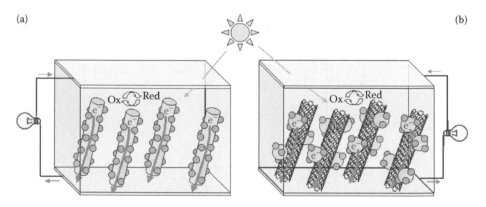

FIGURE 3.6   (See color insert.) Directing the flow of photogenerated electrons across nanostructured semiconductor films: (a) nanotube/nanowires modified with light-absorbing dye molecules and (b) nanotubes as support architecture for anchoring dye modified semiconductor nanoparticles. (From P. Brown, K. Takechi and P. V. Kamat, *Journal of Physical Chemistry C*, 2008, 112, 4776–4782.)

comparing with the nanocrystalline TiO_2 photoanode, and it was also much better than the 1D nanomaterial composite electrode (Figure 3.7). A few references have reported this kind of modification.[74–76]

In a conventional DSSC, the counter electrode is Pt sputtered on conductive glass, where Pt acts as the catalyst for mediator reduction. However, Pt is a noble metal, which contradicts with the main advantage of DSSC, the cheapness. Therefore, many efforts have been made to replace Pt with cheaper carbon materials. Takahashi[77] reported that a new carbon electrode prepared with active carbon was superior to a Pt sputtered electrode as the counter electrode of DSSCs. The photovoltaic performance was largely influenced by the roughness factor of carbon electrode. The open-circuit voltage increased by about 60 mV using the carbon counter electrode compared with the Pt counter electrode because of positive shift of the formal potential for I_3^-/I^- couple (Figure 3.8). CNT,[78–82] graphene,[83,84] and other carbon materials[85–88] are also capable for such tasks. Carbon materials are also the support of other catalysts, such as PEDOT-PSS,[89] TiN,[90] and TiO_2.[55]

The transparent conductive anode is also a very important component in DSSC. An eligible anode should have high enough transparency and conductivity to ensure high efficiency. Conventional anodes are conductive glass (CG) coated with indium tin oxide (ITO) or fluorine doped tin oxide (FTO) film. Disadvantages of the CG include high cost and brittleness. On the other hand, CNT[91] and graphene[92] have successfully been developed as transparent conductive film with much better flexibility than CG. Zhi and Müllen[93] demonstrated the transparent, conductive, and ultrathin graphene films, as an alternative to the ubiquitously employed metal oxides window electrodes for solid-state dye-sensitized solar cells. These graphene films are fabricated from exfoliated graphite oxide, followed by thermal reduction. The obtained films exhibit a high conductivity of 550 S/cm and a transparency of more than 70% over 1000–3000 nm. Furthermore, they show high chemical and thermal stabilities as well as an ultrasmooth surface with tunable wettability (Figure 3.9).

An interesting result was shown by Kim.[94] A mixture of graphene oxide (GO) and TiO_2 nanocomposites was reduced photocatalytically by UV irradiation and

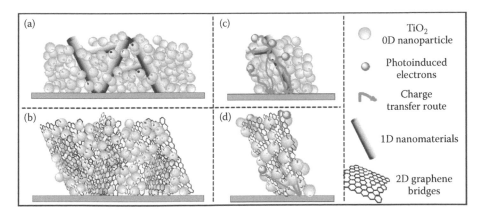

FIGURE 3.7 (See color insert.) Differences between (a, c) 1D and (b, d) 2D nanomaterial composite electrodes. The transfer barrier is larger and the recombination is much easier to happen. (From N. L. Yang, J. Zhai, D. Wang, Y. S. Chen and L. Jiang, *ACS Nano*, 2010, 4, 887–894.)

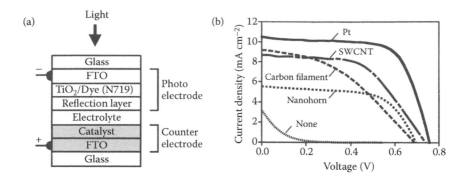

FIGURE 3.8 (a) Schematic diagram of DSSC with carbon catalyst spread on FTO substrate. (b) Photocurrent density–voltage characteristics of the cells under 1.5 AM, 100 mW/cm^2 illumination. (From K. Imoto, K. Takahashi, T. Yamaguchi, T. Komura, J. Nakamura and K. Murata, *Solar Energy Materials and Solar Cells*, 2003, 79, 459–469.)

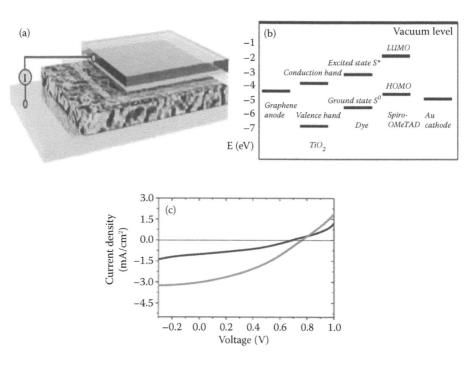

FIGURE 3.9 **(See color insert.)** Illustration and performance of solar cell based on graphene electrodes. (a) Illustration of DSSC using graphene film as electrode, the four layers from bottom to top are Au, dye-sensitized heterojunction, compact TiO$_2$, and graphene film. (b) The energy-level diagram of graphene/TiO$_2$/dye/spiro-OMeTAD/Au device. (c) *I–V* curve of graphene-based cell (black) and the FTO-based cell (red), illuminated under AM solar light (1 sun). (From X. Wang, L. J. Zhi and K. Mullen, *Nano Letters*, 2008, 8, 323–327.)

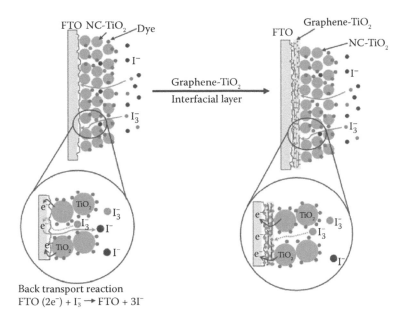

Back transport reaction
FTO $(2e^-) + I_3^- \rightarrow$ FTO $+ 3I^-$

FIGURE 3.10 **(See color insert.)** Schematic representation and mechanism of applied graphene–TiO$_2$ interfacial layer to prevent back-transport reaction of electrons. (From S. R. Kim, M. K. Parvez and M. Chhowalla, *Chemical Physics Letters*, 2009, 483, 124–127.)

applied as an interfacial layer between a FTO layer and a nanocrystalline TiO$_2$ film. Impedance spectra implied a decreased back-transport reaction of electrons. The grapheme–TiO$_2$ interfacial layer effectively reduced the contact between I$_3^-$ ions in the electrolyte and FTO layer, which inhibited back-transport reaction. The introduction of graphene-TiO$_2$ increased V$_{oc}$ by 54 mV and the photoconversion efficiency was improved from 4.89% to 5.26% (Figure 3.10).

3.2.2 WATER-PHOTOLYSIS CELL

Water photolysis, or photoelectrochemical water splitting, is another route to harvest solar energy. A water photolysis cell (WPC) converts solar energy into chemical energy, which is stored in produced H$_2$. The semiconductor in WPC experiences excitation under illumination to give an electron–hole pairs, as dye in DSSC, but the electrons are finally given to water on cathode to produce H$_2$. The holes grab electrons from water to produce O$_2$ on the surface of the semiconductor anode. Carbon has a similar effect as in DSSC of easing the electron–hole separation, which enhances the efficiency of WPC.

A schematic stretch of WPC can be found in Amal's research.[95] BiVO$_4$ is incorporated with reduced graphene oxide (RGO) using a facile single-step photocatalytic reaction to improve its photoresponse in visible light. A remarkable 10-fold enhancement in photoelectrochemical water splitting reaction is observed on BiVO$_4$-RGO composite compared with pure BiVO$_4$ under visible illumination. This improvement is attributed to the longer electron lifetime of excited BiVO$_4$ as the electrons are injected to RGO instantly at the site of generation, leading to a minimized charge recombination.

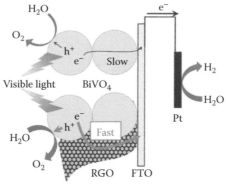

*Reduced graphene oxide (RGO) promotes
electron migration from BiVO₄ to FTO.*

FIGURE 3.11   (See color insert.) Schematic representation and mechanism of applied graphene–BiVO₄ photolysis cell. (From Y. H. Ng, A. Iwase, A. Kudo and R. Amal, *Journal of Physical Chemistry Letters*, 2010, 1, 2607–2612.)

Improved contact between BiVO₄ particles with transparent conducting electrode using RGO scaffold also contributes to this photoresponse enhancement (Figure 3.11).

In Yu's study,[96] a high efficiency of the photocatalytic H₂ production was achieved using graphene nanosheets decorated with CdS clusters as visible-light-driven photocatalysts. These nanosized composites reach a high H₂-production rate of 1.12 mmol h⁻¹ (about 4.87 times higher than that of pure CdS nanoparticles) at graphene content of 1.0 wt% and Pt 0.5 wt% under visible-light irradiation and an apparent quantum efficiency (QE) of 22.5% at wavelength of 420 nm. This high photocatalytic H₂-production activity is attributed predominantly to the presence of graphene, which serves as an electron collector and transporter to efficiently lengthen the lifetime of the photogenerated charge carriers from CdS nanoparticles (Figure 3.12).

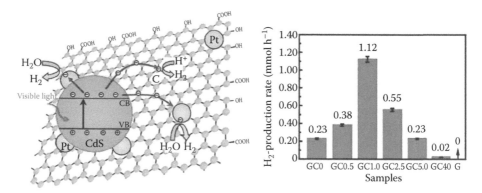

FIGURE 3.12   (See color insert.) Schematic illustration of the charge separation and transfer in the graphene–CdS system under visible light. (From Q. Li, B. D. Guo, J. G. Yu, J. R. Ran, B. H. Zhang, H. J. Yan and J. R. Gong, *Journal of the American Chemical Society*, 2011, 133, 10878–10884.)

3.3 CARBON MATERIALS FOR LITHIUM-ION BATTERIES

The ever-increasing demand for a mobile and heavy-duty energy-storage medium with acceptable cost and volume has attracted broad attention from chemical and material scientists. Due to their relatively high power density and stable charge–discharge properties, lithium ion batteries (LIBs) have been selected as the power sources for most portable electronic devices such as laptops and smart phones.[97]

When discharging, LIB works on the principle of the electrochemical intercalation of Li ion into the cathode and deintercalation of Li ion from the anode as shown in Figure 3.13a. In most of the commercially available single cells, LIBs are composed of anode material that can reversibly store and release Li ions such as graphite and the cathode materials based on Li metal oxide with the general formula Li_xMO_2 where M is Co, Ni, or Mn. The cathode and anode materials are supported on aluminum and copper foil as the current collector, separated, and immersed in a nonaqueous electrolyte (Figure 3.13b).[97] Graphite is still the exclusively used anode material in commercial LIBs due to its high conductivity and a moderate specific energy capacity of ~372 mAh/g when the LiC_6 is formed.[98,99]

However, the power density of current LIBs is still limited by the kinetics of Li ion intercalation and deintercalation into the interlayer spaces of graphite. To improve the kinetics of Li insertion, graphite was prepared with larger interlayer spacing, which also resulted in a Li intercalation on both sides of the graphite plane and a Li_2C_6 intercalation.[100] Moreover, higher Li ion capacity can also be achieved by covalent trapping of Li ions in the micropores of polymer derived carbons.[101] The discovery of novel carbon nanomaterials such as CNTs and graphene also provided an opportunity for increase of the specific capacity and power density of LIBs through different mechanisms of Li ion storage.

3.3.1 CARBON NANOMATERIALS AS ANODE MATERIALS OF LI-ION BATTERIES

Due to the prominent mechanical and electrical properties as well as the unique structures of carbon nanomaterials, they have been studied extensively as the potential substitutes for graphite in LIBs. Both the SWCNTs and the MWCNTs have been

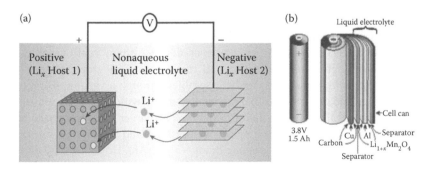

FIGURE 3.13 **(See color insert.)** Schematic operation mechanism of lithium ion battery (a) and a setup of a cylindrical battery (b). (From J. M. Tarascon and M. Armand, *Nature*, 2001, 414, 359–367.)

investigated as the anode materials and a higher capacity/stability has been observed compared with the graphite. As indicated by the first-principles calculation of the energy status and electronic structures of the Li ion intercalated on SWCNTs, the intercalation density of Li ion on an SWCNT is significantly larger than that on graphite and the inner walls of nanotubes are also favorable for the Li ion intercalation.[102] These results predict higher theoretical specific capacity of SWNTs. Besides higher capacity for Li insertion, the deformation of SWCNTs by the Li insertion/deinsertion is also rather small, which is critical for good charge–discharge performance of LIBs.

Using the purified SWCNTs as the anode materials for a LIB, a reversible capacity of 460 mAh/g was obtained, which was 23% larger than the theoretical value for graphite (372 mAh/g).[103] The high-temperature treatment of CNTs is known to enhance the electrical conductivity of carbon materials and to eliminate the surface functional groups.[104] The graphitization degree of CNTs can be significantly improved by heat treatment at higher temperatures from 1600 to 2800°C. CNTs treated at 900°C with lower crystalline degree showed the highest overall capacity of 952 mAh/g, but the irreversible Li ion capacity took up 505 mAh/g of the total value, and the reversible capacity dropped quickly after several cycles from 447 mAh/g to approximately 100 mAh/g due to the continuous formation of the passive layer during the charge and discharge process and possible electrical insulation of the isolated parts of the electrode. On the other hand, CNTs with higher graphite degree treated at high temperatures showed higher stability but low capacity. Another aspect affecting the capacity of the CNT-based anode in LIBs is its morphology.[105] Aligned CNTs prepared by the CVD method on anodic alumina membrane showed more electrochemical active surface area for the Li ion insertion than the disordered CNTs.

Graphene has also been investigated as an anode material for LIBs. The two-dimensional mono/few-layered carbon nanostructure gives distinct advantages over the bulky graphite or the one-dimensional CNTs, providing superior electron conductivity, larger surface area, and mechanical flexibility. Thermally reduced and expanded high-quality powdery graphene with 4 atom layers and specific surface area of 492.5 m^2/g, which was previously used as a catalyst support for a fuel cell,[56,57] showed the initial overall Li storage capacity of 1264 mAh/g. After 40 cycles at the current density of 100 mA/g, the capacity was still as high as 848 mAh/g.[106] These results indicate that high quality graphene possess excellent electrochemical Li ion storage properties when applied as anode in the LIBs.

Apart from the powdery graphene materials used in the above-mentioned research, graphite-like materials with larger interlayer distance than "real" graphite could be prepared through functionalization and reassembling of graphene nanosheets. Chemically converted graphene can be assembled through a vacuum filtration process forming a "bulky" graphite-like graphene paper[107] with an interlayer distance of 0.379 nm compared with 0.336 nm of graphite. This larger interlayer space brought a higher initial Li ion charge/discharge capacity of 680 and 528 mAh/g, respectively, higher than that obtained from graphite being ~372 mAh/. However, only 84 mAh/g was retained in the second circle, indicating the unsuitability of this material for a rechargeable or secondary Li ion battery.

Graphene oxide could also be prepared into a paper-like assembly through a similar process and then be reduced into graphene using hydrazine with the interlayer space of 0.37 nm.[108] By lowering the charge/discharge current density from 50 mA/g to 10 mA/g, the specific reversible capacity increased by 150%, which indicated low Li ion diffusion rate in such material. Increased interlayer spaces (>0.336 nm) can also be obtained through coaxially anchoring graphene outside the walls of CNTs through oxalyl bonding. In this graphene–CNT composite, an interlayer space of 0.378 nm was obtained, and the reversible capacity of 375 mAh/g was measured.[109] Fullerenes, also named as C60, were employed as a layer extending agent for graphene.[110] Graphene@C60 was prepared by hydrazine reduction of the graphene oxide and the acid functionalized C60 in an aqueous suspension followed by vacuum filtration and drying. As-prepared composite had an interlayer distance of approximately 0.42 nm and the reversible capacity as high as 784 mAh/g.

3.3.2 CARBON NANOMATERIAL BASED COMPOSITES FOR LIB ANODE MATERIALS

Apart from electrochemical storage of Li ion in a graphitic structure, Li ion can also be stored by its reversible reactions with other elements including Sn, Si, or Sb, or metal oxides such as SnO_2, MnO_x, TiO_2, or Fe_3O_4.[111–117] For example, the calculated theoretical Li ion storage capacity in Si and Sn is 4200 and 993 mAh/g, respectively, which is considerably higher than the value for graphite.[118,119] However, for these elements or compounds with high capacity, volume changes always take place during the Li ion insertion and extraction, which results in electrode crumbling and poor cycling behavior.[120] At the same time, these materials are poorly electrically conductive, leading to the high internal resistance and low power output of a LIB.[119] To deal with these issues, carbon nanomaterials with superior mechanical and electrical properties have been combined with these materials to form nanocomposites with improved energy storage and cycling stability. One such carbon nanomaterial is graphene. By simply mixing of commercially available nanosized silicon and graphene in a mortar, graphene-wrapped Si nanocomposite could be prepared (Figure 3.14).[121] Such easily

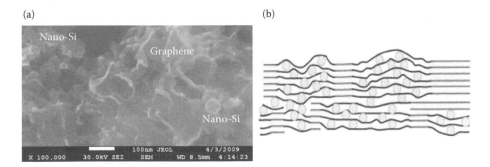

FIGURE 3.14 Microstructures of Si/graphene composites prepared by simple mixing of Si and graphene (a) and a vacuum filtration of Si and graphene followed by reduction in hydrogen (b). (From S.-L. Chou, J.-Z. Wang, M. Choucair, H.-K. Liu, J. A. Stride and S.-X. Dou, *Electrochemistry Communications*, 2009, 12, 303–306; J. K. Lee, K. B. Smith, C. M. Hayner and H. H. Kung, *Chemical Communications*, 2010, 46, 2025–2027.)

prepared composite anode materials showed reversible capacity of 1168 mAh/g after 30 cycles, and the charge transfer resistance was lowered by half compared with the pure nanosized silicon. Another paper-like Si/graphene was prepared via filtration of a suspension of nanosized silicon and graphene, followed by drying in air and reduction in hydrogen.[122] This composite materials comprised oriented graphene sheets with silicon nanoparticles anchored in between them as displayed in Figure 3.14. A high reversible capacity of 2200 mAh/g was measured even after 50 cycles.

In these Si/graphene composites, graphene contributed to the prominent overall Li ion storage performance in several aspects: (1) it connected Si particles forming a three-dimensional high conductive interconnected network with an enhanced electron transfer, so that all Si particles could contribute to the capacity. (2) The highly porous structure of the graphene matrix accommodated large strain and alleviated the volume changes. (3) Highly dispersed Si particles in the porous graphene matrix were easily accessible to the electrolyte compared with the heavily aggregated Si particles.

Sn is another element that can store Li by formation $Li_{4.4}Sn$. Sn/graphene nanocomposite was prepared by the co-reduction of Sn^{2+} from $SnCl_2$ aqueous solution and the graphene oxide using $NaBH_4$ reduction agent.[123] This composite had a similar structure as illustrated in Figure 3.14 with Sn nanoparticles located within the stacked graphene layers, but the Sn particle size was as small as 2–5 nm. The composite showed high reversible Li ion capacity of 795 mAh/g in the second cycle and 508 mAh/g in the 100th cycle, which was higher than both bare graphene and Sn nanoparticles. In this composite, Sn particles not only stored Li ion but also acted as an intercalation agent enlarging the interlayer distance between the graphene layers and thus facilitating the Li ion insertion on both sides of graphene.

Transition metal oxides can also reversibly react with Li ion and potentially be used in a LIB. Fe_3O_4 is a very attractive candidate due to its high Li ion storage capacity, environmental friendliness, and low cost. Recently, a Fe_3O_4/graphene composite was successfully fabricated by hydrolysis of iron chloride in the presence of graphene.[124] As-prepared material had well-organized and flexible tunnel-like structure stacked by graphene and the spindle-like Fe_3O_4 decorated in between the graphene layers. The hollow structure of the composite ensured a good mass transfer during charging and discharging. The confined structure of Fe_3O_4 inside the channels of stacked graphene led to a reversible capacity of 1026 mAh/g after 100 cycles.

SnO_2 is another material that has been intensively investigated. SnO_2/graphene was firstly prepared in a liquid phase, similar with simultaneous reduction of graphene oxide and formation of SnO_2 nanoparticles. A moderate reversible Li ion capacity of 690 mAh/g was obtained.[125] Another similar composite was obtained through hydrolysis of $SnCl_4$ in the presence of a graphene sheet,[126] which achieved a reversible capacity of 810 mAh/g. On the other hand, bare SnO_2 nanoparticles had an initial capacity of 550 mAh/g, but this value quickly dropped to 60 mAh/g after just 15 cycles. The enhanced charge/discharge stability of SnO_2/graphene composites was the result of the dimensional confinement and the volume expansion buffer effect from the graphene. Similar graphene-based transition metal oxide composite materials for LIB anode application include Mn_3O_4/graphene,[127] Fe_3O_4/graphene,[128] TiO_2/graphene,[129] and Co_3O_4/graphene.[130]

TABLE 3.2

Composite Electrode Materials for LIBs Based on CNTs

Electrode Type	Initial Charge Capacity (mAh/g)	Initial Discharge Capacity (mAh/g)	Cycle Number	Residual Reversible Capacity (mAh/g)	Chare Transfer Resistance (Ω)	Ref.
Sn/MWCNT	643	1590	40	627	N/A	111
Sn/MWCNT	N/A	570	30	442	16.4	112
SnNi/MWCNT	N/A	512	30	431	17.3	112
Bi/MWCNT	308	570	50	315	N/A	132
Sb	648	1023	30	115	N/A	113
SnSb$_{0.5}$	726	951	30	171	N/A	113
Sb/MWCNT	462	1266	30	287	N/A	113
SnSb$_{0.5}$/MWCNT	518	1092	30	348	N/A	113
CoSb$_3$/MWCNT	312	915	30	265	7.2	133
Ag/Fe/Sn/MWCNT	530	N/A	300	420	N/A	134
TiO$_2$	52	287	75	21	123	114
TiO$_2$/MWCNT	168	830	75	165	75	114
SnSb/MWCNT	680	1408	50	480	N/A	135
Ag-TiO$_2$/MWCNT	500	250	30	172	15.8	136
SnO$_2$/MWCNT			30	>400		137
SnO$_2$	N/A	728.3	40	126.4	N/A	115
SnO$_2$/MWCNT	N/A	665.1	40	505.9	N/A	115
Li$_4$Ti$_5$O$_{12}$/CNT	145	145	500	142	38	138

CNTs have also been employed as conductive and expansion filler. Some recent works have been reviewed by Liu as shown in Table 3.2.[131]

3.4   ELECTROCHEMICAL CAPACITORS

Electrochemical capacitors (EC), also called supercapacitor or ultracapacitor, is an energy storage device with a very high specific power density and relatively low specific energy storage. EC can be classified in two groups: the electric double-layer capacitor (EDLC) and the pseudocapacitor.

3.4.1   ELECTRIC DOUBLE-LAYER CAPACITOR

EDLC stores electrical energy in the electrical double-layer (EDL) formed at the electrode/electrolyte interface as shown in Figure 3.15.[139]

The capacitance C of an EDLC is calculated as shown in Equation 3.1:

$$C = \varepsilon_r \varepsilon_0 \frac{A}{d} \tag{3.1}$$

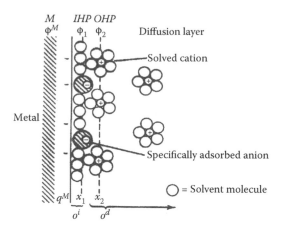

FIGURE 3.15 Scheme of the electrical double layer. (From E. Frackowiak and F. Beguin, *Carbon*, 2001, 39, 937–950.)

where ε_r is the relative permittivity of an electrolyte, ε_0 is the vacuum permittivity (constant), A is the surface area of an electrode, and d is the thickness of EDLC.

The energy density of an EDLC is calculated by Equation 3.2:

$$E = \frac{1}{2}CV^2 \tag{3.2}$$

where C is the capacitance and V is the voltage.

Based on these equations, it can be concluded that the capacitance of an EDLC increased by using the electrode material with the high specific surface area and the energy storage capacity is directly proportional to the capacitance and operational voltage. The voltage of supercapacitor is related to the electrolyte used, which is around 1 V for water-based electrolytes and 2.7 V for organic electrolytes.

Carbon material is the most widely used electrode material in a supercapacitor because of its high specific surface area, high chemical resistance, good electrical conductivity, and relatively low cost. Various activated carbons (ACs) have been investigated as the electrode materials for supercapacitors, and it has been generally accepted that the specific surface area is not the only factor that influences specific capacitance. The pore structure and pore size affect the capacitive performance to a great extent. Shi[140] investigated various kinds of commercial AC materials by nitrogen gas adsorption measurement, alternating current impedance, and constant current discharge techniques. Those ACs with a larger percentage of mesopores were confirmed as more suitable to high-power EC applications because they could deliver high energy at a high rate, although storing less total energy. Simon[141] tested and electrochemically characterized several AC materials obtained from PICA

Company. Galvanostatic cycling of EC cells showed important differences between the activated carbons tested. The PICACTIF SC carbon (an AC material obtained from PICA Company) was found to be particularly interesting, with a specific capacitance of 125 F/g and a series cell resistance of 3.5 Ω cm^2 in 1.7 M organic electrolyte of $N(C_2H_5)_4CH_3SO_3$ salt in acetonitrile. Porosity measurements have then shown that this interesting behavior could be explained by an increase of the mesoporous volume in the carbon that was assumed to facilitate ions accessibility and adsorption on the carbon surface.

The synthesis methods of AC for electrochemical capacitors are also investigated. In Yuan's work,[142] the hydrothermal route using sucrose as a precursor without any catalysts was employed to prepare uniform carbon spheres. The monodispersed 100- to 150-nm carbon spheres were obtained with the activation treatment in a molten KOH. A single electrode of carbon such nanosphere materials showed a very high specific gravimetric capacitance of 328 F g^{-1}, a specific capacitance per surface area of 19.2 μF cm^{-2}, and a volumetric capacitance of 383 F cm^{-3}. Hulicova-Jurcakova[143] reported AC produced from waste coffee grounds activated with $ZnCl_2$. EC electrodes prepared from this coffee grounds carbon exhibited energy densities up to 20 Wh kg^{-1} in 1 M H_2SO_4, and excellent stability at high charge–discharge rates. In a two-electrode cell a specific capacitance as high as 368 F g^{-1} was observed, with rectangular cyclic voltammetry curves and stable performance over 10,000 cycles at a cell potential of 1.2 V and current load of 5 A g^{-1} (Figure 3.16). The good electrochemical performance of the coffee grounds carbon was attributed to a well-developed porosity, with a distribution of micropores, and mesopores 2–4 nm wide,

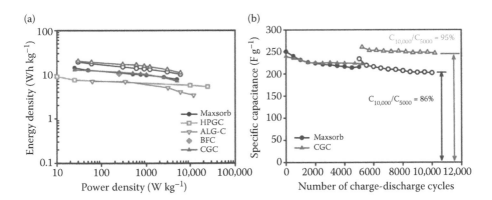

FIGURE 3.16 (a) Ragone plot showing performance of coffee ground carbon (CGC) relative to Maxsorb, hierarchical porous graphitic carbon (HPGC), ALG-C, and BFC. The weight of the cell components is not included in these calculations. (b) Electrochemical stability of CGC and Maxsorb over 5000 cycles at a cell potential of 0–1 V (closed symbols) followed by 5000 cycles with a cell potential of 0–1.2 V (open symbols). (From T. E. Rufford, D. Hulicova-Jurcakova, Z. H. Zhu and G. Q. Lu, *Electrochemistry Communications*, 2008, 10, 1594–1597.)

and the presence of electrochemically active quinone oxygen groups and nitrogen functional groups.

The porous carbon prepared by a template method is another choice for supercapacitor electrodes. Vix-Guterl[144] used the highly ordered MCM-48 mesoporous silica as a hard template to prepare reverse replica carbon materials for ECs. By using different routes and carbon precursors, a variety of carbon materials were synthesized with different porous characteristics that strongly influenced their performance in ECs. Good energy storage characteristics with capacitance values as high as 200 and 110 F/g were obtained in aqueous and organic media, respectively. Quick charge propagation during the supercapacitor performance was observed for the three kinds of carbon materials due to their very unique interconnected porous structure with a well-balanced micro/mesoporosity (Figures 3.16 and 3.17).

Xia and colleagues[145] demonstrated a two-step template approach for preparation of the three-dimensional ordered mesoporous carbon sphere arrays. The ordered macroporous silica skeleton was formed from silicon alkoxide precursor templating around polystyrene (latex) spheres followed by the removal of polystyrene spheres. Such structure was then used as a hard template and infiltrated with a solution mixture of amphiphilic triblock copolymer PEO-PPO-PEO and soluble resol. Using such combination of evaporation-induced surfactant-templating organic resol self-assembly with thermosetting, carbonization, and hydrofluoric acid removal of silica mesoporous carbons with a pore size of 10.4 nm, interconnected window size of about 60 nm, surface area of 601 m^2/g, and pore volume of 1.70 cm^3/g were obtained. They showed impressive electrochemical properties as electrode materials in nonaqueous supercapacitors with the rectangular cyclic voltammograms over a wide range of scan rates even up to 200 mV/s between 0 and 3 V and delivered a large capacitance of 14 $\mu F/cm^2$ (84 F/g), and a good cycling stability with capacitance retention of 93% over 5000 cycles.

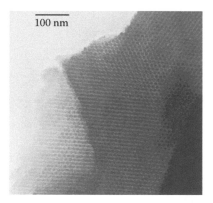

100 nm

FIGURE 3.17 TEM micrograph of the carbon replica. (From C. Vix-Guterl, S. Saadallah, K. Jurewicz, E. Frackowiak, M. Reda, J. Parmentier, J. Patarin and F. Beguin, *Materials Science and Engineering B, Solid-State Materials for Advanced Technology*, 2004, 108, 148–155.)

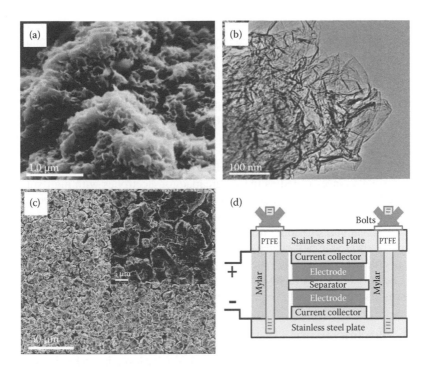

FIGURE 3.18 **(See color insert.)** (a) SEM image of CMG particle surface, (b) TEM image of individual graphene sheets extended from CMG particle surface, (c) low and high (inset) magnification SEM images of CMG particle electrode surface, and (d) a schematic of test cell assembly. (From M. D. Stoller, S. J. Park, Y. W. Zhu, J. H. An and R. S. Ruoff, *Nano Letters*, 2008, 8, 3498–3502.)

Niu and colleagues[146] prepared CNT sheet electrodes from catalytically grown carbon nanotubes with high purity and narrow diameter distribution, centred around 8 nm. Such free-standing mats of entangled nanotubes with an open porous structure, which were almost impossible to obtain with AC or carbon fiber, showed specific capacitances of 102 and 49 F/g at 1 and 100 Hz, respectively, on a single cell device with 38 wt% H_2SO_4 as the electrolyte. The same cell had a power density of >8000 W/kg.

Lee and his team[147] investigated the key factors determining the performance of ECs using single-walled CNT electrodes. They obtained a maximum specific capacitance of 180 F/g with a large power density of 20 kW/kg at an energy density of 6.5 Wh/kg. The heat treatment at high temperature was necessary to increase the capacitance and reduce the CNT-electrode resistance. The increased capacitance was well explained by the enhancement of the specific surface area and the abundant pore distributions at lower pore sizes of 30–50 Å, estimated from the BET (N_2) measurements. Ruoff's group[148] had pioneered a new carbon material that was called a chemically modified graphene (CMG) (Figure 3.18). CMG material was made from 1-atom thick sheets of carbon and further functionalized with functional groups (e.g.,

–OH and –COOH), which showed specific capacitance of 135 and 99 F/g in aqueous and organic electrolyte, respectively. In addition, high electrical conductivity gave these materials consistently good performance over a wide range of voltage scan rates. Chen[149] also investigated graphene materials (GMs) as the EC electrode materials. GMs were prepared from graphite oxide sheets reduced in a gas-phase reaction with hydrazine. A maximum specific capacitance of 205 F/g and a power density of 10 kW/kg at the energy density of 28.5 Wh/kg in an aqueous electrolyte solution were obtained. The supercapacitor manufactured by such carbon exhibited a very good long cycle life.

Other carbon materials including carbide-derived carbons[150] and onion-like carbon[151] were also investigated as the supercapacitor electrodes with high volumetric and specific capacitances in organic and ionic electrolytes. Carbide-derived carbons served as model compounds for computational modeling of the energy storage capacity due to their uniform and narrow pore size distributions. In conclusion, carbon nanomaterials are the best choice of materials as EDLC electrode, not only for their high specific area, but also good chemical stability and low cost. High conductivity can be achieved by using highly graphitic material as CNT or graphene, whereas the pore-size distribution is also controllable.

3.4.2 PSEUDOCAPACITOR

Along with the electric energy stored in electric double layer, it can be also stored and released through the redox reaction, as in the batteries. If the redox process occurs within a few nanometres on the electrode, the performance of the device is similar to that of EDLC and such device is then called the pseudocapacitor. Materials used as pseudocapacitor electrode are usually heteroatom-doped carbons, metal oxides (hydroxides), and conductive polymers.

Cheng and colleagues[152] prepared mesoporous carbon with homogeneous boron dopant by coimpregnation and carbonization of sucrose and boric acid confined in mesopores of SBA-15 silica template. Even at low doping levels, it showed a catalytic effect on oxygen chemisorption at the edge planes and altered the electronic structure of space charge layer of doped mesoporous carbon. As a result, substantial improvement of the interfacial capacitance by 1.5–1.6 times compared with boron-free carbon in alkaline electrolyte (6 M KOH) as well as acid electrolyte (1 M H_2SO_4) was observed (Figure 3.19).

Hulicova-Jurcakova and colleagues[153] showed that phosphorus-rich microporous carbons (P-carbons) prepared by a simple H_3PO_4 activation of three different carbon precursors exhibit enhanced supercapacitive performance in 1 M H_2SO_4 when highly stable performance was achieved at potentials larger than the theoretical decomposition potential of water. This led to the enhancement of the energy density of such supercapacitors capable of delivering 16 Wh/kg compared with 5 Wh/kg for the commercial carbon. An intercept-free multiple linear regression model confirmed the strongest influence of phosphorus on pseudocapacitance together with the micropores 0.65–0.83 nm in width that were most effective in forming the electric double layer.

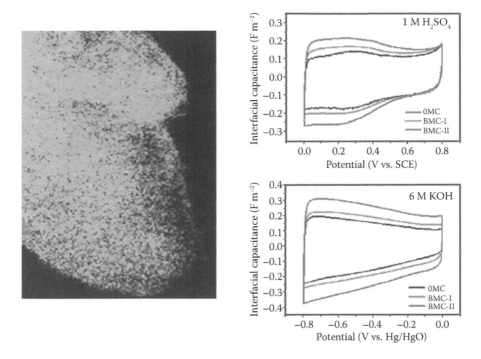

FIGURE 3.19 **(See color insert.)** EELS mapping images of BMC-II sample: boron. Cyclic voltammetry plots of OMC, BMC-I, and BMC-II samples in 1 M H$_2$SO$_4$ electrolyte and 6 M KOH electrolyte recorded at 10 mV s^{-1}. (From D. W. Wang, F. Li, Z. G. Chen, G. Q. Lu and H. M. Cheng, *Chemistry of Materials*, 2008, 20, 7195–7200.)

Wei and his team[154] studied the long cycle performance of amorphous MnO$_2$ (a-MnO$_2$) and SWCNTs composites at a high charge–discharge current of 2 A/g. Such composites were successfully synthesized via a novel room temperature route starting with KMnO$_4$, ethanol, and commercial SWCNTs. All studied composites with different SWCNT loads showed excellent cycling capability, even at high current loads of 2 A/g, with the MnO$_2$:20 wt% SWCNT composite showing the best combination of coulombic efficiency of 75% and specific capacitance of 110 F/g after 750 cycles. On the other hand, the composite with 5 wt% SWCNTs showed the highest specific capacitance during the initial cycles (Figure 3.20). Zhu[155] fabricated a composite of GO supported by needle-like MnO$_2$ nanocrystals (GO-MnO$_2$ nanocomposites) through a simple soft chemical route in a water-isopropyl alcohol system. The formation mechanism of these intriguing nanocomposites investigated by transmission electron microscopy, Raman spectroscopy, and ultraviolet–visible absorption spectroscopy was proposed as an intercalation and adsorption of manganese ions onto the GO sheets, followed by the nucleation and growth of the crystal species in a double solvent system via dissolution–crystallization and oriented attachment mechanisms, which in turn resulted in the exfoliation of GO sheets. It

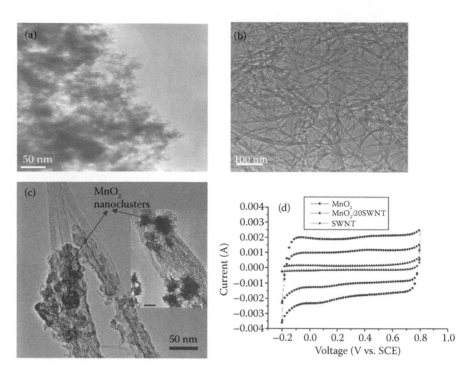

FIGURE 3.20   (a–c) TEM pictures of pure MnO_2, pure SWNT, and MnO_2:20 wt% SWNT composite. Inset of c: high magnification of the composite, scale bar 20 nm. (d) Cyclic voltammograms of the pure MnO_2, the pure SWNT, and the MnO_2:20 wt% SWNT composite at a scan rate of 2 mV/s. (From V. Subramanian, H. W. Zhu and B. Q. Wei, *Electrochemistry Communications*, 2006, 8, 827–832.)

was also found that the electrochemical performance of as-prepared nanocomposites was enhanced by the chemical interaction between the GO and MnO_2.

Lee[156] fabricated a nanocomposite electrode of single-walled SWCNT and polypyrrole (Ppy). The individual nanotubes were uniformly coated with Ppy particles as a result of the in-situ chemical polymerization of pyrrole. Such SWNT-Ppy nanocomposite electrode showed much higher specific capacitance than pure Ppy and as-grown SWNT electrodes, due to the uniform coating of Ppy on the SWNTs and the combination of pseudocapacitance and a double-layer capacitance. Zhao and colleagues[157] prepared chemically modified graphene and polyaniline (PANi) nano-fiber composites by *in situ* polymerization of aniline monomer in the presence of graphene oxide under the acid conditions. The obtained graphene oxide/PANi composites with different mass ratios were reduced to graphene using hydrazine followed by reoxidation and reprotonation of the reduced PANi to give the graphene/PANi nanocomposites. It was found that the chemically modified graphene and the PANi nanofibers formed a uniform nanocomposite with the PANi fibers absorbed on the graphene surface and/or filled between the graphene sheets (Figure 3.21). Such uniform structure together with the observed high conductivities afforded high

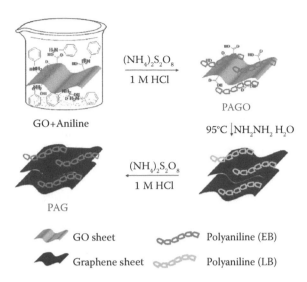

GO+Aniline

PAG

$(NH_4)_2S_2O_8$
1 M HCl

PAGO

95°C ↓$NH_2NH_2·H_2O$

$(NH_4)_2S_2O_8$
1 M HCl

GO sheet   Polyaniline (EB)

Graphene sheet   Polyaniline (LB)

FIGURE 3.21   (See color insert.) Illustration of the process for preparation of graphene–PANi composites. (From K. Zhang, L. L. Zhang, X. S. Zhao and J. S. Wu, *Chemistry of Materials*, 2010, 22, 1392–1401.)

specific capacitance and good cycling stability during the charge–discharge process when used as supercapacitor electrodes. A specific capacitance of as high as 480 F/g at a current density of 0.1 A/g was achieved.

3.5   CURRENT ISSUES AND OUTLOOK

Different sorts of carbons have been vastly applied in energy conversion and storage fields. The performance of these carbon materials is largely dependent on the microstructure, porosity, and surface functionality. Although in some areas, carbons with a certain structure have been used for a long period, e.g., carbon blacks, which have been almost exclusively employed as a current commercial fuel cell catalyst support, the research for a more suitable substitute is still necessary. On the one hand, these conventional carbons still cannot meet all the requirements for the high performance applications, although they are cheap and widely used. On the other hand, the emergence of carbon materials with new structures can always shed light on the current studies.

Novel carbon material includes carbon nanomaterial and porous carbons with controllable pore structures. The former one has shown us a bottom-up way for constructing carbon materials with high crystalline degree, which gives them high mechanical/electrical properties as well as excellent electrochemical corrosion resistance. However, for energy-related applications, it is necessary to modify the surface of such carbon nanomaterials with appropriate functional groups. Methods have been developed, including oxidation treatment, surface functional group grafting, molecular modification or heteroatom doping. These treatments have been applied on various sorts of carbons and have resulted in performance improvement in the corresponding applications. However, to controllably decorate the carbon nanomaterials as well as to construct these materials into desirable porous frame is still necessary for exploring.

Another group of novel carbons is the mesoporous carbon with controllable porosity. The recent development in the template-derived porous carbon has given people significant freedom in choosing porous carbon with suitable pore size and pore structure. The relatively high specific surface area of mesoporous carbon can endow these carbons with more surface active sites, while the large pore size and tailorable pore structure can facilitate the energy exchange by a high rate of mass transfer. However, for this class of material, the amorphous nature makes them vulnerable to the chemical attacks. Although a high-temperature treatment can, to a large extent, enhance their graphitization degree, the structural deterioration cannot be neglected. Direct building of highly crystallized carbon with mesoporous structure should be studied in the future study.

Another issue existing in current research and future commercialization of novel carbons is the high cost of their manufacture. Although some of them have become significantly cheaper compared with the price many years ago, they are still too expensive compared with the commercial carbon blacks, graphites, or active carbon. If this is not taken into consideration, the application of these high-performance materials would be limited to a small range where cost is not a problem and the initial intention of the studying in energy conversion/storage would be hard to realize.

REFERENCES

1. P. Costamagna and S. Srinivasan, *Journal of Power Sources*, 2001, **102**, 242–252.
2. D. A. Landsman and F. J. Luczak, *Catalyst studies and coating technologies*, John Wiley & Sons, Ltd, 2010.
3. E. Antolini, *Applied Catalysis B: Environmental*, 2009, **88**, 1–24.
4. D. Santiago, G. G. Rodriguez-Calero, H. Rivera, D. A. Tryk, M. A. Scibioh and C. R. Cabrera, *Journal of The Electrochemical Society*, 2010, **157**, F189–F195.
5. M. Uchida, Y. Aoyama, M. Tanabe, N. Yanagihara, N. Eda and A. Ohta, *Journal of the Electrochemical Society*, 1995, **142**, 2572–2576.
6. S. Tang, G. Sun, J. Qi, S. Sun, J. Guo, Q. Xin and G. M. Haarberg, *Chinese Journal of Catalysis*, 2010, **31**, 12–17.
7. E. Antolini, R. R. Passos and E. A. Ticianelli, *Journal of Power Sources*, 2002, **109**, 477–482.
8. K. Kanno, K. Eui Yoon, J. J. Fernandez, I. Mochida, F. Fortin and Y. Korai, *Carbon*, 1994, **32**, 801–807.
9. G. Wang, G. Sun, Q. Wang, S. Wang, J. Guo, Y. Gao and Q. Xin, *Journal of Power Sources*, 2008, **180**, 176–180.
10. K. Sakanishi, H.-U. Hasuo, I. Mochida and O. Okuma, *Energy & Fuels*, 1995, **9**, 995–998.
11. J. Yu, Y. Yoshikawa, T. Matsuura, M. N. Islam and M. Hori, *Electrochemical and Solid-State Letters*, 2005, **8**, A152–A155.
12. E. Passalacqua, P. L. Antonucci, M. Vivaldi, A. Patti, V. Antonucci, N. Giordano and K. Kinoshita, *Electrochimica Acta*, 1992, **37**, 2725–2730.
13. D. A. Stevens, M. T. Hicks, G. M. Haugen and J. R. Dahn, *Journal of the Electrochemical Society*, 2005, **152**, A2309–A2315.

14. K. Kinoshita, 1988.
15. D. A. Stevens and J. R. Dahn, *Carbon*, 2005, **43**, 179–188.
16. K. I. Ota, A. Ishihara, S. Mitsushima, K. Lee, Y. Suzuki, N. Horibe, T. Nakagawa and N. Kamiya, *Journal of New Materials for Electrochemical Systems*, 2005, **8**, 25–35.
17. J. M. Thomas, *Carbon*, 1970, **8**, 413–421.
18. C.-C. Hung, P.-Y. Lim, J.-R. Chen and H. C. Shih, *Journal of Power Sources*, 2011, **196**, 140–146.
19. F. Coloma, A. Sepulvedaescribano and F. Rodriguezreinoso, *Journal of Catalysis*, 1995, **154**, 299–305.
20. H.-Q. Li, Y.-G. Wang, C.-X. Wang and Y.-Y. Xia, *Journal of Power Sources*, 2008, **185**, 1557–1562.
21. S. B. Yoon, G. S. Chai, S. K. Kang, J.-S. Yu, K. P. Gierszal and M. Jaroniec, *Journal of the American Chemical Society*, 2005, **127**, 4188–4189.
22. Z. Li, M. Jaroniec, Y.-J. Lee and L. R. Radovic, *Chemical Communications*, 2002, 1346–1347.
23. V. Raghuveer and A. Manthiram, *Electrochemical and Solid-State Letters*, 2004, **7**, A336–A339.
24. D. R. Rolison, *Science*, 2003, **299**, 1698–1701.
25. M. Kim, J.-N. Park, H. Kim, S. Song and W.-H. Lee, *Journal of Power Sources*, 2006, **163**, 93–97.
26. S. Jun, S. H. Joo, R. Ryoo, M. Kruk, M. Jaroniec, Z. Liu, T. Ohsuna and O. Terasaki, *Journal of the American Chemical Society*, 2000, **122**, 10712–10713.
27. S. H. Joo, S. J. Choi, I. Oh, J. Kwak, Z. Liu, O. Terasaki and R. Ryoo, *Nature*, 2001, **412**, 169–172.
28. R. Ryoo, S. H. Joo and S. Jun, *The Journal of Physical Chemistry B*, 1999, **103**, 7743–7746.
29. R. Ryoo, S. H. Joo, M. Kruk and M. Jaroniec, *Advanced Materials*, 2001, **13**, 677–681.
30. J. Lee, S. Yoon, S. M. Oh, C. H. Shin and T. Hyeon, *Advanced Materials*, 2000, **12**, 359–362.
31. F. Kleitz, S. Hei Choi and R. Ryoo, *Chemical Communications*, 2003, 2136–2137.
32. S. Tanaka, N. Nishiyama, Y. Egashira and K. Ueyama, *Chemical Communications*, 2005, 2125–2127.
33. F. Zhang, Y. Meng, D. Gu, Yan, C. Yu, B. Tu and D. Zhao, *Journal of the American Chemical Society*, 2005, **127**, 13508–13509.
34. H. I. Lee, G. D. Stucky, J. H. Kim, C. Pak, H. Chang and J. M. Kim, *Advanced Materials*, 2011, **23**, 2357–2361.
35. Z. Li and M. Jaroniec, *Chemistry of Materials*, 2003, **15**, 1327–1333.
36. Z. Li and M. Jaroniec, *Journal of the American Chemical Society*, 2001, **123**, 9208–9209.
37. C. Moreno-Castilla and F. J. Maldonado-Hodar, *Carbon*, 2005, **43**, 455–465.
38. W. C. Choi, S. I. Woo, M. K. Jeon, J. M. Sohn, M. R. Kim and H. J. Jeon, *Advanced Materials*, 2005, **17**, 446–451.
39. J. Ding, K.-Y. Chan, J. Ren and F.-S. Xiao, *Electrochimica Acta*, 2005, **50**, 3131–3141.
40. H. Yamada, T. Hirai, I. Moriguchi and T. Kudo, *Journal of Power Sources*, 2007, **164**, 538–543.
41. A. K. Geim and K. S. Novoselov, *Nature Materials*, 2007, **6**, 183–191.
42. S. Iijima, *Nature*, 1991, **354**, 56–58.
43. S. Iijima, *Physica B: Condensed Matter*, 2002, **323**, 1–5.
44. X. Wang, W. Li, Z. Chen, M. Waje and Y. Yan, *Journal of Power Sources*, 2006, **158**, 154–159.
45. A. Thess, R. Lee, P. Nikolaev, H. Dai, P. Petit, J. Robert, C. Xu, Y. H. Lee, S. G. Kim, A. G. Rinzler, D. T. Colbert, G. E. Scuseria, D. Tomanek, J. E. Fischer and R. E. Smalley, *Science*, 1996, **273**, 483–487.

46. H. Hiura, T. W. Ebbesen and K. Tanigaki, *Advanced Materials*, 1995, **7**, 275–276.
47. V. Lordi, N. Yao and J. Wei, *Chemistry of Materials*, 2001, **13**, 733–737.
48. J. Prabhuram, T. S. Zhao, Z. K. Tang, R. Chen and Z. X. Liang, *The Journal of Physical Chemistry B*, 2006, **110**, 5245–5252.
49. Z. D. Wei, C. Yan, Y. Tan, L. Li, C. X. Sun, Z. G. Shao, P. K. Shen and H. W. Dong, *The Journal of Physical Chemistry C*, 2008, **112**, 2671–2677.
50. M. S. Saha, R. Li and X. Sun, *Journal of Power Sources*, 2008, **177**, 314–322.
51. Z. He, J. Chen, D. Liu, H. Tang, W. Deng and Y. Kuang, *Materials Chemistry and Physics*, 2004, **85**, 396–401.
52. D.-J. Guo and H.-L. Li, *Electroanalysis*, 2005, **17**, 869–872.
53. Y. Xing, *The Journal of Physical Chemistry B*, 2004, **108**, 19255–19259.
54. C.-C. Chen, C. F. Chen, C.-H. Hsu and I. H. Li, *Diamond and Related Materials*, 2005, **14**, 770–773.
55. P. Joshi, Y. Xie, M. Ropp, D. Galipeau, S. Bailey and Q. Q. Qiao, *Energy & Environmental Science*, 2009, **2**, 426–429.
56. R. Kou, Y. Shao, D. Wang, M. H. Engelhard, J. H. Kwak, J. Wang, V. V. Viswanathan, C. Wang, Y. Lin, Y. Wang, I. A. Aksay and J. Liu, *Electrochemistry Communications*, 2009, **11**, 954–957.
57. M. J. McAllister, J.-L. Li, D. H. Adamson, H. C. Schniepp, A. A. Abdala, J. Liu, M. Herrera-Alonso, D. L. Milius, R. Car, R. K. Prud'homme and I. A. Aksay, *Chemistry of Materials*, 2007, **19**, 4396–4404.
58. Y. Shao, S. Zhang, C. Wang, Z. Nie, J. Liu, Y. Wang and Y. Lin, *Journal of Power Sources*, 2010, **195**, 4600–4605.
59. D.-Q. Yang, J.-F. Rochette and E. Sacher, *The Journal of Physical Chemistry B*, 2005, **109**, 4481–4484.
60. N. Shang, P. Papakonstantinou, P. Wang and S. R. P. Silva, *The Journal of Physical Chemistry C*, 2010, **114**, 15837–15841.
61. M. H. Seo, S. M. Choi, H. J. Kim and W. B. Kim, *Electrochemistry Communications*, 2011, **13**, 182–185.
62. S. Bong, Y.-R. Kim, I. Kim, S. Woo, S. Uhm, J. Lee and H. Kim, *Electrochemistry Communications*, 2010, **12**, 129–131.
63. H. Zhang, X. Xu, P. Gu, C. Li, P. Wu and C. Cai, *Electrochimica Acta*, 2011, **56**, 7064–7070.
64. B. Oregan and M. Gratzel, *Nature*, 1991, **353**, 737–740.
65. M. Gratzel, *Journal of Photochemistry and Photobiology C: Photochemistry Reviews*, 2003, **4**, 145–153.
66. K. H. Jung, J. S. Hong, R. Vittal and K. J. Kim, *Chemistry Letters*, 2002, 864–865.
67. P. Brown, K. Takechi and P. V. Kamat, *Journal of Physical Chemistry C*, 2008, **112**, 4776–4782.
68. K. H. Jung, S. R. Jang, R. Vittal, V. D. Kim and K. J. Kim, *Bulletin of the Korean Chemical Society*, 2003, **24**, 1501–1504.
69. S. L. Kim, S. R. Jang, R. Vittal, J. Lee and K. J. Kim, *Journal of Applied Electrochemistry*, 2006, **36**, 1433–1439.
70. C. Y. Yen, Y. F. Lin, S. H. Liao, C. C. Weng, C. C. Huang, Y. H. Hsiao, C. C. M. Ma, M. C. Chang, H. Shao, M. C. Tsai, C. K. Hsieh, C. H. Tsai and F. B. Weng, *Nanotechnology*, 2008, **19**.
71. J. G. Yu, J. J. Fan and B. Cheng, *Journal of Power Sources*, 2011, **196**, 7891–7898.
72. S. W. Zhang, H. H. Niu, Y. Lan, C. Cheng, J. Z. Xu and X. K. Wang, *Journal of Physical Chemistry C*, 2011, **115**, 22025–22034.
73. N. L. Yang, J. Zhai, D. Wang, Y. S. Chen and L. Jiang, *ACS Nano*, 2010, **4**, 887–894.
74. J. L. Song, Z. Y. Yin, Z. J. Yang, P. Amaladass, S. X. Wu, J. Ye, Y. Zhao, W. Q. Deng, H. Zhang and X. W. Liu, *Chemistry—A European Journal*, 2011, **17**, 10832–10837.

75. S. R. Sun, L. Gao and Y. Q. Liu, *Applied Physics Letters*, 2010, **96**.
76. M. Y. Yen, M. C. Hsiao, S. H. Liao, P. I. Liu, H. M. Tsai, C. C. M. Ma, N. W. Pu and M. D. Ger, *Carbon*, 2011, **49**, 3597–3606.
77. K. Imoto, K. Takahashi, T. Yamaguchi, T. Komura, J. Nakamura and K. Murata, *Solar Energy Materials and Solar Cells*, 2003, **79**, 459–469.
78. S. I. Cha, B. K. Koo, S. H. Seo and D. Y. Lee, *Journal of Materials Chemistry*, 2010, **20**, 659–662.
79. J. Han, H. Kim, D. Y. Kim, S. M. Jo and S. Y. Jang, *ACS Nano*, 2010, **4**, 3503–3509.
80. X. G. Mei, S. J. Cho, B. H. Fan and J. Y. Ouyang, *Nanotechnology*, 2010, **21**.
81. E. Ramasamy, W. J. Lee, D. Y. Lee and J. S. Song, *Electrochemistry Communications*, 2008, **10**, 1087–1089.
82. K. Suzuki, M. Yamaguchi, M. Kumagai and S. Yanagida, *Chemistry Letters*, 2003, **32**, 28–29.
83. L. Kavan, J. H. Yum and M. Gratzel, *ACS Nano*, 2011, **5**, 165–172.
84. J. D. Roy-Mayhew, D. J. Bozym, C. Punckt and I. A. Aksay, *ACS Nano*, 2010, **4**, 6203–6211.
85. J. K. Chen, K. X. Li, Y. H. Luo, X. Z. Guo, D. M. Li, M. H. Deng, S. Q. Huang and Q. B. Meng, *Carbon*, 2009, **47**, 2704–2708.
86. Z. Huang, X. H. Liu, K. X. Li, D. M. Li, Y. H. Luo, H. Li, W. B. Song, L. Q. Chen and Q. B. Meng, *Electrochemistry Communications*, 2007, **9**, 596–598.
87. T. N. Murakami, S. Ito, Q. Wang, M. K. Nazeeruddin, T. Bessho, I. Cesar, P. Liska, R. Humphry-Baker, P. Comte, P. Pechy and M. Gratzel, *Journal of the Electrochemical Society*, 2006, **153**, A2255–A2261.
88. G. Q. Wang, W. Xing and S. P. Zhuo, *Journal of Power Sources*, 2009, **194**, 568–573.
89. W. J. Hong, Y. X. Xu, G. W. Lu, C. Li and G. Q. Shi, *Electrochemistry Communications*, 2008, **10**, 1555–1558.
90. G. R. Li, F. Wang, Q. W. Jiang, X. P. Gao and P. W. Shen, *Angewandte Chemie— International Edition*, 2010, **49**, 3653–3656.
91. Z. C. Wu, Z. H. Chen, X. Du, J. M. Logan, J. Sippel, M. Nikolou, K. Kamaras, J. R. Reynolds, D. B. Tanner, A. F. Hebard and A. G. Rinzler, *Science*, 2004, **305**, 1273–1276.
92. X. S. Li, Y. W. Zhu, W. W. Cai, M. Borysiak, B. Y. Han, D. Chen, R. D. Piner, L. Colombo and R. S. Ruoff, *Nano Letters*, 2009, **9**, 4359–4363.
93. X. Wang, L. J. Zhi and K. Mullen, *Nano Letters*, 2008, **8**, 323–327.
94. S. R. Kim, M. K. Parvez and M. Chhowalla, *Chemical Physics Letters*, 2009, **483**, 124–127.
95. Y. H. Ng, A. Iwase, A. Kudo and R. Amal, *Journal of Physical Chemistry Letters*, 2010, **1**, 2607–2612.
96. Q. Li, B. D. Guo, J. G. Yu, J. R. Ran, B. H. Zhang, H. J. Yan and J. R. Gong, *Journal of the American Chemical Society*, 2011, **133**, 10878–10884.
97. J. M. Tarascon and M. Armand, *Nature*, 2001, **414**, 359–367.
98. N. A. Kaskhedikar and J. Maier, *Advanced Materials*, 2009, **21**, 2664–2680.
99. J. Hou, Y. Shao, M. W. Ellis, R. B. Moore and B. Yi, *Physical Chemistry Chemical Physics*, **13**, 15384–15402.
100. J. R. Dahn, T. Zheng, Y. Liu and J. S. Xue, *Science*, 1995, **270**, 590–593.
101. K. Sato, M. Noguchi, A. Demachi, N. Oki and M. Endo, *Science*, 1994, **264**, 556–558.
102. J. Zhao, A. Buldum, J. Han and J. Ping Lu, *Physical Review Letters*, 2000, **85**, 1706–1709.
103. A. S. Claye, J. E. Fischer, C. B. Huffman, A. G. Rinzler and R. E. Smalley, *Journal of the Electrochemical Society*, 2000, **147**, 2845–2852.
104. E. Frackowiak, S. Gautier, H. Gaucher, S. Bonnamy and F. Beguin, *Carbon*, 1999, **37**, 61–69.
105. G. Che, B. B. Lakshmi, E. R. Fisher and C. R. Martin, *Nature*, 1998, **393**, 346–349.

106. P. Lian, X. Zhu, S. Liang, Z. Li, W. Yang and H. Wang, *Electrochimica Acta*, 2010, **55**, 3909–3914.
107. C. Wang, D. Li, C. O. Too and G. G. Wallace, *Chemistry of Materials*, 2009, **21**, 2604–2606.
108. A. Abouimrane, O. C. Compton, K. Amine and S. T. Nguyen, *The Journal of Physical Chemistry C*, 2010, **114**, 12800–12804.
109. X. Liu, Y.-S. Hu, J.-O. Müller, R. Schlögl, J. Maier and D. S. Su, *ChemSusChem*, 2010, **3**, 261–265.
110. E. Yoo, J. Kim, E. Hosono, H.-S. Zhou, T. Kudo and I. Honma, *Nano Letters*, 2008, **8**, 2277–2282.
111. T. P. Kumar, R. Ramesh, Y. Y. Lin and G. T. K. Fey, *Electrochemistry Communications*, 2004, **6**, 520–525.
112. Z. P. Guo, Z. W. Zhao, H. K. Liu and S. X. Dou, *Carbon*, 2005, **43**, 1392–1399.
113. W. X. Chen, J. Y. Lee and Z. Liu, *Carbon*, 2003, **41**, 959–966.
114. H. Huang, W. K. Zhang, X. P. Gan, C. Wang and L. Zhang, *Materials Letters*, 2007, **61**, 296–299.
115. Y. Fu, R. Ma, Y. Shu, Z. Cao and X. Ma, *Materials Letters*, 2009, **63**, 1946–1948.
116. Z. Wen, Q. Wang, Q. Zhang and J. Li, *Advanced Functional Materials*, 2007, **17**, 2772–2778.
117. Y. He, L. Huang, J. S. Cai, X. M. Zheng and S. G. Sun, *Electrochimica Acta*, **55**, 1140–1144.
118. M. Winter, J. O. Besenhard, M. E. Spahr and P. Novak, *Advanced Materials*, 1998, **10**, 725–763.
119. J. Hou, Y. Shao, M. W. Ellis, R. B. Moore and B. Yi, *Physical Chemistry Chemical Physics*, 2011, **13**, 15384–15402.
120. H. Li, X. Huang, L. Chen, Z. Wu and Y. Liang, *Electrochemical and Solid-State Letters*, 1999, **2**, 547–549.
121. S.-L. Chou, J.-Z. Wang, M. Choucair, H.-K. Liu, J. A. Stride and S.-X. Dou, *Electrochemistry Communications*, 2009, **12**, 303–306.
122. J. K. Lee, K. B. Smith, C. M. Hayner and H. H. Kung, *Chemical Communications*, 2010, **46**, 2025–2027.
123. G. Wang, B. Wang, X. Wang, J. Park, S. Dou, H. Ahn and K. Kim, *Journal of Materials Chemistry*, 2009, **19**, 8378–8384.
124. G. Zhou, D.-W. Wang, F. Li, L. Zhang, N. Li, Z.-S. Wu, L. Wen, G. Q. Lu and H.-M. Cheng, *Chemistry of Materials*, 2010, **22**, 5306–5313.
125. J. Liang, W. Wei, D. Zhong, Q. Yang, L. Li and L. Guo, *ACS Applied Materials & Interfaces*, 2012, **4**, 454–459.
126. S.-M. Paek, E. Yoo and I. Honma, *Nano Letters*, 2008, **9**, 72–75.
127. H. Wang, L. F. Cui, Y. Yang, H. Sanchez Casalongue, J. T. Robinson, Y. Liang, Y. Cui and H. Dai, *Journal of the American Chemical Society*, 2010, **132**, 13978–13980.
128. G. Zhou, D. W. Wang, F. Li, L. Zhang, N. Li, Z. S. Wu, L. Wen, G. Q. Lu and H. M. Cheng, *Chemistry of Materials*, 2010, **22**, 5306–5313.
129. D. Wang, D. Choi, J. Li, Z. Yang, Z. Nie, R. Kou, D. Hu, C. Wang, L. V. Saraf, J. Zhang, I. A. Aksay and J. Liu, *ACS Nano*, 2009, **3**, 907–914.
130. Z.-S. Wu, W. Ren, L. Wen, L. Gao, J. Zhao, Z. Chen, G. Zhou, F. Li and H.-M. Cheng, *ACS Nano*, 2010, **4**, 3187–3194.
131. X.-M. Liu, Z. D. Huang, S. W. Oh, B. Zhang, P.-C. Ma, M. M. F. Yuen and J.-K. Kim, *Composites Science and Technology*, 2012, **72**, 121–144.
132. Y. NuLi, J. Yang and M. Jiang, *Materials Letters*, 2008, **62**, 2092–2095.
133. J. Xie, X. B. Zhao, G. S. Cao and M. J. Zhao, *Electrochimica Acta*, 2005, **50**, 2725–2731.
134. J. Yin, M. Wada, Y. Kitano, S. Tanase, O. Kajita and T. Sakai, *Journal of the Electrochemical Society*, 2005, **152**, A1341–A1346.

135. M. S. Park, S. A. Needham, G. X. Wang, Y. M. Kang, J. S. Park, S. X. Dou and H. K. Liu, *Chemistry of Materials*, 2007, **19**, 2406–2410.
136. J. Yan, H. Song, S. Yang, J. Yan and X. Chen, *Electrochimica Acta*, 2008, **53**, 6351–6355.
137. G. An, N. Na, X. Zhang, Z. Miao, S. Miao, K. Ding and Z. Liu, *Nanotechnology*, 2007, **18**, 435707.
138. J. Huang and Z. Jiang, *Electrochimica Acta*, 2008, **53**, 7756–7759.
139. E. Frackowiak and F. Beguin, *Carbon*, 2001, **39**, 937–950.
140. D. Y. Qu and H. Shi, *Journal of Power Sources*, 1998, **74**, 99–107.
141. J. Gamby, P. L. Taberna, P. Simon, J. F. Fauvarque and M. Chesneau, *Journal of Power Sources*, 2001, **101**, 109–116.
142. D. S. Yuan, J. X. Chen, J. H. Zeng and S. X. Tan, *Electrochemistry Communications*, 2008, **10**, 1067–1070.
143. T. E. Rufford, D. Hulicova-Jurcakova, Z. H. Zhu and G. Q. Lu, *Electrochemistry Communications*, 2008, **10**, 1594–1597.
144. C. Vix-Guterl, S. Saadallah, K. Jurewicz, E. Frackowiak, M. Reda, J. Parmentier, J. Patarin and F. Beguin, *Materials Science and Engineering B, Solid-State Materials for Advanced Technology*, 2004, **108**, 148–155.
145. H. J. Liu, W. J. Cui, L. H. Jin, C. X. Wang and Y. Y. Xia, *Journal of Materials Chemistry*, 2009, **19**, 3661–3667.
146. C. M. Niu, E. K. Sichel, R. Hoch, D. Moy and H. Tennent, *Applied Physics Letters*, 1997, **70**, 1480–1482.
147. K. H. An, W. S. Kim, Y. S. Park, Y. C. Choi, S. M. Lee, D. C. Chung, D. J. Bae, S. C. Lim and Y. H. Lee, *Advanced Materials*, 2001, **13**, 497–+.
148. M. D. Stoller, S. J. Park, Y. W. Zhu, J. H. An and R. S. Ruoff, *Nano Letters*, 2008, **8**, 3498–3502.
149. Y. Wang, Z. Q. Shi, Y. Huang, Y. F. Ma, C. Y. Wang, M. M. Chen and Y. S. Chen, *Journal of Physical Chemistry C*, 2009, **113**, 13103–13107.
150. J. Chmiola, C. Largeot, P. L. Taberna, P. Simon and Y. Gogotsi, *Science*, 2010, **328**, 480–483.
151. D. Pech, M. Brunet, H. Durou, P. H. Huang, V. Mochalin, Y. Gogotsi, P. L. Taberna and P. Simon, *Nature Nanotechnology*, 2010, **5**, 651–654.
152. D. W. Wang, F. Li, Z. G. Chen, G. Q. Lu and H. M. Cheng, *Chemistry of Materials*, 2008, **20**, 7195–7200.
153. D. Hulicova-Jurcakova, A. M. Puziy, O. I. Poddubnaya, F. Suarez-Garcia, J. M. D. Tascon and G. Q. Lu, *Journal of the American Chemical Society*, 2009, **131**, 5026–+.
154. V. Subramanian, H. W. Zhu and B. Q. Wei, *Electrochemistry Communications*, 2006, **8**, 827–832.
155. S. Chen, J. W. Zhu, X. D. Wu, Q. F. Han and X. Wang, *ACS Nano*, 2010, **4**, 2822–2830.
156. K. H. An, K. K. Jeon, J. K. Heo, S. C. Lim, D. J. Bae and Y. H. Lee, *Journal of the Electrochemical Society*, 2002, **149**, A1058–A1062.
157. K. Zhang, L. L. Zhang, X. S. Zhao and J. S. Wu, *Chemistry of Materials*, 2010, **22**, 1392–1401.

4 Solar Energy Storage with Nanomaterials

Nurxat Nuraje, Sarkyt Kudaibergenov, and Ramazan Asmatulu

CONTENTS

4.1 GENERAL INTRODUCTION

Although coal, natural gas, and petroleum-based fossil fuels have been mainly recognized to meet the energy requirements in the world for a while, technological developments, new demands, and environmental and health concerns forced many countries to seek new sources of energy. Thus, new research emphasis has been directed on the utilization of alternative renewable sources of energy. Other alternatives of energy, such as nuclear, hydraulic, biomass, and geothermal are not adequate or have other concerns to meet this huge demand. Besides, the exploitation of the major sources of energy (fossil fuels) has a massive impact on the environment as these fuels are considered to have enormous global warming (Kamat 2007). Considering the population growth, economic development, environmental, and health concerns, and increasing demands for the new energy, the world has been seeking to find alternate energy sources to replace the conventional sources in an economic and environmental ways.

Hydrogen-based energy systems (e.g., fuel cells) are of great interest worldwide because of their environmentally clean nature and high efficiency. Hydrogen and oxygen are usually used to produce electricity in hydrogen fuel cells. The by-product of the hydrogen fuel cells is clean water (Lewis 2007). Hydrogen is the most viable choice of storing solar energy in the absence of sunlight. As shown in Equation 4.1, solar energy is converted and stored in a hydrogen chemical bond. Currently, hydrogen is usually produced from fossil fuels, based on the reaction given in Equation 4.2. During this process, CO_2 is one of the products generated along with hydrogen gas, which causes a greenhouse effect (Kitano and Hara 2010; Kudo and Miseki 2009). This process is not an environmentally friendly technique, so hydrogen is the best source produced using an artificial photosynthesis.

$$H_2O \xrightarrow[\text{catalyst}]{h\gamma} H_2 + \frac{1}{2}O_2 \qquad (4.1)$$

$$CH_4 + 2H_2O \longrightarrow 4H_2 + CO_2 \quad \Delta G^\circ = 131 \text{ kJ mol}^{-1} \qquad (4.2)$$

The sun is recognized to be the major promising energy source for modern society since it is an inexhaustible natural source with a magnitude of 3.0×10^{24} J/year ($\sim 10^5$ TW) (Walter et al. 2010a; Gratzel 2005; Lewis 2007; Hagfeldt et al. 2010; Cook et al. 2010). The current energy consumption of the world is around 4.0×10^{20} J/year (~ 12 TW), corresponding to about 0.01% of solar energy reaching the Earth's surface. Solar energy that reaches the Earth far exceeds the need of the modern society. According to the calculations (Gratzel 2001; Gratzel 2005; Turner 1999), an area of 10^5 km² that is installed with solar cells at 10% working efficiency is enough to provide our energy needs without other alternatives. Even though the sun is an ideal source to meet our energy demand, new initiatives are required to improve the harnessing of incident photons and to improve storage capacity at a great rate of efficiency since solar energy density varies considerably with seasonal changes and locations, such as the Sahara Desert, Amazon region, equatorial region, and North and South Poles (Walter et al. 2010b; Bolton 1977; Boer and Rothwarf 1976; Chen et al. 2010).

One of the main issues for solar energy is that the energy conversion and energy storage rates of the solar energy systems are considerably lower and need to be improved. Nature provides an inspiration to solving these problems through photosynthesis (Kalyanasundaram and Graetzel 2010; Hagfeldt et al. 2010). Nanotechnology is an emerging technology that could provide light-energy harvesting assemblies and an innovative strategy for desired energy conversion devices (Mao and Chen 2007; Fichtner 2005). Nanomaterials, as building blocks for solar energy conversion devices, have been applied in the following three ways (Kutal 1983): (i) the assembly of molecular and clusters of donor–acceptor mimicking photosynthesis, (ii) the production of solar fuel using semiconductor-assisted photocatalysis, and (iii) the use of nanostructured semiconductor materials in solar cells. Among the nanostructured solar energy conversion devices (Chen et al. 2010; Li and Zhang 2010; Walter et al. 2010b), binary and ternary metal oxides are the most widely considered ones.

Although a number of articles (Walter et al. 2010b; Chen et al. 2010; Hagfeldt et al. 2010; Chen 2009; Kudo and Miseki 2009; Li and Zhang 2010; Minggu et al.

2010; Navarro Yerga et al. 2009; Zhu and Zäch 2009) have been published on solar cells and solar fuels in general, only a few articles were published on the metal oxide materials and their applications. Thus, this book chapter summarizes the use of metal oxide-based nanomaterials in solar energy conversion using binary and ternary metal oxides. The basic principle, synthesis methods for metal oxides, and current strategies and future perspectives are discussed in detail.

4.2 METAL OXIDE PHOTOCATALYSTS IN WATER SPLITTING

4.2.1 General Background

The decomposition of water into hydrogen and oxygen using hetereogenous photocatalysts was initiated by the work of Fujishima and Honda (Fujishima and Honda 1972). In this work, overall water splitting was conducted using a photoelectrochemical (PEC) cell consisting of titania as the photoanode and platinum as the counter electrode under UV irradiation and external bias. This work has stimulated the research involving overall water splitting using particulate photocatalysts. Recently, there has been significant progress in developing various photocatalysts under visible light (Kudo and Miseki 2009; Lee 2005; Kitano and Hara 2010; Kudo 2003, 2006; van de Krol et al. 2008; Walter et al. 2010b). The maximum efficiency obtained during the overall water-splitting process is around 5.9%; however, these results still do not satisfy the requirement for the practical application (10%) (Navarro Yerga et al. 2009). Among the options, metal oxide-based photocatalysts possess advantages in comparison with other semiconductor photocatalysts since they have chemical stability, a negative band-gap position for hydrogen generation, and use the natural resources currently available. For instance, CdS series photocatalysts can be used in a special condition because of their instability (Kudo and Miseki 2009; Kitano and Hara 2010; Walter et al. 2010b). This study mainly focuses on the binary and ternary metal oxide photocatalysts and discusses their photocatalytic performance under UV and visible light conditions.

4.2.2 Basic Principle

Overall water splitting, as shown in Equation 4.1, is a thermodynamically unfavorable process and has a positive Gibbs free-energy change (ΔG = +237.2 kJ/mol, 2.46 eV per molecule) (Kudo and Miseki 2009; Navarro Yerga et al. 2009). Thus, the photon energy is required to overcome the large positive change of the Gibbs free energy during the water-decomposition process. As shown in Equation 4.3, 4.4, and 4.5, the decomposition of water into hydrogen and oxygen using an electrochemical cell is a two-electron stepwise process. Semiconductor photocatalysts can absorb photons and generate electrons and holes on their surfaces by absorbing solar energy. The photogenerated electron and hole pairs are able to drive the reduction and oxidization reactions, respectively, of the water molecules.

$$\text{Oxidation: } H_2O + 2h^+ \longrightarrow 2H^+ + \frac{1}{2}O_2 \qquad (4.3)$$

$$\text{Reduction } 2H^+ + 2e^- \longrightarrow H_2 \qquad (4.4)$$

$$\text{Overall reaction: } H_2O \longrightarrow H_2 + \frac{1}{2}O_2 \qquad (4.5)$$

Photocatalysts are utilized for water splitting in two ways: one with photo-electrochemical cells (Youngblood et al. 2009; Gratzel 2001) and the other in a particulate photocatalytic system (Kitano and Hara 2010; Kudo and Miseki 2009). In photoelectrochemical cells, photocatalysts are used to make electrodes and are immersed in an aqueous electrolyte. In the cell, the photocatalyst can be used to transport electrons to the electrode through an external circuit (Youngblood et al. 2009) or it can directly absorb photons without a dye sensitization (Gratzel 2001). In a particulate photocatalytic system (Kudo and Miseki 2009), the photocatalysts are suspended in a solution. Each particle acts as a microphotoelectrode, which conducts the redox reaction of water molecules. This system has some disadvantages in the charge, as well as hydrogen and oxygen separations. Despite these disadvantages, photocatalysts are widely studied because of their simple, scalable, and inexpensive advantages (Kitano and Hara 2010).

In the particulate photocatalytic system (Figure 4.1a), an electron is excited in the conduction band (CB) after a photocatalyst absorbs light. The electron in the CB reduces the water molecules to hydrogen gas on the active sites of the photocatalyst. The positive holes in the valence band are transferred to the active site of the photo-catalyst to oxidize the water molecules into oxygen gas with the help of a co-catalyst.

This process includes three main steps, as shown in Figure 4.1b. In the first step, the semiconductors absorb photons and create electron/hole pairs in the conduction and valence bands, respectively. The semiconductors only absorb the energy of incident light, which is larger than that of the band gap. In the second step, charge separation and migration of photogenerated carriers occur. The crystallinity, shape, size of particles, and crystal structures of the photocatalyst are important in this step. Defects are not desired in the charge separation if they act as a recombination center. The majority of photogenerated electron–hole recombination occurs in bulk defects or on surface defects. Surface defects serve as charge-carrier traps as well as adsorption sites and can inhibit electron–hole recombination (Zhao et al. 2012; Kong et al. 2011). Size is also critical to separating electron/hole pairs. If the particle size is small, then the photogenerated carriers can easily be separated and reach the surface, and the recombination probability of the photogenerated carriers can be decreased. In the third step, chemical reactions (decomposition of water) on the surface of the photocatalyst occur. The important parameters for the decomposition of water are surface active sites (such as a co-catalyst) and the quantity of surface areas. The co-catalyst on the surface of the photocatalyst can perform as electron sinks to separate the excited electrons from the semiconductor band-gap excitation, which retard the charge recombination process. Therefore, co-catalysts, such as Pt and RuO_2 are attached to the surface of the photocatalyst. The smaller the particle size of photocatalyst, the more the active site for water decomposition occurs. The valence of metal oxides consists of a 2p orbital of oxygen, which is sufficient to

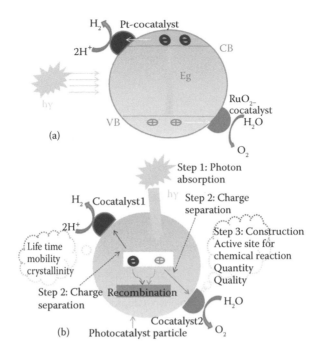

FIGURE 4.1   (See color insert.) (a) Schematic diagram for overall water-splitting reaction on a solid photocatalyst. (b) Main process of electron transport during water-splitting reaction. (Reproduced by permission of The Bentham Science Publisher. Nuraje, N., R. Asmatulu, and S. Kudaibergenov. 2012. Metal oxide-based functional materials for solar energy conversion: a review. *Current Inorganic Chemistry* 2 (2):124–146.)

oxidize the water molecules. Therefore, metal oxides do not require a co-catalyst for the oxidization of water.

In the half reaction of hydrogen or oxygen production, sacrificial agents are used to improve the efficiency of the products. The co-catalyst plays an important role in reducing the overall potential, providing active sites for the reaction, and improving electron transport. In the half reaction of hydrogen production, as shown in Figure 4.2A, the methanol, dieathol amine, triethanol amine, and ethanol can be used as sacrificial agents. For more detail, specific references can be checked to see the specific photocatalysts and co-catalysts (Kudo and Miseki 2009; Chen et al. 2010). Under UV irradiation, metal oxide photocatalysts with a co-catalyst are directly exposed to the irradiation. After the photocatalyst absorbs the photon and generates an electron/hole pair, the electron will be responsible for hydrogen production at the surface of the co-catalyst. The sacrificial agents, the electron donor element (e.g., methanol), provide an electron to the positive hole in the valence band of the photocatalyst. Since most metal oxide band gaps are in the UV region, dyes are used to absorb the visible light and transport the electron to the conduction band gap of the metal oxides under the visible-light irradiation. The separated electrons are transported to the co-catalyst surface of the metal oxide where water molecules are reduced. The reaction

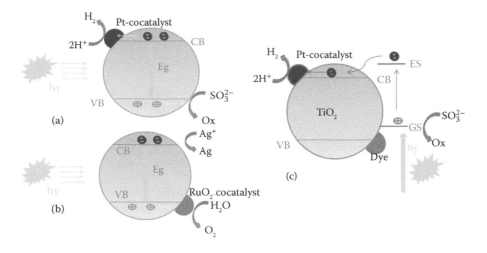

FIGURE 4.2 (**See color insert.**) (a) Half reaction for hydrogen production using photo-catalyst. (b) Half reaction for oxygen evolution using photocatalyst. (c) Dye-sensitized metal oxide particulate system for hydrogen evolution. (Reproduced by permission of The Bentham Science Publisher. Nuraje, N., R. Asmatulu, and S. Kudaibergenov. 2012. Metal oxide-based functional materials for solar energy conversion: a review. *Current Inorganic Chemistry* 2 (2):124–146.)

continues until the exhaustion of sacrificial agents, as shown in Figure 4.2c. In the half reaction of water oxidation, an electron acceptor, such as silver nitrate or iron (III) chloride, is applied to take photogenerated electrons from the conduction band of the photocatalysts (Figure 4.2b). The water oxidization reaction takes place at the surface of the co-catalysts, such as RuO_2 and IrO_2, and positive holes in the valence band of the metal oxides oxidize the water to produce oxygen under UV or visible irradiation. The process totally depends on the band gap of the photocatalysts.

For the overall water splitting under visible-light irradiation, the following two strategies are applied. Figure 4.3a shows the dual-bed configuration, which does not require an external circuit or wire to transport electrons. In the dual-bed (or Z-scheme), electron transport depends on the redox mediator. The second case is the dye-sensitized photoelectrochemical cell, which requires an external circuit or wire to transport the electron from one electrode to other one (Figure 4.3b). In most cases, dye is required to sensitize the porous electrodes made of metal oxide nanoparticles. In the dual photocatalyst system (Figure 4.3a), two different semiconductors with band gap and band positions can be combined to form a photocatalyst system for the overall water splitting. One type of semiconductor is usually used for the oxygen evolution. Another is used for hydrogen production. They have been chosen for an overall water splitting based on the band positions. The photocatalyst (catalyst 2 in Figure 4.3a) in the conduction band position, which is more negative than the reduction potential of water, is used to reduce the water molecules. The other photocatalyst (catalyst 1 in Figure 4.3a), which has a more positive valence band in comparison to the oxidation potential of water molecules, is chosen to oxidize the water molecules.

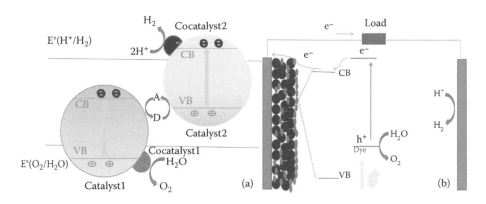

FIGURE 4.3 **(See color insert.)** (a) Schematic diagram of dual-bed configuration for overall water-splitting reaction on two different solid photocatalysts, and (b) dye-sensitized photoelectrochemical cell for overall water splitting. (Reproduced by permission of The Bentham Science Publisher. Nuraje, N., R. Asmatulu, and S. Kudaibergenov. 2012. Metal oxide-based functional materials for solar energy conversion: a review. *Current Inorganic Chemistry* 2 (2):124–146.)

In the hydrogen evolution photocatalyst, after water is reduced to H_2 by photoexcited electrons, the electron donor (D in Figure 4.3a) is oxidized through holes to its electron-acceptor form (A in Figure 4.3a). In the oxygen evolution photocatalyst, the photoexcited electrons reduce the electron acceptor (A) back to its electron-donor form (D), as holes oxidize the water to oxygen.

In a dye-sensitized photoelectrochemical cell (Figure 4.3b), a metal oxide nanoparticle employed to form a photoelectrode is loaded with dye molecules. Here, the main idea is very similar to the DSSC. Upon visible-light irradiation, the dye absorbs photons in order to excite the electron, which in turn is injected into the conduction band of the metal oxide electrode. The electron reaches the counter electrode through an external wire where the electron reduces water to hydrogen gas. With the help of co-catalysts, such as IrO_2 or RuO_2, the oxidization reaction of water occurs at the surface of the co-catalyst. The dye are regenerated after taking the electrons from the water oxidization reactions (Youngblood et al. 2009).

The photocatalytic activity of the water-splitting system can be evaluated in two common ways. The first is the direct measurement of the amounts of hydrogen, while the second is an indirect method, whereby the electron is transported from the semiconductor to the water within a certain time period under light irradiation. It is difficult to directly compare the results from different research groups and the photocatalytic hydrogen generation systems for the same catalyst, even if they test the same photocatalyst, due to the different experimental setups/procedures. Usually, the rate of gas (O_2 and H_2) evolution (units such as $\mu mol/h^{-1}$ and $\mu mol/h/g$ catalyst) is applied to make the relative comparison between different photocatalysts under similar experimental conditions. Hence, the quantum yield is always used to make a direct comparison. A thermopile or Si photodiode can be utilized to determine the incident photons. The real amount of absorbed photons is difficult to measure because of dispersion system scattering. The real quantum yield, Equation 4.6 is larger than the apparent quantum yield, Equation 4.7, because the number of absorbed photons is usually smaller than that of the incident light; however, the apparent quantum yield is usually reported. The number of reacted electrons is calculated from the amount of produced hydrogen gas. In this case, the quantum yield is different from the solar energy conversion, as seen in Equation 4.8.

$$\text{Overall quantum yeild } (\%) = \frac{\text{Number of reacted electrons}}{\text{Number of absorbed photons}} \times 100\% \quad (4.6)$$

$$\text{Apparent quantum yeild } (\%) = \frac{\text{Number of reacted electrons}}{\text{Number of incident photons}} \times 100\% \quad (4.7)$$

$$\text{Solar energy conversion } (\%) = \frac{\text{Output energy as } H_2}{\text{Energy of incident solar light}} \times 100\% \quad (4.8)$$

4.2.3 METAL OXIDE PHOTOCATALYST (UV-ACTIVE)

First, due to the simplicity of the molecular formula of the binary metal oxide and the common photocatalyst, it is convenient to introduce the photocatalysts in the particulate system including the synthesis method and hydrogen production at different conditions. On the basis of the binary metal oxide, ternary metal oxide photocatalysts can be readily understood. With the introduction of metal oxides, their photocatalytic activity can be reasonably compared, although all of these photocatalytic evaluation conditions for various photocatalysts, including light source (intensity), different co-catalysts, and sacrificial chemicals, are different. Since Fujishima and Honda (Fujishima and Honda 1972) reported photocatalytic activity of titania in 1972, water splitting has been under intensive study. Many oxide photocatalyst (Chen et al. 2010) materials have been investigated, so far. They have been introduced and divided into the following three metal-oxide groups: d^0 (Ti^{4+}, Zr^{4+}, Nb^{5+}, Ta^{5+}, W^{6+}, and Mo^{6+}), d^{10} (In^{3+}, Ga^{3+}, Ge^{4+}, Sn^{4+}, and Sb^{5+}), and f^0 (Ce^{4+}). Therefore, the introduction of binary photocatalysts is begun using UV and visible light conditions.

4.2.3.1 Binary Metal Oxide Photocatalysts (UV-Active)

The synthesis methods for the binary metal oxide photocatalysts is tabulated in reference (Nuraje et al. 2012). Titanium dioxide is usually synthesized via sol–gel methods, although there are many other ways, such as solid-state reactions, polymerizable complex method, hydrothermal method, etc. Usually particles synthesized by soft methods, such as a polymerizable complex and sol–gel method provide higher performance than those of a synthesized solid state reaction due to the small particle size and good crystallinity (Figure 4.1b, step 2).

The band gaps of metal oxide with d^0 metal ions are usually formed from an O 2p orbital and nd orbitals of metal cations, which are more negative than the zero potential of hydrogen ions. The band gaps of the metal oxides are usually in the UV range. Powdered titania photocatalysts cannot split water without any modifications, such as a co-catalyst (Pt) (Kudo and Miseki 2009). Hydrogen production experiments have been conducted using a TiO_2 photocatalyst (band gap: 3.2 eV; crystal structure: anatase) under different conditions, including pure water, vapor, and an aqueous solution including an electron donor with the assistance of a co-catalyst (Duonghong et al. 1981; Sayama and Arakawa 1997; Shi et al. 2006; Zhang et al. 2008; Yamaguti and Sato 1985; Kudo et al. 1987; Tabata et al. 1995). The addition of NaOH or Na_2CO_3 was used to split water with a loaded Pt. (Sayama and Arakawa 1994; Kudo et al. 1987). Under UV irradiation, the efficiency of titania doped with metal ions (Jing et al. 2005; Sasikala et al. 2008; Zalas and Laniecki 2005) improved. Mesoporous titania structures of MCM-41 and MCM-48 (Zhao et al. 2010) showed high photocatalytic activity over bulk titania under UV irradiation since they have high surface areas.

ZrO_2 (Band gap: 5.0 eV) is a photocatalyst that can split water without a co-catalyst (Sayama and Arakawa 1994, 1993; Sayama and Arakawa 1996; Reddy et al. 2003) under UV irradiation because of the high conduction band position. Photocatalytic activity of ZrO_2 decreased when it was loaded with co-catalysts such as Pt, Au, Cu, and RuO_2. It is likely that the large electronic barrier height of the semiconductor band metal hurdled electron transport. However, the photocatalytic activity improved with the addition of Na_2CO_3.

Nb_2O_5 (Band gap: 3.4 eV) (Soares et al. 2011) is not active for water splitting without any modification under UV irradiation (Sayama et al. 1996). It decomposes water efficiently in a mixture of water and methanol after being loaded with a platinum co-catalyst (Chen et al. 2007). Its higher photocatalytic activity under UV irradiation was observed as assembled mesoporous Nb_2O_5 (Chen et al. 2007). Ta_2O_5 with a band gap of 4.0 eV is a well-known photocatalyst. It can produce a small amount of hydrogen and no oxygen without any modification (Sayama et al. 1996). Ta_2O_5 loaded with NiO and RuO_2 shows great photocatalytic activity for generating both hydrogen and oxygen (Kato and Kudo 1998). Addition of Na_2CO_3 and a mesoporous structure of the catalyst show enhanced photocatalytic activity (Takahara et al. 2001; Sayama and Arakawa 1994). The narrow band gap metal oxide WO_3, which has a relatively positive conduction band, proved dissatisfactory for hydrogen production and created major impediments for the efficient performance of these two photocatalysts in visible-light–driven water splitting (Kudo and Miseki 2009).

Nanostructured VO_2 with a body-centered cubic (bcc) structure and a large optical band gap of 2.7 eV demonstrated excellent photocatalytic activity in hydrogen production from a solution of water and ethanol under UV irradiation (Wang et al. 2008), and it exhibited a high quantum efficiency of 38.7%. All of the metal oxides with d^{10} metal ions (Zn^{2+}, In^{3+}, Ga^{3+}, Ge^{4+}, Sn^{4+}, Sb^{5+}) are effective photochemical water-splitting catalysts under UV irradiation. Among the d^{10} metal oxides, ZnO (band gap: 3.2 eV) (Thorat et al. 2011) and In_2O_3 (band gap: 3.6 eV) (Pentyala et al. 2011) are not photocatalysts for water splitting because of their photo instability and low conduction band level (Kudo and Miseki 2009). Ni-loaded Ga_2O_3 (Band gap: 4.6 eV) (Yanagida et al. 2004; Kudo and Miseki 2009) showed decent photocatalytic performance for overall water splitting. The addition of Ca, Cr, Zn, Sr, Ba, and Ta ions to the photocatalyst enhanced photocatalytic activity. In particular, Zn ion-doped Ga_2O_3 showed remarkable photocatalytic activity when Ni was used as a co-catalyst, with 20% of an apparent quantum yield (Sakata et al. 2008). The f^0 block metal oxide (3.19 eV) (Goharshadi et al. 2011), CeO_2, is not active for splitting water under UV irradiation. However, CeO_2 doped with Sr shows reasonable photocatalytic activity (Kadowaki et al. 2007).

4.2.3.2   UV-Active Ternary Metal Oxide Photocatalysts

Although binary metal oxides with d^0, d^{10}, and f^0 metal ions show efficient photocatalytic activity, their ternary oxides have been widely studied and also proven to have the same effect. For example, $SrTiO_3$ (band gap: 3.2 eV) and $KTaO_3$ (band gap: 3.6 eV) (Kudo and Miseki 2009) photoelectrodes with a perovskite structure can split water without an external bias because of their high conduction band (Chen et al. 2010). These materials can be used as powdered photocatalysts. Domen and coworkers studied the photocatalytic performance of NiO-loaded $SrTiO_3$ powder for water splitting (Domen et al. 1980; Domen, Naito, Onishi, Tamaru, et al. 1982; Domen et al. 1982; Domen et al. 1986; Kudo et al. 1988; Domen, Kudo, Onishi, et al. 1986). A reduction in H_2 is responsible for the activation of the NiO co-catalyst for H_2 evolution. Then, subsequent O_2 oxidation to form a NiO/Ni double-layer structure provides a further path for the electron migration from a photocatalyst substrate to a co-catalyst surface. The NiO co-catalyst prevents the back reaction between H_2 and O_2, which is totally different for Pt. The NiO co-catalyst has often been applied to many photocatalysts for water splitting.

The $SrTiO_3$ photocatalyst with an Rh co-catalyst was studied (J.M. Lehn 1980). The enhanced photocatalytic activity of $SrTiO_3$ was also reported through a new modified preparation method (Liu et al. 2008) or suitable metal–cation doping (such as La^{3+} (Qin et al. 2007), Ga^{3+} (Takata and Domen 2009), and Na^+ (Takata and Domen 2009)). Many ternary titanates are efficient photocatalysts for water splitting under UV irradiation. Shibata et al. (1987), studied the H_2 evolution of photocatalysts of $Na_2Ti_3O_7$ (crystal structure: layered structure), $K_2Ti_2O_5$ (crystal structure: layered structure), and $K_2Ti_4O_9$ (crystal structure: layered structure) from aqueous methanol solutions in the absence of a Pt co-catalyst. The quantum yield of the materials studied for H^+-exchanged $K_2Ti_2O_5$ reaches 10%. The method of catalyst preparation also shows a different activity.

$BaTiO_3$ (band gap: 3.22 eV; crystal structure: perovskite) prepared with a polymerized complex method has high photocatalytic activity in comparison with materials prepared by the traditional method (Yamashita et al. 1998) because of the smaller size and larger surface area. $CaTiO_3$ (band gap: 3.5 eV; crystal structure: perovskite) loaded with Pt showed a good photocatalytic activity under UV irradiation (Mizoguchi et al. 2002). The activity of $CaTiO_3$ doped with a Zr^{4+} solid solution was further increased. Quantum yields were reported to be up to 1.91% and 13.3% for H_2 evolution from pure water and an aqueous ethanol solution, respectively (Sun et al. 2007). Kim and coworkers investigated a series of perovskites including La_2TiO_5, $La_2Ti_3O_9$, and $La_2Ti_2O_7$ with layered structures and reported much higher photocatalytic activities under UV irradiation than bulk $LaTiO_3$ (Kim et al. 2001; Kim et al. 2002; Kim et al. 2004). The photoactivities of $La_2Ti_2O_7$ doped with Ba, Sr, and Ca was improved sufficiently (Kim et al. 2005). $La_2Ti_2O_7$ (band gap: 3.8 eV) prepared using a polymerized approach showed higher photoactivity than the traditional solid-state method (Li, Chen, et al. 2008).

A niobate photocatalyst (Domen, Kudo, Shibata, et al. 1986; Domen, Kudo, Shinozaki, et al. 1986), $K_4Nb_6O_{17}$ (band gap: 3.4 eV; crystal structure: layered structure) was recently reported as a high and stable material for H_2 evolution from an aqueous methanol solution without any co-catalysts. Potassium ions are located in two different kinds of interlayers of this niobate, which are composed of layers of niobium oxide sheets. The potassium ions between the niobium oxide layers can be replaced by many other cations, such as transition metal ions. The catalysts exchanged with H^+, Cr^{3+}, and Fe^{3+} ions showed higher activity than the original $K_4Nb_6O_{17}$. In particular, the H^+-exchanged $K_4Nb_6O_{17}$ showed the highest activity for the H_2 evolution from an aqueous methanol solution (quantum yield, ca. 50% at 330 nm) (Kudo et al. 1988). After loading with co-catalysts, such as NiO (Ikeda et al. 1997; Kudo et al. 1989; Domen et al. 1990; Sayama et al. 1990), Au (Iwase et al. 2006), Pt (Sayama et al. 1991; Sayama et al. 1998), and Cs (Chung and Park 1998), $K_4Nb_6O_{17}$ became more efficient for overall water splitting (Sayama et al. 1996). Other alkaline-metal niobates loaded with co-catalysts, such as $LiNbO_3$, $NaNbO_3$, $KNbO_3$, and $Cs_2Nb_4O_{11}$ (band gap: 3.7 eV; crystal structure: pyrochlore-like), catalyzed water under UV irradiation (Li, Kako, et al. 2008; Ding et al. 2008; Zielinska, Borowiak-Palen, and Kalenzuk 2008; Miseki, Kato, and Kudo 2005). A new NiO-loaded $ZnNb_2O_6$ photocatalyst (band gap: 4.0 eV; crystal structure: columbite) showed a high activity under UV irradiation (Kudo, Nakagawa, and Kato 1999).

Kato and Kudo (Kato and Kudo 2001; Kudo and Miseki 2009) studied the photocatalytic water splitting of $LiTaO_3$ (band gap: 4.7 eV; crystal structure: ilmenite), $NaTaO_3$ (band gap: 4.0 eV; crystal structure: perovskite), and $KTaO_3$ (band gap: 3.6 eV; crystal structure: perovskite) and reported higher activities under UV irradiation. Among the tantalates, NiO-loaded $NaTaO_3$ demonstrated the highest activity. As

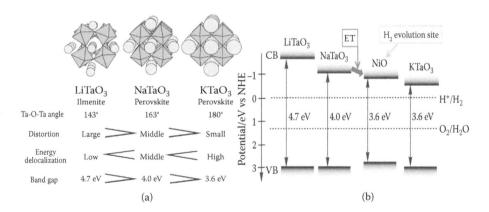

FIGURE 4.4 (See color insert.) (a) Crystal and energy structures of alkali tantalate and (b) band structures of alkali tantalates with NiO co-catalyst. (From Kato, H., and A. Kudo. 2001. Water splitting into H_2 and O_2 on alkali tantalate photocatalysts $ATaO_3$ (A = Li, Na, and K). *J. Phys. Chem. B* 105 (19):4285–4292. doi: 10.1021/jp004386b. Reprinted with permission of AAAS.)

shown in Figure 4.4a, the photocatalytic activity of tantalates depends on alkaline cations, since the bond angles of Ta–O–Ta are different. When the bond angles reach 180 degrees, excited electron–hole pairs are transported easily, and band gap becomes smaller. However, in NiO-loaded tantalates, the conduction band of $NaTaO_3$ is higher than that of NiO, as shown in Figure 4.4b, and the electron is easy to migrate from the conduction band to NiO under this condition. Although the electron transfer is easier in $KTaO_3$, the photocatalytic activity is low because of mismatching between the conduction band of $KTaO_3$ and NiO. This finding is also used to explain another photocatalytic system in which $Sr_2Ta_2O_7$ (band gap: 4.6 eV; crystal structure: layered perovskite) has a higher photocatalytic activity than $Sr_2Nb_2O_7$ (band gap: 4.0 eV; crystal structure: layered perovskite) (Kudo et al. 1999). The lanthanum ions used as dopants greatly increase the photocatalytic activity of $NiO/NaTaO_3$ (Kato et al. 2003). The study showed that the optimized photocatalyst (NiO (0.2 wt%)/$NaTaO_3$:La (2%)) provided a high activity with yield of 56% for water splitting. This high photocatalytic activity continued for more than 400 hrs under the mercury lamp of high irritation (400 W). The reason for high photocatalytic activity of La-doped $NaTaO_3$ is that the particle size of the material has decreased and an ordered surface structure has formed.

The small number of heterogeneous photocatalysts based on either tungstates or molybdates for H_2 or O_2 evolution were found to be active for water splitting only under UV irradiation even though they showed an optical absorption in the visible region. Inoue and coworkers (Kadowaki et al. 2006; Saito et al. 2004) reported a high and stable photocatalytic activity of $PbWO_4$ (band gap: 3.9 eV; crystal structure: scheelite) possessing a WO_4 tetrahedron for the overall splitting of water. $PbMoO_4$ (band gap: 3.31 eV; crystal structure: scheelite) produced hydrogen from aqueous methanol solution and oxygen evolution from aqueous silver nitrate solution under UV irradiation (Kudo et al. 1990). Kudo and coworkers (Kudo et al. 1990; Kudo and Kato 1997; Kato, Matsudo, and Kudo 2004) have broadly extended the study of the photocatalytic activities of tungstates and molybdates. Both $Na_2W_4O_{13}$ (band gap: 3.1 eV; crystal structure: layered structure) and $Bi_2W_2O_9$ (bandgap: 3.0 eV; crystal structure: aurivillius-like) (Kudo and Hijii 1999) showed photocatalytic activity for the hydrogen evolution in the presence of Pt as co-catalyst and oxygen evolution. However, low catalytic activity of Bi_2MoO_6 (band gap: 3.0 eV; crystal structure: aurivillius structure) (Kudo and Hijii 1999) was reported in the $AgNO_3$ aqueous solution. Sato and coworkers (Sato, Kobayashi, et al. 2003; Sato, Saito, et al. 2003; Sato, Kobayashi, and Inoue 2003; Sato et al. 2001a; Sato et al. 2001c) reported that the photocatalytic properties of indates with the octahedrally In^{3+} d^{10} configuration ion could be utilized for the water decompositions.

The large photocatalytic activity for water decomposition under UV irradiation was reported for RuO_2-dispersed $CaIn_2O_4$ (crystalline structure: tunnel structure), $SrIn_2O_4$ (band gap: 3.6 eV; crystalline structure: tunnel structure), and $Sr_{0.93}Ba_{0.07}In_2O_4$ but was very poor for RuO_2-dispersed $LiInO_2$, $NaInO_2$ (band gap: 3.9 eV; crystal structure: layered structure) and $LaInO_3$ (band gap: 4.1 eV), $NdInO_3$. Distorted InO_6 octahedra is responsible for the photocatalytic activity of indates because the dipole moments provide internal fields for a charge separation in the early stage of photoexcitation. Metal oxides with d^{10} configurations and distorted octahedral and/or tetrahedral structures were reported to have stable activity in decomposing water to H_2 and O_2 under UV irradiation when combined with RuO_2 or Pt as co-catalysts (Sato et al. 2001b; Sato et al. 2001c; Sato et al. 2002; Ikarashi et al. 2002; Sato et al. 2004; Kadowaki et al. 2005). For example, they are the distorted SbO_6 octahedra in $Ca_2Sb_2O_7$ (band gap: 3.9 eV; crystal structure: weberite), $Sr_2Sb_2O_7$ (band gap: 4.0 eV; crystal structure: weberite), $CaSb_2O_6$ (band gap: 3.6 eV; crystal structure: layered structure), and $NaSbO_3$ (band gap: 3.6 eV; crystal structure: ilmenite); the distorted GeO_4 tetrahedra in Zn_2GeO_4 (band gap: 4.6 eV; crystal structure: willemite); and the distorted InO_6 octahedra and GeO_4 tetrahedra in $LiInGeO_4$. Photocatalytic activity of the f^0 metal oxide photocatalyst $BaCeO_3$ (band gap: 3.2 eV; crystal structure: perovskite) was reported by Yuan et al. (2008), where overall water splitting with the aid of RuO_2 loading was observed under UV irradiation.

4.2.4 VISIBLE LIGHT-SENSITIVE METAL OXIDE PHOTOCATALYST

WO_3 (band gap: 2.8 eV) is widely studied for O_2 evolution in the presence of sacrificial reagents such as Ag^+ and Fe^{3+} under visible light (Darwent and Mills 1982; Erbs et al. 1984; Miseki et al. 2010). Both Bi_2WO_6 (band gap: 2.8 eV; crystal structure: aurivillius structure) and Bi_2MoO_6 (band gap: 2.7 eV; crystal structure: aurivillius structure) are active for an O_2 evolution reaction and not active for H_2 evolution because of the low conduction band level (Kudo and Miseki 2009). α-Fe_2O_3 has a band gap of 2.2 eV and a low-cost semiconductor. The major drawback of this catalyst is its high resistivity and high recombination rate of photogenerated charge carriers (Satsangi et al. 2008; Ingler, Baltrus, and Khan 2004; Aroutiounian et al. 2002). Most research has focused on increasing its photoconductivity and reducing the recombination rates of the charge carriers.

To make photocatalysts of a visible-light active for water splitting, the following approaches have been applied (Kudo and Miseki 2009; Chen et al. 2010): (1) metal or/and nonmetal ion doping, (2) dye sensitization of photocatalysts, (3) development of a novel single phase for active photocatalysts through band-gap engineering, and (4) development of solid solutions to control the band structure. In the next section, three of these approaches are discussed. The last case is not included because it is a very effective method for sulfide-based semiconductors.

4.2.4.1 Doped Metal Oxide Photocatalyst (Visible Light)

The engineering band gap of metal oxides with metal ions or nonmetal ions is the most common technique to prepare visible light-driven photocatalysts. Usually, titanium photocatalyst is a host material for many doping processes. With the doping of transition metal cations, in general, the photocatalytic behavior considerably reduced due to the recombination formations between photogenerated electrons and holes. This can even happen under a band-gap excitation.

However, the meaning of transition metal doping is to develop visible light-sensitive photocatalysts when a suitable dopant is chosen. Although some semiconductors such as CdS (Matsumura, Saho, and Tsubomura 1983; Reber and Rusek 1986) have a narrow band gap for visible light absorption, serious photocorrosion of CdS has been observed in the photocatalytic reaction. Numerous researches were performed on the modification of wide–band-gap photocatalysts using metal-ion doping including doped TiO_2, doped $SrTiO_3$, and doped $La_2Ti_2O_7$. Borgarello et al. (Chen and Mao 2007; Ji et al. 2010; Leung et al. 2010; Kato and Kudo 2002; Liu et al. 2006), produced hydrogen and oxygen via sustained water cleavage under visible-light (400–550 nm) irradiation with Cr^{5+}-doped TiO_2 as a photocatalyst (Borgarello et al. 1982). Previously, a number of different metal ions have been reported to dope TiO_2 to improve the visible-light absorption and photocatalytic activities (Chen and Mao 2007; Ji et al. 2010; Leung et al. 2010; Kato and Kudo 2002; Liu et al. 2006).

Doping not only allows metal oxides to use solar irradiation more effectively, but it also initiates effective photocatalytic reactions under both UV and visible-light irradiation. Pt^{4+}- and Ag^+-doped TiO_2 nanoparticles (Kim, Hwang, and Choi 2005; Rengaraj and Li 2006) proved the improvement of photocatalytic activities under visible-light or UV irradiation. In this case, these metal ions used as dopants contribute the visible light absorption and also serves as a recombination inhibitor by the electrons or holes. However, in some cases, the metal-ion dopants are responsible for low photocatalytic activity even under UV irradiation because of serving as recombination sites for photoinduced charges. Nonmetal ion (e.g., C, N, and S)-doped TiO_2 was studied for optical and photocatalytic properties. The absorption spectra are red-shifted to longer wavelengths and enhanced photocatalytic behaviors (Burda et al. 2003; Choi, Umebayashi, and Yoshikawa 2004; Ohno, Mitsui, and Matsumura 2003). Attention has also been paid to other oxide semiconductors as host photocatalysts for metal-ion doping. Wang et al. (Wang et al. 2004; Wang et al. 2009) investigated the behaviors of $SrTiO_3$ doped with Cr cations, including photophysical and photocatalytic.

The recombination sites of the dopant in the catalyst are formed at a certain degree. In this case, the efficiency level of the metal ions is generally separate. Thus, the formation of valence band in the oxide photocatalysts is critical to design visible light-driven photocatalysts. For this reason, orbitals of Pb 6s in Pb^{2+}, Bi 6s in Bi^{3+}, Sn 5s in Sn^{2+}, and Ag 4d in Ag^+ were used to build the valence bands above the valence band consisting of O 2p orbitals in the new metal oxide photocatalysts (Kudo and Miseki 2009; Kitano and Hara 2010). The conduction band of $BiVO_4$ (band gap: 2.4 eV) is composed of V 3d as in other d^0 oxide photocatalysts. The valence band formed with Bi 6s orbitals possesses the potential for water oxidation to form O_2.

4.2.4.2 Dye-Sensitized Metal Oxide Photocatalyst (Half Reaction)

Dye-sensitized photocatalysts for water splitting have been studied under visible-light irradiation by sensitization of wide–band-gap semiconductor photocatalysts, as shown. TiO_2 and $K_4Nb_6O_{17}$ loaded with dyes preceded H_2 evolution as shown in Figure 4.2c. Since the electrons are excited from the highest occupied molecular orbital (HOMO) to the lowest occupied molecular orbital (LUMO) of a dye using visible light, the electrons are introduced into the conduction band. During which process, H_2 gas is produced on the wide–band-gap of the photocatalysts. Ru $(bpy)_3^{2+}/K_4Nb_6O_{17}$ thin-film electrode is applied in the sensitization process to provide a photocurrent responding to visible light. Photocatalytic hydrogen production systems in which ruthenium (II) complex dyes that are sensitize to wide–band-gap semiconductors to visible light have been the focus of intensive research for many years. Duonghong and coworkers (Duonghong, Borgarello, and Graetzel 1981; Borgarello, Kiwi, Pelizzetti, Visca, and Graetzel 1981; Borgarello, Kiwi, Pelizzetti,

Visca, and Gratzel 1981; Duonghond, Serpone, and Grätzel 1984) proved Pt/RuO_2-loaded TiO_2 particles as an efficient photocatalysts for the water-splitting process with visible light using Ru $(bpy)_3^{2+}$ and its amphiphilic derivatives as sensitizers. Pt/TiO_2 sensitized with a polymer-pendant Ru $(bpy)_3^{2+}$ complex was reported for H_2 evolution in the presence of the sacrificial donor (e.g., EDTA) under the visible-light irradiation (Nakahira et al. 1988).

The dynamics of photoexcited Ru $(bpy)_3^{2+}$ intercalated into the $K_4Nb_6O_{17}$ interlayers were studied (Furube et al. 2002). They observed the fast and efficient electron transfer between Ru $(bpy)_3^{2+}$ and $K_4Nb_6O_{17}$ and showed fast and nonexponential decay of the transient bleaching of the Ru $(bpy)_3^{2+}$ band showed. The platinized $H_4Nb_6O_{17}$ nanoscrolls show better electron transfer mediator than the acid-restacked $HCa_2Nb_3O_{10}$ nanosheets. The apparent quantum yield of photocatalytic hydrogen production of $Pt/H_4Nb_6O_{17}$ nanoscrolls was <25% at 450 nm (Maeda et al. 2009). Since some transition-metal complexes (Gurunathan et al. 1997), especially Ru complexes, are quite expensive, metal free dyes (including but not limited to) porphine, xanthene, melocyanine, and coumarin, have been well studied. Some xanthene dye-sensitized particulate Pt-loaded semiconductor catalysts were studied by Shimidzu et al. (1985). They reported high quantum yields of hydrogen production for sensitized platinized semiconductors using heavy-halogenated xanthene dyes, e.g., rose bengal, erythrosine, and eosine bluish.

4.2.5 PHOTOCATALYST SYSTEMS FOR OVERALL WATER SPLITTING UNDER VISIBLE LIGHT

Two types of photocatalyst systems for water splitting under the visible-light irradiation are the biomimetic Z-scheme water-splitting system, which is a two-photon system and a dye-sensitized photoelectrochemical cell, as shown in Figure 4.3a. The biomimetic Z-scheme system mechanism using reversible redox mediators has been investigated with a view to constructing a photocatalytic system for overall water splitting into H_2 and O_2. A Z-scheme constructed from both $Pt/SrTiO_3$:Cr/Ta (band gap: 2.3 eV) for the H_2 evolution photocatalyst and Pt/WO_3 for the O_2 evolution photocatalyst decomposes water into H_2 and O_2 in stoichiometric amounts under the visible-light irradiation (Sayama et al. 2001). A Z-scheme system was made of $Pt/SrTiO_3$:Rh (band gap: 2.3 eV) and $BiVO_4$ or Bi_2MoO_6 and show overall water splitting capability in the presence of an Fe^{3+}/Fe^{2+} redox couple (Kato et al. 2004). The energy and band gaps of the system of $Pt/SrTiO_3$:Rh and $BiVO_4$ were shown in Figure 4.3a. Other combined systems are listed in reference (Nuraje et al. 2012).

Youngblood et al. (2009) recently demonstrated photoelectrochemical cell (PEC) water splitting under visible light using a dye-sensitized photoelectrochemical cell (Figure 4.3b), in which hydrated iridium oxide ($IrO_2 \cdot nH_2O$) nanoparticles are attached to Ru complex sensitizer molecules to serve as an O_2 evolution catalyst. An electron is excited from HOMO to LUMO by visible-light absorption, injected into the conduction band of TiO_2, and then transferred to a counter electrode (Pt) through the outer circuit to reduce water to H_2. The oxidized state of the dye is regenerated to the ground state by accepting an electron from the $IrO_2 \cdot nH_2O$ nanoparticle where water oxidation to O_2 proceeds. There are various studies (Zhou et al. 2009; Wolcott et al. 2009; Park et al. 2005) on carbon-doped titania tubes for photoelectrochemical water splitting. Park et al. (2005) discussed the higher photocurrent densities and more efficient water splitting of vertically grown carbon-doped TiO_2 nanotube arrays with high aspect ratios (~3 μm length) under the visible-light illumination than those of pure titania nanotube arrays. The difference between this system and using the dye-sensitized photoelectrochemical cell for water splitting is that carbon-doped titania tubes directly used as both photoelectrode and light collector absorbed the visible light without any help from the dye.

4.3 CONCLUSIONS

In the water splitting process, five different approaches were used to produce hydrogen and oxygen, including full reaction under UV and visible-light irradiation. This study also explains the synthesis methods, hydrogen and oxygen evolution rate, and quantum yield of binary and ternary metal oxides for water splitting. Metal oxides have been studied extensively in recent years since they have an appropriate band gap, chemical stability, and availability. A powdered photocatalyst of $NiO/NaTaO_3$:La was reported to have 56% apparent quantum yield at 270 nm. Since most metal oxides have a wide band gap, their applications are limited under the UV condition. To extend their application in the visible-light region, several approaches, including metal/nonmetal ion doping, dye sensitization, novel single-phase through–band-gap engineering, and the addition of solid solutions to control the band structure, have been applied. Although more metal oxide-based photocatalysts have been considered in recent years, the maximum efficiency obtained during the overall water-splitting process is still below 6.0%. As discussed in the basic principles of water splitting, three steps need to be studied in detail. For example, in water splitting, size and crystallinity of the photo catalysts are very important to produce a charge separation. Nanotechnology is a very important technology to develop new materials and methods to study the mechanism of water splitting from synthesizing particles to developing new water-splitting devices. To improve photocatalytic activity of overall water splitting, novel redox mediators also need to be addressed.

REFERENCES

Aroutiounian, V. M., V. M. Arakelyan, G. E. Shahnazaryan, G. M. Stepanyan, J. A. Turner, and O. Khaselev. 2002. Investigation of ceramic Fe_2O_3 < Ta > photoelectrodes for solar energy photoelectrochemical converters. *Int. J. Hydrogen Energy* 27 (1):33–38.
Boer, K. W., and A. Rothwarf. 1976. Materials for solar photovoltaic energy conversion. *Annu. Rev. Mater. Sci.* 6 (1):303–333. doi: 10.1146/annurev.ms.06.080176.001511.
Bolton, J. R. 1977. Photochemical conversion and storage of solar energy. *J. Solid State Chem.* 22 (1):3–8. doi: 10.1016/0022-4596(77)90183-9.
Borgarello, E., J. Kiwi, E. Pelizzetti, M. Visca, and M. Graetzel. 1981. Sustained water cleavage by visible light. *J. Am. Chem. Soc.* 103 (21):6324–6329. doi: 10.1021/ja00411a010.
Borgarello, E., J. Kiwi, M. Graetzel, E. Pelizzetti, and M. Visca. 1982. Visible light induced water cleavage in colloidal solutions of chromium-doped titanium dioxide particles. *J. Am. Chem. Soc.* 104 (11):2996–3002. doi: 10.1021/ja00375a010.

Borgarello, E., J. Kiwi, E. Pelizzetti, M. Visca, and M. Gratzel. 1981. Photochemical cleavage of water by photocatalysis. *Nature* 289 (5794):158–160.

Burda, C., Y. Lou, X. Chen, A. C. S. Samia, J. Stout, and J. L. Gole. 2003. Enhanced nitrogen doping in TiO2 nanoparticles. *Nano Lett.* 3 (8):1049–1051. doi: 10.1021/nl034332o.

Chen, X. 2009. Titanium dioxide nanomaterials and their energy applications. *Chinese J. Catal.* 30 (8):839–851. doi: 10.1016/s1872-2067(08)60126-6.

Chen, X., and S. S. Mao. 2007. Titanium dioxide nanomaterials: synthesis, properties, modifications, and applications. *Chem. Rev.* 107 (7):2891–2959. doi: 10.1021/cr0500535.

Chen, X., S. Shen, L. Guo, and S. S. Mao. 2010. Semiconductor-based photocatalytic hydrogen generation. *Chem. Rev.* 110 (11):6503–6570. doi: 10.1021/cr1001645.

Chen, X., T. Yu, X. Fan, H. Zhang, Z. Li, J. Ye, and Z. Zou. 2007. Enhanced activity of mesoporous Nb2O5 for photocatalytic hydrogen production. *Appl. Surf. Sci.* 253 (20):8500–8506. doi: 10.1016/j.apsusc.2007.04.035.

Choi, Y., T. Umebayashi, and M. Yoshikawa. 2004. Fabrication and characterization of C-doped anatase TiO2 photocatalysts. *J. Mater. Sci.* 39 (5):1837–1839. doi: 10.1023/b:jmsc.0000016198.73153.31.

Chung, K.-H., and D.-C. Park. 1998. Photocatalytic decomposition of water over cesium-loaded potassium niobate photocatalysts. *J. Mol. Catal. A: Chemical* 129 (1):53–59. doi: 10.1016/s1381-1169(97)00130-1.

Cook, T. R., D. K. Dogutan, S. Y. Reece, Y. Surendranath, T. S. Teets, and D. G. Nocera. 2010. Solar energy supply and storage for the legacy and nonlegacy worlds. *Chem. Rev.* 110 (11):6474–6502. doi: 10.1021/cr100246c.

Darwent, J. R., and A. Mills. 1982. Photo-oxidation of water sensitized by WO3 powder. *J. Chem. Soc., Faraday Trans. 2: Molecular and Chemical Physics* 78 (2):359–367.

Ding, Q.-P., Y.-P. Yuan, X. Xiong, R.-P. Li, H.-B. Huang, Z.-S. Li, T. Yu, Z.-G. Zou, and S.-G. Yang. 2008. Enhanced photocatalytic water splitting properties of KNbO3 nanowires synthesized through hydrothermal method. *J. Phys. Chem. C* 112 (48):18846–18848. doi: 10.1021/jp8042768.

Domen, K., S. Naito, T. Onishi, and K. Tamaru. 1982. Photocatalytic decomposition of liquid water on a NiO–SrTiO3 catalyst. *Chem. Phys. Lett.* 92 (4):433–434. doi: 10.1016/0009-2614(82)83443-x.

Domen, K., A. Kudo, and T. Onishi. 1986. Mechanism of photocatalytic decomposition of water into H2 and O2 over NiO–SrTiO3. *J. Catal.* 102 (1):92–98. doi: 10.1016/0021-9517(86)90143-0.

Domen, K., A. Kudo, T. Onishi, N. Kosugi, and H. Kuroda. 1986. Photocatalytic decomposition of water into hydrogen and oxygen over nickel(II) oxide-strontium titanate (SrTiO3) powder. 1. Structure of the catalysts. *J. Phys. Chem.* 90 (2):292–295. doi: 10.1021/j100274a018.

Domen, K., A. Kudo, M. Shibata, A. Tanaka, K.-I. Maruya, and T. Onishi. 1986. Novel photocatalysts, ion-exchanged K4Nb6O17, with a layer structure. *J. Chem. Soc., Chem. Commun.* 23:1706–1707.

Domen, K., A. Kudo, A. Shinozaki, A. Tanaka, K.-I. Maruya, and T. Onishi. 1986. Photodecomposition of water and hydrogen evolution from aqueous methanol solution over novel niobate photocatalysts. *J. Chem. Soc., Chem. Commun.* (4):356–357.

Domen, K., A. Kudo, A. Tanaka, and T. Onishi. 1990. Overall photodecomposition of water on a layered niobiate catalyst. *Catal. Today* 8 (1):77–84. doi: 10.1016/0920-5861(90)87009-r.

Domen, K., S. Naito, T. Onishi, K. Tamaru, and M. Soma. 1982. Study of the photocatalytic decomposition of water vapor over a nickel(II) oxide-strontium titanate (SrTiO3) catalyst. *J. Phys. Chem.* 86 (18):3657–3661. doi: 10.1021/j100215a032.

Domen, K., S. Naito, M. Soma, T. Onishi, and K. Tamaru. 1980. Photocatalytic decomposition of water vapour on an NiO–SrTiO$_3$ catalyst. *J. Chem. Soc., Chem. Commun.* 12:543–544.

Duonghond, D., N. Serpone, and M. Grätzel. 1984. Integrated systems for water cleavage by visible light; sensitization of TiO$_2$ particles by surface derivatization with ruthenium complexes. *Helvetica Chimica Acta* 67 (4):1012–1018. doi: 10.1002/hlca.19840670413.

Duonghong, D., E. Borgarello, and M. Graetzel. 1981. Dynamics of light-induced water cleavage in colloidal systems. *J. Am. Chem. Soc.* 103 (16):4685–4690. doi: 10.1021/ja00406a004.

Erbs, W., J. Desilvestro, E. Borgarello, and M. Graetzel. 1984. Visible-light-induced oxygen generation from aqueous dispersions of tungsten(VI) oxide. *J. Phys. Chem.* 88 (18):4001–4006. doi: 10.1021/j150662a028.

Fichtner, M. 2005. Nanotechnological aspects in materials for hydrogen storage. *Adv. Eng. Mater.* 7 (6):443–455. doi: 10.1002/adem.200500022.

Fujishima, A., and K. Honda. 1972. Electrochemical photolysis of water at a semiconductor electrode. *Nature* 238 (5358):37–38.

Furube, A., T. Shiozawa, A. Ishikawa, A. Wada, K. Domen, and C. Hirose. 2002. Femtosecond transient absorption spectroscopy on photocatalysts: K$_4$Nb$_6$O$_{17}$ and Ru(bpy)$_3{}^{2+}$–intercalated K$_4$Nb$_6$O$_{17}$ thin films. *J. Phys. Chem. B* 106 (12):3065–3072. doi: 10.1021/jp011083o.

Goharshadi, E. K., S. Samiee, and P. Nancarrow. 2011. Fabrication of cerium oxide nanoparticles: characterization and optical properties. *J. Colloid Interface Sci.* 356 (2):473–480. doi: 10.1016/j.jcis.2011.01.063.

Gratzel, M. 2005. Mesoscopic solar cells for electricity and hydrogen production from sunlight. *Chem. Lett.* 34 (1):8–13.

Gratzel, M. 2001. Photoelectrochemical cells. *Nature* 414 (6861):338–344.

Gurunathan, K., P. Maruthamuthu, and M. V. C. Sastri. 1997. Photocatalytic hydrogen production by dye-sensitized Pt/SnO$_2$ and Pt/SnO$_2$/RuO$_2$ in aqueous methyl viologen solution. *Int. J. Hydrogen Energy* 22 (1):57–62. doi: 10.1016/s0360-3199(96)00075-4.

Hagfeldt, A., G. Boschloo, L. C. Sun, L. Kloo, and H. Pettersson. 2010. Dye-sensitized solar cells. *Chem. Rev.* 110 (11):6595–6663. doi: 10.1021/cr900356p.

Ikarashi, K., J. Sato, H. Kobayashi, N. Saito, H. Nishiyama, and Y. Inoue. 2002. Photocatalysis for water decomposition by RuO$_2$-dispersed ZnGa$_2$O$_4$ with d^{10} configuration. *J. Phys. Chem. B* 106 (35):9048–9053. doi: 10.1021/jp020539e.

Ikeda, S., A. Tanaka, K. Shinohara, M. Hara, J. N. Kondo, K.-I. Maruya, and K. Domen. 1997. Effect of the particle size for photocatalytic decomposition of water on Ni-loaded K$_4$Nb$_6$O$_{17}$. *Microporous Mater.* 9 (5–6):253–258. doi: 10.1016/s0927-6513(96)00112-5.

Ingler, J. P. Baltrus, and S. U. M. Khan. 2004. Photoresponse of p-type zinc-doped iron(III) oxide thin films. *J. Am. Chem. Soc.* 126 (33):10238–10239. doi: 10.1021/ja048461y.

Iwase, A., H. Kato, and A. Kudo. 2006. Nanosized Au particles as an efficient cocatalyst for photocatalytic overall water splitting. *Catal. Lett.* 108 (1):7–10. doi: 10.1007/s10562-006-0030-1.

Ji, P., M. Takeuchi, T.-M. Cuong, J. Zhang, M. Matsuoka, and M. Anpo. 2010. Recent advances in visible light-responsive titanium oxide-based photocatalysts. *Res. Chem. Intermed.* 36 (4):327–347. doi: 10.1007/s11164-010-0142-5.

Jing, D., Y. Zhang, and L. Guo. 2005. Study on the synthesis of Ni doped mesoporous TiO$_2$ and its photocatalytic activity for hydrogen evolution in aqueous methanol solution. *Chem. Phys. Lett.* 415 (1–3):74–78. doi: 10.1016/j.cplett.2005.08.080.

Kadowaki, H., J. Sato, H. Kobayashi, N. Saito, H. Nishiyama, Y. Simodaira, and Y. Inoue. 2005. Photocatalytic activity of the RuO$_2$-dispersed composite p-block metal oxide LiInGeO$_4$ with d^{10}–d^{10} configuration for water decomposition. *J. Phys. Chem. B* 109 (48):22995–23000. doi: 10.1021/jp0544686.

Kadowaki, H., N. Saito, H. Nishiyama, and Y. Inoue. 2007. RuO_2-loaded Sr^{2+}-doped CeO_2 with d^0 electronic configuration as a new photocatalyst for overall water splitting. *Chem. Lett.* 36 (3):440–441.

Kadowaki, H., N. Saito, H. Nishiyama, H. Kobayashi, Y. Shimodaira, and Y. Inoue. 2006. Overall splitting of water by RuO_2-loaded $PbWO_4$ photocatalyst with $d^{10}s^2$–d^0 configuration. *J. Phys. Chem. C* 111 (1):439–444. doi: 10.1021/jp065655m.

Kalyanasundaram, K., and M. Graetzel. 2010. Artificial photosynthesis: biomimetic approaches to solar energy conversion and storage. *Curr. Opin. Biotechnol.* 21 (3):298–310. doi: 10.1016/j.copbio.2010.03.021.

Kamat, P. V. 2007. Meeting the clean energy demand: nanostructure architectures for solar energy conversion. *J. Phys. Chem. C* 111 (7):2834–2860. doi: 10.1021/jp066952u.

Kato, H., M. Hori, R. Konta, Y. Shimodaira, and A. Kudo. 2004. *Chem. Lett.* 33:1348.

Kato, H., K. Asakura, and A. Kudo. 2003. Highly efficient water splitting into H_2 and O_2 over lanthanum-doped $NaTaO_3$ photocatalysts with high crystallinity and surface nanostructure. *J. Am. Chem. Soc.* 125 (10):3082–3089. doi: 10.1021/ja027751g.

Kato, H., and A. Kudo. 1998. New tantalate photocatalysts for water decomposition into H_2 and O_2. *Chem. Phys. Lett.* 295 (5–6):487–492. doi: 10.1016/s0009-2614(98)01001-x.

Kato, H., and A. Kudo. 2001. Water splitting into H_2 and O_2 on alkali tantalate photocatalysts $ATaO_3$ (A = Li, Na, and K). *J. Phys. Chem. B* 105 (19):4285–4292. doi: 10.1021/jp004386b.

Kato, H., and A. Kudo. 2002. Visible-light-response and photocatalytic activities of TiO_2 and $SrTiO_3$ photocatalysts codoped with antimony and chromium. *J. Phys. Chem. B* 106 (19):5029–5034. doi: 10.1021/jp0255482.

Kato, H., N. Matsudo, and A. Kudo. 2004. Photophysical and photocatalytic properties of molybdates and tungstates with a scheelite structure. *Chem. Lett.* 33 (9):1216–1217.

Kim, H., S. Ji, J. Jang, S. Bae, and J. Lee. 2004. Formation of $La_2Ti_2O_7$ crystals from amorphous La_2O_3–TiO_2 powders synthesized by the polymerized complex method. *Korean J. Chem. Eng.* 21 (5):970–975. doi: 10.1007/bf02705579.

Kim, J., D. Hwang, S. Bae, Y. Kim, and J. Lee. 2001. Effect of precursors on the morphology and the photocatalytic water-splitting activity of layered perovskite $La_2Ti_2O_7$. *Korean J. Chem. Eng.* 18 (6):941–947. doi: 10.1007/bf02705623.

Kim, J., D. Hwang, H. Kim, S. Bae, J. Lee, W. Li, and S. Oh. 2005. Highly efficient overall water splitting through optimization of preparation and operation conditions of layered perovskite photocatalysts. *Top. Catal.* 35 (3):295–303. doi: 10.1007/s11244-005-3837-x.

Kim, J., D. W. Hwang, H.-G. Kim, S. W. Bae, S. M. Ji, and J. S. Lee. 2002. Nickel-loaded $La_2Ti_2O_7$ as a bifunctional photocatalyst. *Chem. Commun.* 21:2488–2489.

Kim, S., S.-J. Hwang, and W. Choi. 2005. Visible light active platinum-ion-doped tio_2 photocatalyst. *J. Phys. Chem. B.* 109 (51):24260–24267. doi: 10.1021/jp055278y.

Kitano, M., and M. Hara. 2010. Heterogeneous photocatalytic cleavage of water. *J. Mater. Chem.* 20 (4):627–641. doi: 10.1039/b910180b.

Kong, M., Y. Li, X. Chen, T. Tian, P. Fang, F. Zheng, and X. Zhao. 2011. Tuning the relative concentration ratio of bulk defects to surface defects in TiO_2 nanocrystals leads to high photocatalytic efficiency. *J. Am. Chem. Soc.* 133 (41):16414–16417. doi: 10.1021/ja207826q.

Kudo, A., K. Sayama, A. Tanaka, K. Asakura, K. Domen, K. Maruya, and T. Onishi. 1989. Nickel-loaded $K_4Nb_6O_{17}$ photocatalyst in the decomposition of H_2O into H_2 and O_2: Structure and reaction mechanism. *J. Catal.* 120 (2):337–352. doi: 10.1016/0021-9517(89)90274-1.

Kudo, A. 2003. Photocatalyst materials for water splitting. *Catal. Surv. Asia* 7 (1):31–38. doi: 10.1023/a:1023480507710.

Kudo, A. 2006. Development of photocatalyst materials for water splitting. *Int. J. Hydrogen Energy* 31 (2):197–202. doi: 10.1016/j.ijhydene.2005.04.050.

Kudo, A., K. Domen, K.-I. Maruya, and T. Onishi. 1987. Photocatalytic activities of TiO_2 loaded with NiO. *Chem. Phys. Lett.* 133 (6):517–519. doi: 10.1016/0009-2614 (87)80070-2.

Kudo, A., and S. Hijii. 1999. H_2, or O_2 evolution from aqueous solutions on layered oxide photocatalysts consisting of Bi^{3+} with $6s^2$ configuration and d^0 transition metal ions. *Chem. Lett.* 28 (10):1103–1104.

Kudo, A., and H. Kato. 1997. Photocatalytic activities of $Na_2W_4O_{13}$ with layered structure. *Chem. Lett.* 26 (5):421–422.

Kudo, A., H. Kato, and S. Nakagawa. 1999. Water splitting into H_2 and O_2 on new Sr2M2O7 (M = Nb and Ta) photocatalysts with layered perovskite structures: factors affecting the photocatalytic activity. *J. Phys. Chem. B* 104 (3):571–575. doi: 10.1021/jp9919056.

Kudo, A., and Y. Miseki. 2009. Heterogeneous photocatalyst materials for water splitting. *Chem. Soc. Rev.* 38 (1):253–278.

Kudo, A., S. Nakagawa, and H. Kato. 1999. Overall water splitting into H_2 and O_2 under UV irradiation on NiO-loaded $ZnNb_2O_6$ photocatalysts consisting of d^{10} and d^0 ions. *Chem. Lett.* 28 (11):1197–1198.

Kudo, A., M. Steinberg, A. J. Bard, A. Campion, M. A. Fox, T. E. Mallouk, S. E. Webber, and J. M. White. 1990. Photoactivity of ternary lead-group IVB oxides for hydrogen and oxygen evolution. *Catal. Lett.* 5 (1):61–66. doi: 10.1007/bf00772094.

Kudo, A., A. Tanaka, K. Domen, K.-I. Maruya, K.-I. Aika, and T. Onishi. 1988. Photocatalytic decomposition of water over $NiO-K_4Nb_6O_{17}$ catalyst. *J. Catal.* 111 (1):67–76. doi: 10.1016/0021-9517(88)90066-8.

Kudo, A., A. Tanaka, K. Domen, and T. Onishi. 1988. The effects of the calcination temperature of $SrTiO_3$ powder on photocatalytic activities. *J. Catal.* 111 (2):296–301. doi: 10.1016/0021-9517 (88)90088-7.

Kutal, C. 1983. Photochemical conversion and storage of solar-energy. *J. Chem. Educ.* 60 (10):882–887.

Lehn, J. M., J. P. Sauvage, and R. Ziesel. 1980. *Nouv. J. Chim* 4:623.

Lee, J. 2005. Photocatalytic water splitting under visible light with particulate semiconductor catalysts. *Catal. Surv. Asia* 9 (4):217–227. doi: 10.1007/s10563-005-9157-0.

Leung, D. Y. C., X. Fu, C. Wang, M. Ni, M. K. H. Leung, X. Wang, and X. Fu. 2010. Hydrogen production over titania-based photocatalysts. *ChemSusChem* 3 (6):681–694. doi: 10.1002/cssc.201000014.

Lewis, N. S. 2007. Powering the Planet. *MRS Bull.* 32:808–820.

Li, G., T. Kako, D. Wang, Z. Zou, and J. Ye. 2008. Synthesis and enhanced photocatalytic activity of $NaNbO_3$ prepared by hydrothermal and polymerized complex methods. *J. Phys. Chem. Solids* 69 (10):2487–2491. doi: 10.1016/j.jpcs.2008.05.001.

Li, Y., and J. Z. Zhang. 2010. Hydrogen generation from photoelectrochemical water splitting based on nanomaterials. *Laser Photon. Rev.* 4 (4):517–528. doi: 10.1002/lpor.200910025.

Li, Z., G. Chen, X. Tian, and Y. Li. 2008. Photocatalytic property of $La_2Ti_2O_7$ synthesized by the mineralization polymerizable complex method. *Mater. Res. Bull.* 43 (7):1781–1788. doi: 10.1016/j.materresbull.2007.07.010.

Liu, J. W., G. Chen, Z. H. Li, and Z. G. Zhang. 2006. Electronic structure and visible light photocatalysis water splitting property of chromium-doped $SrTiO_3$. *J. Solid State Chem.* 179 (12):3704–3708. doi: 10.1016/j.jssc.2006.08.014.

Liu, Y., L. Xie, Y. Li, R. Yang, J. Qu, Y. Li, and X. Li. 2008. Synthesis and high photocatalytic hydrogen production of $SrTiO_3$ nanoparticles from water splitting under UV irradiation. *J. Power Sources* 183 (2):701–707. doi: 10.1016/j.jpowsour.2008.05.057.

Maeda, K., M. Eguchi, S.-H. A. Lee, W. J. Youngblood, H. Hata, and T. E. Mallouk. 2009. Photocatalytic hydrogen evolution from hexaniobate nanoscrolls and calcium niobate nanosheets sensitized by ruthenium(II) bipyridyl complexes. *J. Phys. Chem. C* 113 (18):7962–7969. doi: 10.1021/jp900842e.

Mao, S. S., and X. Chen. 2007. Selected nanotechnologies for renewable energy applications. *Int. J. Energy Research* 31 (6–7):619–636. doi: 10.1002/er.1283.

Matsumura, M., Y. Saho, and H. Tsubomura. 1983. Photocatalytic hydrogen production from solutions of sulfite using platinized cadmium sulfide powder. *J. Phys. Chem.* 87 (20):3807–3808. doi: 10.1021/j100243a005.

Minggu, L. J., W. R. W. Daud, and M. B. Kassim. 2010. An overview of photocells and photoreactors for photoelectrochemical water splitting. *Int. J. Hydrogen Energy* 35 (11):5233–5244. doi: 10.1016/j.ijhydene.2010.02.133.

Miseki, Y., H. Kato, and A. Kudo. 2005. Water splitting into H_2 and O_2 over $Cs_2Nb_4O_{11}$ photocatalyst. *Chem. Lett.* 34 (1):54–55.

Miseki, Y., H. Kusama, H. Sugihara, and K. Sayama. 2010. Cs-modified WO_3 photocatalyst showing efficient solar energy conversion for O_2 production and Fe (III) ion reduction under visible light. *J. Phys. Chem. Lett.* 1 (8):1196–1200. doi: 10.1021/jz100233w.

Mizoguchi, H., K. Ueda, M. Orita, S.-C. Moon, K. Kajihara, M. Hirano, and H. Hosono. 2002. Decomposition of water by a $CaTiO_3$ photocatalyst under UV light irradiation. *Mater. Res. Bull.* 37 (15):2401–2406. doi: 10.1016/s0025-5408(02)00974-1.

Nakahira, T., Y. Inoue, K. Iwasaki, H. Tanigawa, Y. Kouda, S. Iwabuchi, K. Kojima, and M. Grätzel. 1988. Visible light sensitization of platinized TiO_2 photocatalyst by surface-coated polymers derivatized with ruthenium tris (bipyridyl). *Die Makromolekulare Chemie, Rapid Communications* 9 (1):13–17. doi: 10.1002/marc.1988.030090103.

Navarro Yerga, R. M., M. C. Álvarez Galván, F. del Valle, J. A. Villoria de la Mano, and J. L. G. Fierro. 2009. Water splitting on semiconductor catalysts under visible-light irradiation. *ChemSusChem.* 2 (6):471–485. doi: 10.1002/cssc.200900018.

Nocera, D. G. 2009. Personalized energy: the home as a solar power station and solar gas station. *ChemSusChem.* 2 (5):387–390. doi: 10.1002/cssc.200900040.

Nozik, A. J. 1978. Photoelectrochemistry: applications to solar energy conversion. *Annu. Rev. Phys. Chem.* 29 (1):189–222. doi:10.1146/annurev.pc.29.100178.001201.

Nuraje, N., R. Asmatulu, and S. Kudaibergenov. 2012. Metal oxide-based functional materials for solar energy conversion: a review. *Current Inorganic Chemistry* 2 (2):124–146.

Ohno, T., T. Mitsui, and M. Matsumura. 2003. Photocatalytic activity of S-doped TiO_2 photocatalyst under visible light. *Chem. Lett.* 32 (4):364–365.

Park, J. H., S. Kim, and A. J. Bard. 2005. Novel carbon-doped TiO_2 nanotube arrays with high aspect ratios for efficient solar water splitting. *Nano Lett.* 6 (1):24–28. doi: 10.1021/nl051807y.

Pentyala, N., R. K. Guduru, E. M. Shnerpunas, and P. S. Mohanty. 2011. Synthesis of ultrafine single crystals and nanostructured coatings of indium oxide from solution precursor. *Appl. Surf. Sci.* 257 (15):6850–6857. doi: 10.1016/j.apsusc.2011.03.018.

Qin, Y., G. Wang, and Y. Wang. 2007. Study on the photocatalytic property of La-doped CoO/$SrTiO_3$ for water decomposition to hydrogen. *Catal. Commun.* 8 (6):926–930. doi: 10.1016/j.catcom.2006.11.025.

Reber, J. F., and M. Rusek. 1986. Photochemical hydrogen production with platinized suspensions of cadmium sulfide and cadmium zinc sulfide modified by silver sulfide. *J. Phys. Chem.* 90 (5):824–834. doi: 10.1021/j100277a024.

Reddy, V. R., D. W. Hwang, and J. S. Lee. 2003. Photocatalytic water splitting over ZrO_2 prepared by precipitation method. *Korean J. Chem. Eng.* 20 (6):1026–1029.

Rengaraj, S., and X. Z. Li. 2006. Enhanced photocatalytic activity of TiO_2 by doping with Ag for degradation of 2,4,6-trichlorophenol in aqueous suspension. *J. Mol. Catal. A: Chem.* 243 (1):60–67. doi: 10.1016/j.molcata.2005.08.010.

Saito, N., H. Kadowaki, H. Kobayashi, K. Ikarashi, H. Nishiyama, and Y. Inoue. 2004. A new photocatalyst of RuO_2-loaded $PbWO_4$ for overall splitting of water. *Chem. Lett.* 33 (11):1452–1453.

Sakata, Y., Y. Matsuda, T. Yanagida, K. Hirata, H. Imamura, and K. Teramura. 2008. Effect of metal ion addition in a Ni supported Ga_2O_3 photocatalyst on the photocatalytic overall splitting of H_2O. *Catal. Lett.* 125 (1):22–26. doi: 10.1007/s10562-008-9557-7.

Sasikala, R., V. Sudarsan, C. Sudakar, R. Naik, T. Sakuntala, and S. R. Bharadwaj. 2008. Enhanced photocatalytic hydrogen evolution over nanometer sized Sn and Eu doped titanium oxide. *Int. J. Hydrogen Energy* 33 (19):4966–4973. doi: 10.1016/j.ijhydene.2008.07.080.

Sato, J., H. Kobayashi, K. Ikarashi, N. Saito, H. Nishiyama, and Y. Inoue. 2004. Photocatalytic activity for water decomposition of RuO_2-dispersed Zn_2GeO_4 with d^{10} configuration. *J. Phys. Chem. B* 108 (14):4369–4375. doi: 10.1021/jp0373189.

Sato, J., H. Kobayashi, and Y. Inoue. 2003. Photocatalytic activity for water decomposition of indates with octahedrally coordinated d^{10} configuration. II. Roles of geometric and electronic structures. *J. Phys. Chem. B* 107 (31):7970–7975. doi: 10.1021/jp030021q.

Sato, J., H. Kobayashi, N. Saito, H. Nishiyama, and Y. Inoue. 2003. Photocatalytic activities for water decomposition of RuO_2-loaded $AlnO_2$ (A = Li, Na) with d^{10} configuration. *J. Photochem. Photobiol. A: Chem.* 158 (2–3):139–144. doi: 10.1016/s1010-6030(03)00028-5.

Sato, J., N. Saito, H. Nishiyama, and Y. Inoue. 2001a. New photocatalyst group for water decomposition of RuO_2-loaded p-block metal (In, Sn, and Sb) oxides with d^{10} configuration. *J. Phys. Chem. B* 105 (26):6061–6063. doi: 10.1021/jp010794j.

Sato, J., N. Saito, H. Nishiyama, and Y. Inoue. 2001b. New photocatalyst group for water decomposition of RuO_2-loaded p-block metal (In, Sn, and Sb) oxides with d^{10} configuration. *J. Phys. Chem. B* 105 (26):6061–6063. doi: 10.1021/jp010794j.

Sato, J., N. Saito, H. Nishiyama, and Y. Inoue. 2002. Photocatalytic water decomposition by RuO_2-loaded antimonates, $M_2Sb_2O_7$ (M = Ca, Sr), $CaSb_2O_6$ and $NaSbO_3$, with d^{10} configuration. *J. Photochem. Photobiol. A: Chem.* 148 (1–3):85–89. doi: 10.1016/s1010-6030(02)00076-x.

Sato, J., N. Saito, H. Nishiyama, and Y. Inoue. 2003. Photocatalytic activity for water decomposition of indates with octahedrally coordinated d^{10} configuration. I. Influences of preparation conditions on activity. *J. Phys. Chem. B* 107 (31):7965–7969. doi: 10.1021/jp030020y.

Sato, J., N. Saito, H. Nishiyama, and Y. Inoue. 2001c. Photocatalytic activity for water decomposition of RuO_2-loaded $SrIn_2O_4$ with d^{10} configuration. *Chem. Lett.* 30 (9):868–869.

Satsangi, V. R., S. Kumari, A. P. Singh, R. Shrivastav, and S. Dass. 2008. Nanostructured hematite for photoelectrochemical generation of hydrogen. *Int. J. Hydrogen Energy* 33 (1):312–318. doi: 10.1016/j.ijhydene.2007.07.034.

Sayama, K., and H. Arakawa. 1993. Photocatalytic decomposition of water and photocatalytic reduction of carbon dioxide over zirconia catalyst. *J. Phys. Chem.* 97 (3):531–533. doi: 10.1021/j100105a001.

Sayama, K., and H. Arakawa. 1994. Effect of Na_2CO_3 addition on photocatalytic decomposition of liquid water over various semiconductor catalysis. *J. Photochem. Photobiol. A: Chemistry* 77 (2–3):243–247. doi: 10.1016/1010-6030(94)80049-9.

Sayama, K., H. Arakawa, and K. Domen. 1996. Photocatalytic water splitting on nickel intercalated $A_4Ta_xNb_{6-x}O_{17}$ (A = K, Rb). *Catal. Today* 28 (1–2):175–182. doi: 10.1016/0920-5861 (95)00224-3.

Sayama, K., A. Tanaka, K. Domen, K. Maruya, and T. Onishi. 1990. Improvement of nickel-loaded $K_4Nb_6O_{17}$ photocatalyst for the decomposition of H_2O. *Catal. Lett.* 4 (3):217–222. doi: 10.1007/bf00765937.

Sayama, K., A. Tanaka, K. Domen, K. Maruya, and T. Onishi. 1991. Photocatalytic decomposition of water over platinum-intercalated potassium niobate ($K_4Nb_6O_{17}$). *J. Phys. Chem.* 95 (3):1345–1348. doi: 10.1021/j100156a058.

Sayama, K., K. Yase, H. Arakawa, K. Asakura, A. Tanaka, K. Domen, and T. Onishi. 1998. Photocatalytic activity and reaction mechanism of Pt-intercalated $K_4Nb_6O_{17}$ catalyst on the water splitting in carbonate salt aqueous solution. *J. Photochem. Photobiol. A: Chem.* 114 (2):125–135. doi: 10.1016/s1010-6030(98)00202-0.

Sayama, K., and H. Arakawa. 1996. Effect of carbonate addition on the photocatalytic decomposition of liquid water over a ZrO_2 catalyst. *J. Photochem. Photobiol. A: Chemistry* 94 (1):67–76. doi: 10.1016/1010–6030 (95)04204-0.

Sayama, K., and H. Arakawa. 1997. Effect of carbonate salt addition on the photocatalytic decomposition of liquid water over $Pt–TiO_2$ catalyst. *J. Chem. Soc., Faraday Transactions* 93 (8):1647–1654.

Sayama, K., K. Mukasa, R. Abe, Y. Abe, and H. Arakawa. 2001. Stoichiometric water splitting into H and O using a mixture of two different photocatalysts and an IO/I shuttle redox mediator under visible light irradiation. *Chem. Commun.* (23):2416–2417.

Shi, J., J. Chen, Z. Feng, T. Chen, Y. Lian, X. Wang, and C. Li. 2006. Photoluminescence characteristics of TiO_2 and their relationship to the photoassisted reaction of water/methanol mixture. *J. Phys. Chem. C* 111 (2):693–699. doi: 10.1021/jp065744z.

Shibata, M., A. Kudo, A. Tanaka, K. Domen, K.-I. Maruya, and T. Onishi. 1987. Photocatalytic activities of layered titanium compounds and their derivatives for H_2 evolution from aqueous methanol solution. *Chem. Lett.* 16:1017–1018.

Shimidzu, T., T. Iyoda, and Y. Koide. 1985. An advanced visible-light-induced water reduction with dye-sensitized semiconductor powder catalyst. *J. Am. Chem. Soc.* 107 (1):35–41. doi: 10.1021/ja00287a007.

Soares, M. R. N., S. Leite, C. Nico, M. Peres, A. J. S. Fernandes, M. P. F. Graça, M. Matos, R. Monteiro, T. Monteiro, and F. M. Costa. 2011. Effect of processing method on physical properties of Nb_2O_5. *J. Eur. Ceram. Soc.* 31 (4):501–506. doi: 10.1016/j.jeurceramsoc.2010.10.024.

Sun, W., S. Zhang, C. Wang, Z. Liu, and Z. Mao. 2007. Enhanced photocatalytic hydrogen evolution over $CaTi_{1-x}Zr_xO_3$ composites synthesized by polymerized complex method. *Catal. Lett.* 119 (1):148–153. doi: 10.1007/s10562-007-9212-8.

Tabata, S., H. Nishida, Y. Masaki, and K. Tabata. 1995. Stoichiometric photocatalytic decomposition of pure water in Pt/TiO_2 aqueous suspension system. *Catal. Lett.* 34 (1):245–249. doi: 10.1007/bf00808339.

Takahara, Y., J. N. Kondo, T. Takata, D. Lu, and K. Domen. 2001. Mesoporous tantalum oxide. 1. Characterization and photocatalytic activity for the overall water decomposition. *Chem. Mater.* 13 (4):1194–1199. doi: 10.1021/cm000572i.

Takata, T., and K. Domen. 2009. Defect engineering of photocatalysts by doping of aliovalent metal cations for efficient water splitting. *J. Phys. Chem. C* 113 (45):19386–19388. doi: 10.1021/jp908621e.

Thorat, J., K. Kanade, L. Nikam, P. Chaudhari, and B. Kale. 2011. Nanostructured ZnO hexagons and optical properties. *J. Mater. Sci.: Mater. Electron.* 22 (4):394–399. doi: 10.1007/s10854-010-0149-0.

Turner, J. A. 1999. A Realizable Renewable Energy Future. *Science* 285 (5428):687–689. doi: 10.1126/science.285.5428.687.

van de Krol, R., Y. Liang, and J. Schoonman. 2008. Solar hydrogen production with nanostructured metal oxides. *J. Mater. Chem.* 18 (20):2311–2320.

Walter, M. G., E. L. Warren, J. R. McKone, S. W. Boettcher, Q. Mi, E. A. Santori, and N. S. Lewis. 2010a. Solar water splitting cells. *Chem. Rev.* 110 (11):6446–6473. doi: 10.1021/cr1002326.

Wang, J., H. Li, H. Li, S. Yin, and T. Sato. 2009. Preparation and photocatalytic activity of visible light-active sulfur and nitrogen co-doped $SrTiO_3$. *Solid State Sciences* 11 (1):182–188. doi: 10.1016/j.solidstatesciences.2008.04.010.

Wang, J., S. Yin, M. Komatsu, Q. Zhang, F. Saito, and T. Sato. 2004. Preparation and characterization of nitrogen doped $SrTiO_3$ photocatalyst. *J. Photochem. Photobiol. A: Chem.* 165 (1–3):149–156. doi: 10.1016/j.jphotochem.2004.02.022.

Wang, Y., Z. Zhang, Y. Zhu, Z. Li, R. Vajtai, L. Ci, and P. M. Ajayan. 2008. Nanostructured VO_2 photocatalysts for hydrogen production. *ACS Nano* 2 (7):1492–1496. doi: 10.1021/nn800223s.

Wolcott, A., W. A. Smith, T. R. Kuykendall, Y. Zhao, and J. Z. Zhang. 2009. Photoelectrochemical water splitting using dense and aligned TiO_2 nanorod arrays. *Small* 5 (1):104–111. doi: 10.1002/smll.200800902.

Yamaguti, K., and S. Sato. 1985. Photolysis of water over metallized powdered titanium dioxide. *J.Chem. Soc., Faraday Trans. 1: Phys. Chem. Condens. Phases* 81 (5):1237–1246.

Yamashita, Y., K. Yoshida, M. Kakihana, S. Uchida, and T. Sato. 1998. Polymerizable complex synthesis of $RuO_2/BaTi_4O_9$ photocatalysts at reduced temperatures: factors affecting the photocatalytic activity for decomposition of water. *Chem. Mater.* 11 (1):61–66. doi: 10.1021/cm9804012.

Yanagida, T., Y. Sakata, and H. Imamura. 2004. Photocatalytic decomposition of H_2O into H_2 and O_2 over Ga_2O_3 loaded with NiO. *Chem. Lett.* 33 (6):726–727.

Youngblood, W. J., S. H. A. Lee, K. Maeda, and T. E. Mallouk. 2009. Visible light water splitting using dye-sensitized oxide semiconductors. *Acc. Chem. Res.* 42 (12):1966–1973. doi: 10.1021/ar9002398.

Yuan, Y., J. Zheng, X. Zhang, Z. Li, T. Yu, J. Ye, and Z. Zou. 2008. $BaCeO_3$ as a novel photocatalyst with 4f electronic configuration for water splitting. *Solid State Ionics* 178 (33–34):1711–1713. doi: 10.1016/j.ssi.2007.11.012.

Zalas, M., and M. Laniecki. 2005. Photocatalytic hydrogen generation over lanthanides-doped titania. *Sol. Energy Mater. Sol. Cells* 89 (2–3):287–296. doi: 10.1016/j.solmat.2005.02.014.

Zhang, J., Q. Xu, Z. Feng, M. Li, and C. Li. 2008. Importance of the Relationship between surface phases and photocatalytic activity of TiO_2. *Angew. Chem., Int. Ed.* 47 (9):1766–1769. doi: 10.1002/anie.200704788.

Zhao, D., S. Budhi, A. Rodriguez, and R. T. Koodali. 2010. Rapid and facile synthesis of Ti-MCM-48 mesoporous material and the photocatalytic performance for hydrogen evolution. *Int. J. Hydrogen Energy* 35 (11):5276–5283. doi: 10.1016/j.ijhydene.2010.03.087.

Zhao, Y., P. Chen, B. Zhang, D. S. Su, S. Zhang, L. Tian, J. Lu, Z. Li, X. Cao, B. Wang, M. Wei, D. G. Evans, and X. Duan. 2012. Highly dispersed TiO_6 units in a layered double hydroxide for water splitting. *Chemistry* 18 (38):11949–11958. doi: 10.1002/chem.201201065.

Zhou, B., M. Schulz, H. Y. Lin, S. Ismat Shah, J. Qu, and C. P. Huang. 2009. Photoeletrochemical generation of hydrogen over carbon-doped TiO_2 photoanode. *Appl. Catal. B: Environ.* 92 (1–2):41–49. doi: 10.1016/j.apcatb.2009.07.026.

Zhu, J., and M. Zäch. 2009. Nanostructured materials for photocatalytic hydrogen production. *Curr. Opin. Colloid Interface Sci.* 14 (4):260–269. doi: 10.1016/j.cocis.2009.05.003.

Zielinska, B., E. Borowiak-Palen, and R. J. Kalenzuk. 2008. Preparation and characterization of lithium niobate as a novel photocatalyst in hydrogen generation. *J. Phys. Chem. Solids* 69 (1):236–242. doi: 10.1016/j.jpcs.2007.09.001.

Section II

Biofuels from Biomass Valorization Using Nanomaterials

5 Catalytic Reforming of Biogas into Syngas Using Supported Noble-Metal and Transition-Metal Catalysts

Albin Pintar, Petar Djinović, Boštjan Erjavec, and Ilja Gasan Osojnik Črnivec

CONTENTS

5.1 OVERVIEW OF CURRENT MAINSTREAM BIOFUEL PRODUCTION TECHNOLOGIES

Exponential growth of Earth's population projects itself on quickly increasing consumption of fossil fuel resources (natural gas, oil, and coal), which act as a foundation for production of fuels and numerous chemicals, such as fertilizers, plastics, engineering materials, and medicine, that enable a considerable improvement in the quality of living, compared with the standards of several decades ago. Nevertheless, fossil fuel reserves are limited, and their exploitation at today's rate will lead to their extinction, resulting in drastic changes of life standards globally. Consequently, fossil fuel alternatives are being very intensively investigated, and a steady progress toward their practical implementation is being accomplished.

For example, first-generation biofuel production in large-scale biodiesel facilities is already globally operational (2009 world production was estimated at the amount of 16 million ton). Fast expansion of this technology demands cultivating enormous areas of land, which started the competition between fuel and food production. Biodiesel cost is currently higher than the cost of diesel produced from petroleum, therefore improvements in the following fields are crucial: (i) process intensification by reducing the reactor volume and consequently reduction of investment costs, (ii) the use of advanced heterogeneous catalysts, in particular bifunctional acid catalysts, enabling the promotion of both esterification and trans-esterification occurring together in a continuous-flow reactor, (iii) enabling the possibility of using nonrefined oils and oils not competing with food, such as *Jathropa curcas* or algae, and (iv) exploration of new uses for by-product glycerol for fuel additives and/or commodities that can make it a very desirable alternative [1].

Production of another first-generation biofuel, i.e. bioethanol, from corn and sugar cane in quantities exceeding 100 million tons in America and Europe to act as a gasoline additive only, increased the food vs. fuel rivalry. Biochemical conversion of lignocellulosic materials through saccharification and fermentation could represent a major pathway for bioethanol production from biomass, circumventing competition with food production and leading to sustainable and effective utilization of biomass, which would otherwise represent an organic waste. To accomplish this, the following issues need to be solved: (i) biomass consists of many different polymeric molecules, such as cellulose, lignin, lignocellulose, etc., which are hard to depolymerize, separate, transform selectively and efficiently by employing various chemical, thermal or catalytic pathways, (ii) diversity of sugars, which are released when the hemicellulose and cellulose polymers are cleaved and the need to find or genetically engineer organisms to efficiently ferment these sugars, (iii) costs for collection and storage of low density lignocellosic materials [2]. Production of first-generation biofuels relies mainly on unit operations such as milling, extraction, distillation, and use of homogeneous catalysis and biochemical transformation with close to zero involvement of heterogeneous catalysis.

As can be seen, there is a lot of improvement necessary to be made toward more efficient exploitation of biomass, either to act as a fuel, source of platform chemicals, or use in numerous other applications.

Rapid expansion of established processes for biogas production (different cultures of microorganisms under anaerobic conditions decompose organic matter into a mixture of methane and carbon dioxide) resulted in its exponential production in Europe in the last decade, with 10 billion m^3 of biomethane produced in the year 2009, and projected 46 billion m^3 of biomethane primary energy potential in 2020 [3]. This technology is globally widespread, and currently, the final aim of biogas utilization is primarily its combustion for production of electricity and thermal energy in modern co-generation or tri-generation power plants. This could represent an important alternative to the utilization of natural gas, thus decreasing a dependence of today's energy demanding population and economy on fossil fuels [4]. Biogas acts as a source of methane and carbon dioxide of ever increasing importance, since different redundant organic materials (waste streams originating from agricultural activity as well as municipal origin) or purpose-grown energy crops are being dedicated for transformation into this energy-rich gas mixture [5]. Biogas can be used directly as an individual source of either biomethane or carbon dioxide, naturally, after appropriate purification (adsorption of trace impurities and pressure swing adsorption for CO_2 removal) [6].

5.2 CATALYTIC VALORIZATION OF (BIO)METHANE

Methane (originating either from natural gas or separated from biogas) can be catalytically transformed into syngas (a mixture of carbon monoxide and hydrogen) by steam reforming or partial oxidation processes [7,8]. These well-established technologies, with the total of 15 GW_{th} syngas capacity in 2010 [9], represent the basis for synthesis of numerous chemicals and fuels, even though both have their advantages and drawbacks [10]. For example, partial oxidation reaction (Equation 5.1) produces syngas with a H_2/CO ratio around 2 and the energy balance is very favorable due to exothermicity of the reaction. A substantial increase of temperature at the reactor inlet due to methane oxidation and gas-phase channeling can cause the generation of hot spots, which can cause severe catalyst deactivation. Since pure oxygen acts as an oxidant (oxidation with air would result in a substantial dilution of the reaction gas mixture), a cryogenic air separation unit for oxygen production is required.

$$2CH_4 + O_2 \rightarrow 2CO + 4H_2 \quad \Delta H^0_{298\,K} = -36 \text{ kJ/mol} \quad (5.1)$$

Methane steam reforming reaction (Equation 5.2) on the other hand requires high energy input and sophisticated reactor design due to high endothermicity of the reaction. The process faces steam corrosion issues, but produces syngas with a H_2/CO ratio of roughly 3, which is very beneficial for the production of pure hydrogen [11].

$$CH_4 + H_2O \leftrightarrow CO + 3H_2 \quad \Delta H^0_{298\,K} = 206 \text{ kJ/mol} \quad (5.2)$$

Methane dry reforming (MDR) reaction (Equation 5.3) is yet to be widely employed for syngas production. This reaction uses CO_2, which is present in biogas in concentrations up to 50 vol% as an oxidizing agent, much like O_2 or H_2O in partial oxidation and steam reforming reactions, respectively [11]. Consequently, purified biogas can be used directly as a feedstock for the MDR reaction without additional oxidants, thus performing its task as a part of a complementary strategy to reduce the dependence of our society on fossil fuels [12].

5.3 METHANE DRY REFORMING: REACTION CONDITIONS AND THERMODYNAMICS

Methane dry reforming reaction (Equation 5.3) is strongly endothermic [13], and requires high temperatures to achieve noticeable CH_4 and CO_2 conversions and a beneficial H_2/CO molar ratio, which can in the ideal case reach unity. By adding water to the reactant stream (in the form of humidity, which is always present in biogas and depends on the anaerobic digestion temperature), this ratio can be increased to the value of around 1.5.

$$CH_4 + CO_2 \leftrightarrow 2H_2 + 2CO \quad \Delta H^0_{298\,K} = 247 \text{ kJ/mol} \quad (5.3)$$

The products of dry reforming are therefore considered as an attractive prospect, as a low ratio of H_2/CO input is often preferred in many synthesis pathways, e.g., in the production of alkenes via Fischer–Tropsch reactions, in the synthesis of methanol, dimethyl ether (DME), formaldehyde, methyl formate, ethanol, propanol, etc. Comprehensive reviews of such oxo-syntheses and syntheses of olefins are widely available in the literature [14,15]. Hydrogen yield is strongly influenced by the co-occurrence of reverse water gas shift (RWGS) reaction (Equation 5.4). This reaction is quasi-equilibrated in a wide range of operating conditions, which results in water and CO production and considerable decrease of the H_2 yield [16]. The occurrence of the RWGS reaction is also the primary reason for higher CO_2 conversion compared to methane conversion, as observed in numerous studies [13]. Beside RWGS reaction, methane dehydrogenation reaction (Equation 5.5), CO disproportionation (Equation 5.6), and carbon gasification reactions (Equation 5.7) also contribute to the potpourri of possible side reactions, occurring under the relevant operating conditions.

$$H_2 + CO_2 \leftrightarrow H_2O + CO \quad \Delta H^0_{298\,K} = 41 \text{ kJ/mol} \tag{5.4}$$

$$CH_4 \leftrightarrow C + 2H_2 \quad \Delta H^0_{298\,K} = 75 \text{ kJ/mol} \tag{5.5}$$

$$2CO \leftrightarrow C + CO_2 \quad \Delta H^0_{298\,K} = -171 \text{ kJ/mol} \tag{5.6}$$

$$H_2 + CO \leftrightarrow H_2O + C \quad \Delta H^0_{298\,K} = -131 \text{ kJ/mol} \tag{5.7}$$

To minimize the contribution of the RWGS reaction and to improve the H_2 yield, high temperatures (>800°C), and/or a CH_4/CO_2 ratio well above 1 are required accordingly to thermodynamic calculations. This is in agreement with our results, depicted in Figure 5.1a, which show the produced H_2/CO ratios as a function of reaction temperature for different ratios of methane and carbon dioxide in the feed stream. The proximity to the RWGS equilibrium of gas-phase composition during methane dry reforming reaction, calculated on the basis of H_2, CO, H_2O, and CO_2 partial pressures, is presented in Figure 5.1b. As can be seen, RWGS reaction is indeed quasi-equilibrated at temperatures between 600°C and 700°C when a CH_4/CO_2 ratio of 0.67 was used. Higher CO_2 concentration helps to shift the RWGS reaction toward its equilibrium composition, confirming the dominance of this reaction [17]. These results are in very good agreement with the previous findings of Wei et al. [18], who report mechanistic equivalence of methane steam and dry reforming reactions with quasi equilibrated reaction steps leading to RWGS reaction over supported Rh catalysts at 600°C.

High-temperature operation requires the use of thermally stable catalysts and special materials for reactor design, not to mention soaring energy consumption for achieving the desired reaction conditions. On the other hand, by operating in the

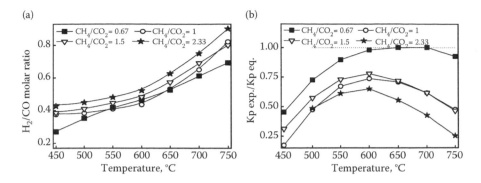

FIGURE 5.1 (a) H_2/CO molar ratio as a function of reaction temperature for different CH_4/CO_2 molar ratios in the feed stream; (b) proximity to RWGS equilibrium, calculated on the basis of partial pressures of H_2, CO, CO_2, and H_2O. Reaction conditions: 75 mg of 3% Ru/γ-Al_2O_3 catalyst. Total volumetric flow rate equals 100 Nml/min.

methane rich regime (CH_4/CO_2 molar ratio in biogas is usually between 1 and 3), carbon deposition on the catalyst surface is strongly enhanced, which renders traditional transition metal catalysts useless, as a result of their fast carbon accumulation [13] via methane dehydrogenation reaction (Equation 5.5). Amounts of accumulated carbon over transition metal catalysts (Figure 5.2a) and noble metal catalysts (Figure 5.2b), when increasing the methane content in the feed, reveals much better resistance of the noble metal catalysts due to the better balance of carbon accumulation and carbon gasification reactions [19]. Coke accumulation and consequent catalyst failure, either due to deactivation or reactor plugging, represents the main challenge obstructing widespread use of transition metal catalysts in the methane dry reforming reaction. Another drawback of using a large surplus of methane in the feed is the need to separate and recycle the gas stream discharged from the reactor, since the

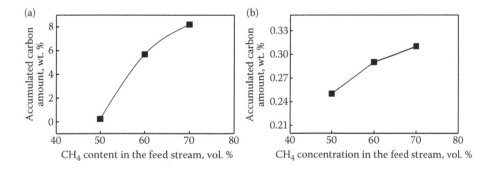

FIGURE 5.2 Amount of accumulated carbon over (a) 3% NiCo/CeZr catalyst and (b) 3% Ru/γ-Al_2O_3 catalyst. Test conditions: time on stream: 80h, isothermal tests at 750°C using undiluted feed stream of CH_4 and CO_2 at different ratios: 50:50, 60:40, and 70:30, and total volumetric flow equal to 100 Nml/min. Catalyst mass: 500 mg, diluted with 2850 mg of SiC.

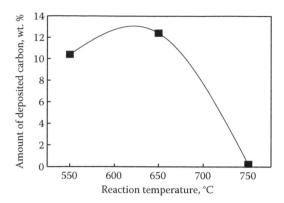

FIGURE 5.3 Amount of deposited carbon over 3% NiCo/CeZr catalyst as a function of reaction temperature. Reaction conditions: time on stream: 80 h, feed gas composition: CH_4 and CO_2 50 Nml/min each. Catalyst mass: 500 mg, diluted with 2850 mg of SiC.

limiting reactant (CO_2) will be consumed, with a substantial fraction of methane still remaining unconverted.

Carbon accumulation over the catalyst, originating from CO disproportionation into carbon and carbon dioxide (Equation 5.6) can also be prominent [20], especially at temperatures below 700°C, where the equilibrium of this reaction is favored. Consequently, the operating reaction temperature for methane dry reforming reaction should be high enough (above 700°C) in order to minimize the negative contribution of this reaction [16]. In Figure 5.3, the amount of accumulated carbon as a function of reaction temperature is presented. As can be seen, carbon accumulation can be indeed eliminated by carrying out the process at sufficiently high temperatures, where the contribution of Equation 5.6 is negligible.

5.4 METHANE DRY REFORMING CATALYSTS BASED ON NOBLE METALS

Active catalysts for the methane dry reforming reaction can be divided into two groups: transition metal catalysts and noble metal catalysts, both exhibiting specific benefits and drawbacks.

An early work of Rostrup-Nielsen [21] compared the activity of Ni, Ru, Rh, Pd, Ir, and Pt and discovered that Ru and Rh exhibited the highest activity and carbon-free operation. This was later confirmed also by Richardson [22], who identified rhodium as the best catalyst in terms of activity and stability in the methane dry reforming reaction. Liao et al. [23] compared the dissociation of methane over various metals and reported the lowest dissociation enthalpies over Rh, Ru, and Ni surfaces. On the other hand, IB metals (Cu, Ag, and Au) are reported to have much higher methane dissociation enthalpies and are consequently very poorly active in methane activation [23]. Methane dissociation over kink and step sites of active metal nanoclusters is energetically less demanding compared with close-packed crystalline planes. This elementary step is regarded as rate determining in the methane dry reforming reaction over Ni and other noble metals (Pt, Pd, Ru, and Rh) catalysts as reported by Wei and Iglesia [18] and references therein. As a result, by decreasing the active metal cluster size, coordinatively unsaturated surface atoms prevail, relative to densely packed low index crystalline planes, which results in increasing turnover rates.

Supported Ru and Rh catalysts exhibit high activity, stability (Rh > Ru), and very low tendency for carbon accumulation in the methane dry reforming reaction over a wide range of operating conditions. Slow deactivation of Ru containing catalysts was ascribed to active metal sintering, as was clearly noticeable from SEM characterization (Figure 5.4) [24].

To achieve high dispersion of noble metal particles as well as to improve their thermal stability in typical reaction conditions, and of course, due to economic considerations, active catalytic phase (up to 5 wt%) is deposited over various oxides, such as ZrO_2, SiO_2, MgO, Al_2O_3, YSZ, CeO_2, perovskites, etc.

Along with an active metal, the catalyst support also plays an important role in methane dry reforming reaction. By dispersing noble (Rh, Ru, Pt, Re, and Pd) and transition (Ni, Fe, Co) metals over various inert and reducible supports (Al_2O_3, SiO_2, ZrO_2, La_2O_3, various molecular sieves, and active carbon) [25], it has been discovered that metals supported over inert materials (e.g., silica) show much lower activity in terms of turnover numbers, compared with metals dispersed over Al_2O_3, TiO_2, ZrO_2, or CeO_2. Mechanistic studies indicate the participation of surface diffusion and spill-over processes of hydrogen and oxygen species from the metal to the support and vice versa, justifying the above mentioned observations. Bifunctional mechanisms in which methane is activated on metallic surface and CO_2 on the support, are reported for various oxidic supports, such as ZrO_2 [26], TiO_2 [27], γ-Al_2O_3 [28], and CeO_2 [29]. Besides modifying the mechanism of methane dry reforming reaction through the enhanced CO_2 dissociation over oxygen vacancy sites in ceria [29], addition of CeO_2 is reported to prevent metal sintering due to strong metal–support interactions (SMSIs) and improving catalyst thermal stability [30].

To demonstrate the involvement of the oxide support on the CH_4-CO_2 reforming activity, pure γ-Al_2O_3 and CeO_2 with BET specific surface area between 120 and 160 m^2/g were tested in the methane dry reforming reaction by feeding an equimolar

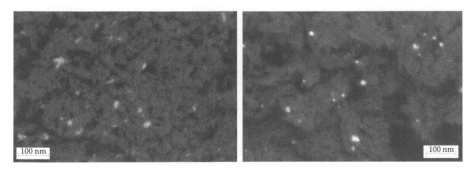

FIGURE 5.4   SEM images of 3% Ru/γ-Al_2O_3 catalyst (a) before and (b) after 70 h time on stream. CH_4-CO_2 reforming test was conducted at 750°C. The brighter areas highlight ruthenium clusters, recorded with the energy selective backscatter detector.

stream of CH_4/CO_2 at 750°C. Pure γ-Al_2O_3 exhibited negligible catalytic activity in terms of CH_4–CO_2 reforming, since the measured H_2 and CO concentrations reached 0.16 and 0.36 vol%, respectively. The amount of deposited carbon on the surface of pure γ-Al_2O_3 after the test was 8.81 wt%. However, pure CeO_2 produced 0.15 and 1.5 vol% of H_2 and CO at 750°C, respectively. After the reaction, 1.38 wt% of carbon was identified on the surface of pure CeO_2. The major difference in the performance of supports was a fivefold increase of CO production, which confirms the positive role of reducible support and oxygen vacancies on the CO_2 activation. Pronounced redox properties of this catalyst, compared with irreducible γ-Al_2O_3, also positively influence the carbon scavenging and oxidation by active oxygen species, originating from the dissociative CO_2 adsorption over oxygen vacancy sites in the CeO_2 lattice, created as a result of exposure to H_2 reducing atmosphere at elevated temperatures [24].

Coking of a catalyst is identified in the literature as the key reason for premature catalyst deactivation in the investigated process due to thermodynamically preferred carbon formation [25].

Coke deposition (Figure 5.5) and subsequent deactivation of the catalyst and clogging of the reformer is influenced mostly by the reaction mechanism and the secondary catalytic effect of the catalyst support. Fereira-Aparicio et al. [31] discovered that, for the Ru/γ-Al_2O_3 catalyst, the alumina support is involved in CO_2 activation by producing formates on its surface that subsequently decompose to CO. Likewise, Zhang et al. [32] tested the Rh/γ-Al_2O_3 catalyst and have identified a similar route of carbon to CO disproportionation. A very recent work of Ocsachoque et al. [33] clearly showed that a substantial increase of methane dry reforming activity as well as improved carbon resistance can be achieved by CeO_2 addition to a Rh/α-Al_2O_3 catalyst. During CH_4 dissociation, the electron transfer takes place, and Ce^{4+}/Ce^{3+} and $Rh^0/Rh^{\delta+}$ redox couples are generated. The $Rh^0/Rh^{\delta+}$ couple is believed to be involved in adsorption of CH_4 and the C–H bond activation, whereas the Ce^{4+}/Ce^{3+} couple promotes CO_2 dissociation to CO. Furthermore, CeO_2 has the extraordinary ability to act as an oxygen reservoir under oxidative and reductive atmosphere, which enables the oxidation of deposited carbon, thus

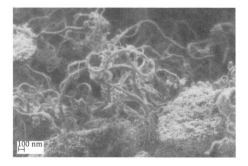

FIGURE 5.5 SEM of carbon nanofibers, accumulated over the catalyst surface during methane dry reforming reaction.

maintaining stable catalytic activity [29]. This promotion is believed to originate from the metal-ceria interaction, redox behavior of Ce^{4+}/Ce^{3+}, and electronic effect on metal [34]. In the case of Rh-based catalysts, CO_2 activation occurs mainly over oxygen vacancy surface sites of CeO_2 [27,29]. Therefore, extensive formation of oxygen vacancies, as reported above, generates many active sites for CO_2 dissociation, which forms CO and replenishes one oxygen vacancy. Methane dehydrogenation over Rh surface constantly supplies CH_x species ($x < 4$) along with H_2. The balance between CH_4 dehydrogenation and CO_2 disproportionation enables continuous partial oxidation of CH_x species, which regenerates oxygen vacancies in CeO_2 and maintains the oxidation/reduction cycle and constant catalytic activity. By preventing the accumulation of CH_x species, which can polymerize and cover the catalyst surface with carbon, a premature activity decline can be omitted. Finally, the higher the extent of ceria reduction (which translates into higher density of oxygen vacancies), the more CO_2 can be dissociated to replenish a higher quantity of oxygen, which accelerates elimination of eventually deposited carbon on the catalyst surface [29].

5.5 METHANE DRY REFORMING CATALYSTS BASED ON TRANSITION METALS

Despite their excellent catalytic performance and stability, noble metal catalysts are less suitable for large-scale applications due to their price, which is roughly three orders of magnitude higher compared with transition metals, such as nickel, copper, cobalt, iron, tin, and manganese [35]. Consequently, great economic potential and scientific challenge lie in the systematic development and precise tailoring of an active, stable, and carbon-resistant transition metal catalyst that is capable of dry reforming real biogas streams into syngas.

Transition metal catalysts show high initial activity, which is strongly affected by extensive carbon formation originating either from methane dehydrogenation (Equation 5.5) or CO disproportionation (Equation 5.6). By appropriate tailoring of transition metal catalysts, coke accumulation can be greatly diminished, which importantly contributes to the overall catalytic performance.

A lot of research work has been performed in the field of optimization of transition metal catalysts, especially nickel, which is generally accepted as the most active among transition metals for methane activation [36]. It has been discovered that to diminish carbon accumulation over the transition metal catalyst, both chemical composition and catalyst morphology play a fundamental role.

When supported, nickel catalysts are used for methane dry reforming reaction, carbon nanofiber formation is prominent, leading to pressure buildup and reactor plugging. This can be to some extent prevented over very small nickel particles, usually below 20 nm [37]. Such small nickel particles are reported to effectively slow down carbon nanofiber growth due to minimal size of nickel ensembles, required for carbon nucleation (Figure 5.6).

This represents only a short-term solution, since the exposure to reaction temperatures above 700°C and hydrothermal conditions (water vapor presence due to

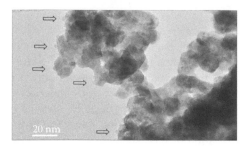

FIGURE 5.6   Nickel and cobalt bimetallic particles roughly 6 nm diameter in size (emphasized with arrows), dispersed over ceria zirconia support.

RWGS reaction) will cause sintering of the catalyst and growth of active particle size. Consequently, it can be expected that nickel critical size will sooner or later be exceeded, leading to pronounced carbon growth [38].

A very creative approach (commercialized as the SPARG process [39]) was developed some years ago, which employs the promotion-by-poisoning effect over nickel alumina catalysts for CO-rich syngas manufacture. The principle behind its operation is passivation of nickel surface sites with sulfur, which decreases the apparent nickel ensemble size, resulting in a higher thermodynamic potential for carbon formation. Since carbon accumulation requires a larger ensemble of surface sites, compared with methane reforming reaction, the former process is effectively stopped [40].

Beside passivation with sulfur, utilization of bimetallic nickel catalysts represents another effective approach toward minimizing carbon accumulation during methane dry reforming reaction.

Addition of tin, for example, causes the rate determining methane activation step to shift from most active low-coordinated to less active well-coordinated Ni sites, which increases the activation barriers for methane dissociation. It also weakens the binding strength of carbon atoms to the most active low-coordinated sites, which is a requirement for carbon nucleation at these sites [41]. Because nickel surface is decorated with tin, active sites are partially blocked, which in turn decreases the catalytic activity. Hou et al. [36] attributed higher dispersion of nickel, improvement of nickel reducibility and better resistance to catalyst sintering, to the presence of tin.

Addition of Cr and Mn significantly inhibited carbide formation, which is widely reported as a precursor for carbon formation [42,43]. Alloying nickel with cobalt breaks the integrity of surface nickel ensembles, which decreases the apparent nickel particle size to the extent that it is too small for carbon nucleation and growth. Consequently, by alloy formation and dilution of nickel, its tendency for carbon dissolution can be reduced, leading to further increase of resistance toward carbon accumulation. In Figure 5.7, the amount of accumulated carbon during methane dry reforming over supported bimetallic NiCo catalysts is presented for gradually increasing average bimetallic particle size (6, 17, 47, and 65 nm, respectively).

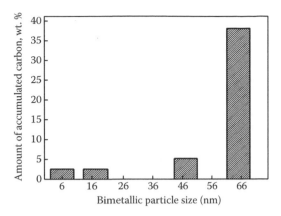

FIGURE 5.7 Amount of accumulated carbon over NiCo/CeZr bimetallic catalysts as a function of active particle size. Test conditions: 150 mg catalyst, diluted with 850 mg of inert SiC, feed stream: CH_4 and CO_2 at 50 Nml/min each, 20 h time on stream.

Zhang et al. [16] assigned improved carbon resistance of supported bimetallic NiCo catalysts, compared with monometallic nickel ones, to good metal dispersion, high metallic surface area, solid solution formation, and SMSIs.

The specific carbon accumulation rates over supported monometallic and bimetallic Ni/CeZr, Co/CeZr, and NiCo/CeZr catalysts containing the total of 12 wt% metals are compared in Figure 5.8. As can be seen, alloying nickel and cobalt in a 40:60 molar ratio leads to a substantial decrease of carbon accumulation rate.

In summary, considerable improvement of catalyst performance in terms of stability and minimization of carbon accumulation can be obtained using bimetallic transition metal catalysts.

The use of catalyst support, which exhibits high oxygen mobility (such as CeO_2, CeO_2-ZrO_2, YSZ, and TiO_2) can act as an efficient way to minimize carbon

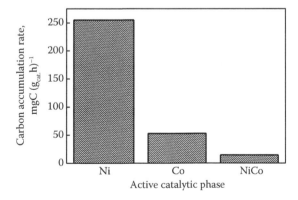

FIGURE 5.8 Carbon accumulation rate during reforming of equimolar methane and carbon dioxide streams using supported 12% Ni/CeZr, 12% Co/CeZr, and 12% NiCo/CeZr catalysts. Test conditions: 150 mg catalyst, diluted with 850 mg of inert SiC, feed stream: CH_4 and CO_2 at 50 Nml/min each, 20 h time on stream.

accumulation also over transition metal catalysts. High dispersion of active metal phase is a prerequisite for good metal–support interaction and high oxygen mobility is the prevailing mechanism for effective oxidation of carbon deposits, formed over active metal surface via oxygen spillover [36,44,45]. To further improve the oxygen mobility and thermal stability of ceria, different synthesis pathways for CeO_2-ZrO_2 solid solution manufacture are studied, such as modified Pechini route [46], reduction in ethylene glycol [47], coprecipitation/digestion [48], etc. Incorporation of isovalent nonreducible elements such as Zr^{4+} into the CeO_2 lattice positively influences the oxygen mobility and oxygen storage capacity in the crystalline lattice of the CeO_2-ZrO_2 solid solution. Deformation of metal–oxygen bonds (due to different Ce^{4+} and Zr^{4+} ionic radii, which causes lattice distortion and favors structural defects) results in highly mobile lattice oxygen species, so that the reduced zone is not confined to the surface layers, but extends deep into the bulk [49]. Mamontov et al. [50] suggested that incorporation of the smaller Zr^{4+} ions gives rise to strain in the structure of mixed oxide, which can be alleviated by the promotion of Ce^{4+} to Ce^{3+} ions (ionic radius of 97 vs. 128 pm, respectively).

By successful substitution of Ce^{4+} with Zr^{4+} in the synthesized CeO_2-ZrO_2 solid solution, up to 67% transformation of Ce^{4+} to Ce^{3+} could be achieved during the performed temperature programmed reduction experiments. On the contrary, by altering the synthesis method, the zirconium substitution was not efficient, allowing only up to 50% Ce^{4+} to Ce^{3+} transformation [19].

By taking into account the following modifications: (i) appropriate synthesis of CeO_2-ZrO_2 solid solution to act as the catalyst support (which will provide ample mobile oxygen species for instantaneous carbon oxidation), (ii) deposition of nickel and cobalt bimetallic particles (with their size in an appropriate nanometer scale in order to minimize carbon nucleation and growth) to provide the methane dissociation activity, and finally (iii) conducting the process in an appropriate temperature interval to decrease the thermodynamic driving force for carbon generation, long-term methane dry reforming performance with minimal carbon accumulation can be achieved (Figure 5.9).

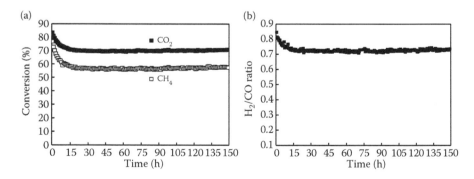

FIGURE 5.9 (a) Methane and carbon dioxide conversion over $NiCo/CeO_2$-ZrO_2 catalyst and (b) the corresponding H_2/CO ratio obtained. Test conditions: 500 mg catalyst, diluted with 2850 mg of inert SiC, feed stream: CH_4 and CO_2 at 50 Nml/min each, 150 h time on stream, reaction temperature: 750°C.

REFERENCES

1. Santacesaria, E., Martinez Vicente, G., Di Serio, M., Tesser, R. 2012. Main technologies in biodiesel production: State of the art and future challenges. *Catal. Today* (article in press), http://dx.doi.org/10.1016/j.cattod.2012.04.057.
2. Balat, M. 2011. Production of bioethanol from lignocellulosic materials via the biochemical pathway: A review. *Energy Convers. Manage.* 52:858–75.
3. 2011 AEBIOM Annual Statistical Report.
4. Thrän, D., Seiffert, M., Müller-Langer, F., Plättner, A., Vogel, A. 2007. Möglichkeiten einer Europäischen Einspeisungsstrategie, Leipzig, Germany, Deublein, D., Steinhauser, A. 2008. *Biogas from Waste and Renewable Resources*, Wiley, Weinheim.
5. Santosh, Y., Sreekrishnan, T.R., Kohli, S., Rana, V. 2004. Enhancement of biogas production from solid substrates using different techniques—A review. *Bioresource Technol.* 95:1–10.
6. Cavenati, S., Grande, C.A., Rodrigues, A.E. 2006. Separation of $CH_4/CO_2/N_2$ mixtures by layered pressure swing adsorption for upgrade of natural gas. *Chem. Eng. Sci.* 61:3893–106.
7. Rostrup-Nielsen, J., Christiansen, L.J. 2011. *Concepts in Syngas Manufacture*, Imperial College Press, London.
8. Ashcroft, A.T., Cheetham, A.K., Foord, J.S., Green, M.L.H., Grey, C.P., Murrell, A.J., Vernon, P.D.F. 1990. *Nature* 344:319–21.
9. The National Energy Technology Laboratory (NETL), NETL: 2010 Worldwide Gasification Database, http://www.netl.doe.gov/technologies/coalpower/gasification/world database/2010_Worldwide_Gasification_Database.pdf, accessed June 21, 2013.
10. Satterfield, C.N. 1996. *Heterogeneous catalysis in industrial practice*, second edition, Krieger Publishing Company, Malabar, FL.
11. Wilhelm, D., Simbeck, D., Karp, A., Dickenson, R. 2001. Syngas production for gas-to-liquids applications: Technologies, issues and outlook. *Fuel Proc. Tech.* 71:139–48.
12. Deublein, D., Steinhauser, A. 2008. *Biogas from Waste and Renewable Resources*, Wiley, Weinheim.
13. Bradford, M.C.J., Vannice, M.A. 1999. CO_2 reforming of CH_4. *Catal. Rev. Sci. Eng.* 41:1–42.
14. Storch, H.H., Golumbic, N., Anderson, R.B. 1951. *The Fischer–Tropsch and Related Syntheses: Including a Summary of Theoretical and Applied Contact Catalysis*, Wiley, Chapman & Hall, New York and London.
15. Liu, K., Song, C., Subramani, V. 2010. *Hydrogen and Syngas Production and Purification Technologies*, Wiley, New York.
16. Zhang, J., Wang, H., Dalai, A.K. 2007. Development of stable bimetallic catalysts for carbon dioxide reforming of methane. *J. Catal.* 249:300–10.
17. Djinović, P., Batista, J., Pintar, A. 2012. Efficient catalytic abatement of greenhouse gases: Methane reforming with CO_2 using a novel and thermally stable $Rh-CeO_2$ catalyst. *Int. J. Hydrogen Energy* 37:2699–707.
18. Wei, J., Iglesia, E. 2004. Structural requirements and reaction pathways in methane activation and chemical conversion catalyzed by rhodium. *J. Catal.* 225:116–127.
19. Djinović, P., Osojnik Črnivec, I.G., Erjavec, B., Pintar, A. 2012. Influence of active metal loading and oxygen mobility on coke-free dry reforming of Ni–Co bimetallic catalysts. *Appl. Catal. B* 125:259–70.
20. Efstathiou, A.M., Kladi, A., Tsipouriari, V.A., Verykios, X.E. 1996. Reforming of methane with carbon dioxide to synthesis gas over supported rhodium catalysts: II. A steady-state tracing analysis: Mechanistic aspects of the carbon and oxygen reaction pathways to form CO. *J. Catal.* 158:64–75.
21. Rostrup-Nielsen, J.R. 1994. Aspects of CO_2-reforming of methane. *Stud. Surf. Sci. Catal.* 81:25–41.

22. Richardson, J.T., Garrait, M., Hung, J.K. 2003. Carbon dioxide reforming with Rh and Pt–Re catalysts dispersed on ceramic foam supports. *Appl. Catal. A* 255:69–82.
23. Liao, M.S., Zhang, Q.E. 1998. Dissociation of methane on different transition metals. *J Mol. Catal. A* 136:185–194.
24. Djinović, P., Črnivec Osojnik, I.G., Batista, J., Levec, J., Pintar, A. 2011. Catalytic syngas production from greenhouse gasses: Performance comparison of Ru–Al_2O_3 and Rh–CeO_2 catalysts. *Chem. Eng. Process.* 50:1054–62.
25. Liu, D., Lau, R., Borgna, A., Yang, Y. 2009. Carbon dioxide reforming of methane to synthesis gas over Ni-MCM-41 catalysts. *Appl. Catal. A* 358:110–18.
26. Bitter, J.H., Seshan, K., Lercher, J.A. 1997. The state of zirconia supported platinum catalysts for CO_2/CH_4 reforming. *J. Catal.* 171:279–86.
27. Ferreira-Aparicio, P., Marquez-Alvarez, C., Rodriguez-Ramos, I., Schuurman, Y., Guerrero-Ruiz, A., Mirodatos, C. 1999. A transient kinetic study of the carbon dioxide reforming of methane over supported Ru catalysts. *J. Catal.* 184:202–212.
28. Nakamura, J., Aikawa, K., Sato, K., Uchijima, T. 1994. Role of support in reforming of CH_4 with CO_2 over Rh catalysts. *Catal. Lett.* 25:265–70.
29. Wang, R., Xu, H., Liu, X., Ge, Q., Li, W. 2006. Role of redox couples of Rh^0/$Rh^{\delta+}$ and Ce^{4+}/Ce^{3+} in CH_4/CO_2 reforming over Rh–CeO_2/Al_2O_3 catalyst. *Appl. Catal. A* 305:204–10.
30. Wang, S., Lu, G.Q. 1998. Role of CeO_2 in Ni/CeO_2–Al_2O_3 catalysts for carbon dioxide reforming of methane. *Appl. Catal. B* 19:267–77.
31. Ferreira-Aparicio, P., Rodriguez-Ramos, I., Anderson, J.A., Guerrero-Ruiz, A. 2000. Mechanistic aspects of the dry reforming of methane over ruthenium catalysts. *Appl. Catal. A* 202:183–196.
32. Zhang, Z.L., Tsipouriari, V.A., Efstathiou, A.M., Verykios, X.E. 1996. Reforming of methane with carbon dioxide to synthesis gas over supported rhodium catalysts: I. Effects of support and metal crystallite size on reaction activity and deactivation characteristics. *J. Catal.* 158:51–63.
33. Ocsachoque, M., Bengoa, J., Gazzoli, D., González, M.G. 2011. Role of CeO_2 in Rh/α-Al_2O_3 catalysts for CO_2 reforming of methane. *Catal. Lett.* 141:1643–50.
34. Jin, T., Okuhara, T., Mains, G.J., White, J.M. 1987. An important role for lattice oxygen in CO oxidation. *J. Phys. Chem.* 91:3310–15.
35. www.lme.com, www.platinum.matthey.com, accesses October 23, 2012.
36. Hou, Z., Yokota, O., Tanaka, T., Yashima, T. 2004. Surface properties of a coke-free Sn doped nickel catalyst for the CO_2 reforming of methane. *Appl. Surf. Sci.* 233:58–68.
37. Zhang, J., Wang, H., Dalai, A.K. 2008. Effects of metal content on activity and stability of Ni–Co bimetallic catalysts for CO_2 reforming of CH_4. *Appl. Catal. A* 339:121–9.
38. Juan-Juan, J., Román-Martínez, M.C., Illán-Gómez, M.J. 2009. Nickel catalyst activation in the carbon dioxide reforming of methane: Effect of pretreatments. *Appl. Catal. A* 355:27–32.
39. Rostrup-Nielsen, J.R. 1984. Sulfur-passivated nickel catalysts for carbon-free steam reforming of methane. *J. Catal.* 85:31–43.
40. Andersen, N.T., Topsøe, F., Alstrup, I., Rostrup-Nielsen, J.R. 1987. Statistical models for ensemble control by alloying and poisoning of catalysts: I. Mathematical assumptions and derivations. *J. Catal.* 104:454–65.
41. Takanabe, K., Nagaoka, K., Nariai, K., Aika, K.I. 2005. Titania-supported cobalt and nickel bimetallic catalysts for carbon dioxide reforming of methane. *J. Catal.* 232:268–75.
42. Chen, P., Zhang, H.B., Lin, G.D., Tsai, K.R. 1998. Development of coking-resistant Ni-based catalyst for partial oxidation and CO_2-reforming of methane to syngas. *Appl. Catal. A* 166:343–50.

43. Choi, J.S., Moon, K.I., Kim, Y.G., Lee, J.S., Kim, C.H., Trimm, D.L. 1998. Stable carbon dioxide reforming of methane over modified Ni/Al₂O₃ catalysts. *Catal. Lett.* 52:43–7.
44. Nikolla, E., Schwank, J., Linic, S. 2009. Comparative study of the kinetics of methane steam reforming on supported Ni and Sn/Ni alloy catalysts: The impact of the formation of Ni alloy on chemistry. *J. Catal.* 263:220–7.
45. Rezaei, M., Alavi, S.M., Sahebdelfar, S., Bai, P., Liu, X., Yan, Z.F. 2008. CO_2 reforming of CH_4 over nanocrystalline zirconia-supported nickel catalysts. *Appl. Catal. B* 77:346–54.
46. Sadykov, V., Muzykantov, V., Bobin, A., Mezentseva, N., Alikina, G., Sazonova N., Sadovskaya, E., Gubanova, L., Lukashevich, A., Mirodatos, C. 2010. Oxygen mobility of Pt-promoted doped CeO_2–ZrO_2 solid solutions: Characterization and effect on catalytic performance in syngas generation by fuels oxidation/reforming. *Catal. Today* 157:55–60.
47. Osojnik Črnivec, I.G., Djinović, P., Erjavec, B., Pintar, A. 2012. Effect of synthesis parameters on morphology and activity of bimetallic catalysts in CO_2–CH_4 reforming. *Chem. Eng. J.* 207–8:299–307.
48. Potdar, H.S., Roh, H.S., Jun, K.W., Ji, M., Liu, Z.W. 2002. Carbon dioxide reforming of methane over co-precipitated Ni–Ce–ZrO_2 catalysts. *Catal. Lett.* 84:95–100.
49. Vlaic, G., Di Monte, R., Fornasiero, P., Fonda, E., Kašpar, J., Graziani, M. 1999. Redox property–local structure relationships in the Rh-loaded CeO_2–ZrO_2 mixed oxides. *J. Catal.* 182:378–89.
50. Mamontov, E., Egami, T., Brezny, R., Koranne, M., Tyagi, S. 2000. Lattice defects and oxygen storage capacity of nanocrystalline ceria and ceria–zirconia. *J. Phys. Chem. B* 104:11110–6.

6 Sulfated Inorganic Oxides for Methyl Esters Production
Traditional and Ultrasound-Assisted Techniques

*Daria C. Boffito, Carlo Pirola, Claudia L. Bianchi,
Giuseppina Cerrato, Sara Morandi, and
Muthupandian Ashokkumar*

CONTENTS

6.1 INTRODUCTION

Biodiesel is a liquid biofuel that is defined as a fatty acid methyl ester matching certain standards. The two most common standards setting the characteristic of biodiesel are EN 14214 (European Commission) and ASTM 6751 (American Society for Testing and Materials).

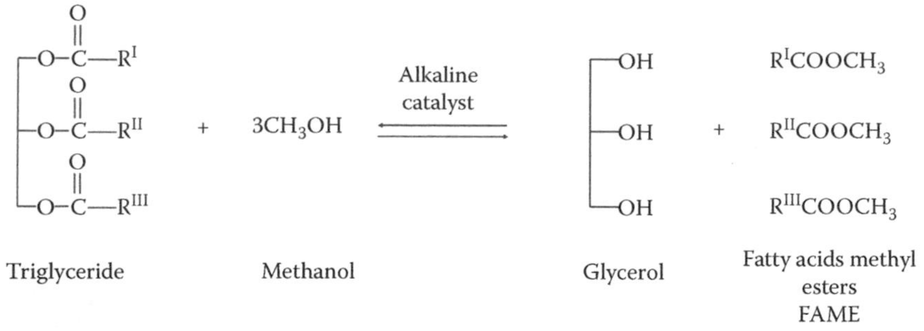

SCHEME 6.1   Transesterification reaction.

Biodiesel is obtained by the transesterification or alcoholysis of natural triacilglycerols (TAG), also known as triglycerides, contained in vegetable oils, animal fats, waste fats and greases, waste cooking oils, or side-stream products of refined edible oil production with short-chain alcohols. The most common alcohol used in the transesterification is methanol, followed by ethanol, which is used in a few cases (Perego and Ricci, 2012; Okoronkwo et al., 2012; Veljković et al., 2012; Ganesan et al., 2009). A representation of transesterification is given in Scheme 6.1.

Nowadays, most biodiesel (BD) is produced from edible oils with methanol and in the presence of an homogeneous alkaline catalysis. NaOH, KOH, and CH_3ONa are the most common catalysts (Santori et al., 2012; Arzamendi et al., 2008; Lotero et al., 2005; Ma and Hanna, 1999).

The presence of free fatty acids (FFAs) in the feedstock, which is very common in the case of unrefined oils, causes the formation of soaps due to the reaction with alkaline catalysts ($RCOOH + NaOH \rightarrow RCOONa + H_2O$). Soaps distribute among the two phases, hindering the contact between reagents, and eventually, the products separating (Perego and Ricci, 2012; Boffito et al., 2012, 2013a,b; Bianchi et al., 2010; Pirola et al., 2010). As a consequence, the process can only accept a restricted range of feedstock, typically with less than 0.5% of FFAs and 0.2% of moisture (Perego and Ricci, 2012; Boffito et al., 2012, 2013ab; Bianchi et al., 2010; Pirola et al., 2010). Another drawback given by the use of alkaline homogeneous catalysts lies in the need of neutralization (and wastewater disposal) and the nonrecovery of the catalyst. A solution could be represented by the use of acid catalysts. Acid catalysts, besides being tolerant to the FFA, also catalyze their esterification, according to Scheme 6.2.

Sulfuric acid, hydrochloric acid, and *p*-toluene sulfonic acids (Bourges and Diaz, 2012) are usually adopted as catalysts in the FFA esterification. However, the use of these catalysts results in some obvious drawbacks, such as corrosion, impossibility

$$RCOOH + CH_3OH \xrightleftharpoons[]{\text{Acid catalyst}} RCOOH + H_2O$$

SCHEME 6.2   FFAs esterification reaction.

of catalyst recovery and safety issues. A clear challenge is the development of heterogeneous catalysts with high efficiency and sustainable costs of manufacturing. Solid acid catalysts have the strong potential to replace liquid acids in the transesterification reaction as they can eliminate separation, corrosion, and environmental problems associated with liquid acid transesterification, besides also promoting the FFA esterification.

Moreover, the FFA esterification reaction may be also an alternative process to transesterification reaction when low-quality oils and fats are used as feedstock. For this reason, many efforts regarding the use of solid acid catalysts for the FFA esterification were recently carried out (Boffito et al., 2012, 2013a,b; Bianchi et al., 2010; Pirola et al., 2010; Marchetti and Errazu, 2008).

Among various solid acid catalysts available, many studies have highlighted the application of sulfated metal oxides (Lotero et al., 2005; Jacobson et al., 2008; Jitputti et al., 2006; Boffito et al., 2012, 2013a).

Zirconium sulfate has been widely studied for the FFA esterification reaction (Boffito et al., 2012a; Juan et al., 2007) due to its high activity. Sulfated zirconia is widely studied for other kinds of reactions, such as isomerizations, alkylations, and esterification of other carboxylic acids (Ardizzone et al., 2004), but its whereas its use for FFA esterification directly performed in the oil is reported in a few studies (Dijs et al., 2003). Nevertheless, SO_4^{2-}/ZrO_2 is reported to exhibit high activity in the biodiesel production through tranesterification (Lotero et al., 2005; Jacobson et al., 2008; Jiputti et al., 2006).

Sulfated tin oxide (SO_4^{2-}/SnO_2) is another potential catalyst for transesterification reaction due to its high surface acidity that is reported to be stronger than SO_4^{2-}/ZrO_2 (Matsuhashi et al., 1990; Lam et al., 2009). Nevertheless, study concerning the usage of SO_4^{2-}/SnO_2 catalyst in biodiesel production is still very limited.

These materials are active because of the presence of both Brønsted acidic centers and Lewis acid sites (i.e., coordinatively unsaturated (cus) Zr^{4+} cations) at the surface, as evidenced by Morterra et al. (2001).

In this chapter, we describe the preparation of different sulfated catalysts such as SO_4^{2-}/ZrO_2, SO_4^{2-}/ZrO_2-TiO_2, SO_4^{2-}/SnO_2, and SO_4^{2-}/SnO_2-TiO_2 and the results of their activity in the FFA esterification reaction.

All the catalysts were characterized with special regard to the surface acidity. Moreover, since catalyst durability is a crucial issue from an industrial standpoint, much attention was devoted to catalyst reuse.

A large part of this chapter is devoted to the description of the ultrasonic enhancement of the acidity, surface area, and FFA esterification catalytic activity of sulfated ZrO_2-TiO_2 systems, using an innovative ultrasound-assisted sol–gel synthesis.

6.2 MATERIALS AND METHODS

6.2.1 TRADITIONAL SYNTHESIS OF SULFATED ZR-BASED SYSTEMS

In Table 6.1 the sulfated Zr-based catalysts prepared using traditional techniques, i.e., sol–gel, or impregnation methods are listed along with the method used for their

TABLE 6.1
Sulfated Zr-Based Systems

	Sample	Composition	Preparation	Precursors	T calc.	SSA (m²g⁻¹)	Vₚ (cm³g⁻¹)	meq H⁺g⁻¹
1	SZ1	SO_4^{2-}/ZrO_2	One-pot sol–gel	ZTNP[a] $(NH_4)_2SO_4$	893 K O_2, 6 h	107	0.09	0.90
2a	SZ2a	SO_4^{2-}/ZrO_2	Two-pots sol–gel	ZTNP H_2SO_4	893 K static, 3 h	102	0.10	0.11
2b	SZ2b	SO_4^{2-}/ZrO_2	Two-pots sol–gel	ZTNP H_2SO_4	653 K static, 3 h	110	0.10	0.12
3	SZ3	SO_4^{2-}/ZrO_2	Solvent-free	$ZrOCl_2$. $8H_2O$ $(NH_4)_2SO_4$	873 K static	81	0.11	1.3
4	SZ4	$Zr(SO_4)_2/SiO_2$	Impregnation	$Zr(SO_4)_2$. $4H_2OSiO_2$	873 K static, 3 h	331	0.08	1.4
5	SZ5	$Zr(SO_4)_2/Al_2O_3$	Impregnation	$Zr(SO_4)_2$. $4H_2OAl_2O_3$	873 K static, 3 h	151	0.09	0.67
6	ZS	$Zr(SO_4)_2.4H_2O$ (commercial)	—	—	—	13	0.12	9.6

[a] Zr-tetra-*n*-propoxide, $Zr(OC_3H_7)_4$.

preparation (entries 1 to 5). These catalysts are referred as SZ. Commercial zirconium sulfate, referred to as ZS in Table 6.1 (entry 6), was characterized and tested in the FFAs esterification for the sake of comparison with the other catalysts.

Catalysts from entries 1 to 3 in Table 6.1, i.e., SZ1, SZ2a, and SZ2b (SO_4^{2-}/ZrO_2), were synthesized using one pot sol-gel technique under acid hydrolysis conditions (HNO_3). Zirconium tetra-*n*-propoxide (ZTNP) and i-PrOH were used for all the sol–gel syntheses as precursor and solvent, respectively, adopting a molar ratio i-PrOH: ZTNP = 15 (Ardizzone et al., 2004). The molar ratio between water and the ZTNP was kept constant at 30, and the amount of sulfates, either from sulfuric acid or from $(NH_4)_2SO_4$, was the same for all the preparations (molar ratio $SO_4/Zr = 0.15$) (Ardizzone et al., 2009). In a typical synthesis, an aqueous solution containing HNO_3, and $(NH_4)_2SO_4$ in the case of one-pot sol–gel synthesis, was added to the precursor solution under stirring at a rate of 1–2 drops per minute. The solution was then kept under stirring for 90 min. The obtained product was dried at 353 K in static air obtaining a xerogel. The xerogel was ground in a mortar and sulfates were added by impregnation through incipient wetness of the zirconia hydrous precursor in the case of samples SZ2a and SZ2b. The xerogels were then calcined in the conditions reported in Table 6.1. All the samples listed in Table 6.1 from entry 1 to 5 were calcined with a heating rate of 2.5 K/min. The choice of the low calcination temperature of the sample SZ2b, along with the slow heating rate and duration of the treatment, was dictated by the need to preserve the largest number of sulfate groups after the calcination, as suggested by Dijs et al. (2003). SZ3 was prepared by a solvent free synthesis (Sun et al., 2005). In this synthesis, zirconyl chloride and $(NH_4)_2SO_4$ were grounded in a mortar at room temperature for 20 min and then left ageing at room temperature for 18 h. Calcination was performed in static air at 873 K (heating rate: 2.5 K/min). Also in this case, the molar ratio SO_4/Zr was kept constant at 0.15.

Samples SZ4 and SZ5, i.e., $Zr(SO_4)_2/SiO_2$ and $Zr(SO_4)_2/Al_2O_3$, respectively, were prepared by the impregnation of 3 g of support with 25 ml of a MeOH solution 0.4M of $Zr(SO_4)_2$. The solvent was then evaporated at 323 K under vacuum, and the samples calcined at 873 K in static air (heating rate: 2.5 K/min), as reported in Table 6.1.

$Zr(SO_4)_2.4H_2O$ (98% purity + metal basis) was an Alfa Aesar product.

6.2.2 CONVENTIONAL SYNTHESIS OF SULFATED SnO_2-TiO_2 SYSTEMS

The sulfated SnO_2-TiO_2 systems are listed in Table 6.2. Different TiO_2 weight amounts were added to the SnO_2. As an example, STTO5 in Table 6.2 indicates a sample containing 95% of SnO_2 and 5% of TiO_2 by weight. Sulfated SnO_2-TiO_2 samples were all prepared grounding SnO_2 (Sigma Aldrich, ~325 mesh) with TiO_2 P25 (Degussa, SSA = 52 m^2g^{-1}) in a mortar for 20 min at room temperature, followed by impregnation by incipient wetness of a water solution. In a typical synthesis, 3 g of compound were impregnated with 25 ml of water solution containing sulfates (H_2SO_4) with a ratio $SO_4^{2-}/(Sn + Ti) = 0.15$ (Ardizzone et al., 2004). Water was then evaporated at 338 K under vacuum, and the samples calcined in static air at 773 K for 3 h (heating rate: 2.5 K/min), as indicated in Table 6.2.

All the chemicals used for the syntheses were Fluka products and were used without further purification.

TABLE 6.2
Sulfated SnO_2-TiO_2 Systems

	Sample	Composition	Preparation	T calc.	SSA ($m^2 g^{-1}$)	V_p ($cm^3 g^{-1}$)	meq $H^+ g^{-1}$
1	STTO0	SO_4^{2-}/SnO_2	Physical	773 K	16.8	0.10	3.15
2	STTO5	$SO_4^{2-}/95\%SnO_2$-$5\%TiO_2$	mixing +	static	15.9	0.11	3.43
3	STTO10	$SO_4^{2-}/90\%SnO_2$-$10\%TiO_2$	impregnation		16.5	0.09	5.07
4	STTO15	$SO_4^{2-}/85\%SnO_2$-$15\%TiO_2$			14.9	0.11	7.13
5	STTO20	$SO_4^{2-}/80\%SnO_2$-$20\%TiO_2$			16.9	0.09	7.33

6.2.3 ULTRASOUND-ASSISTED SYNTHESIS OF THE SULFATED ZrO$_2$-TiO$_2$ SYSTEMS (BOFFITO ET AL., 2013A)

A list of all the Zr/Ti-based catalysts is given in Table 6.3. SO_4^{2-}/ZrO_2 (hereafter referred to as SZ) was synthesized using a conventional sol–gel method, while $SO_4^{2-}/80\%ZrO_2$-20%TiO$_2$ (hereafter referred to as SZT) was synthesized using both conventional and US-assisted sol–gel techniques. Samples termed USZT refer to US-obtained sulfated 80%ZrO$_2$-20%TiO$_2$. The name is followed by the US power, the length of US pulses and the molar ratio of water over precursors, as reported in Table 6.3. For example, USZT_40_0.1_30 indicates a sample $SO_4^{2-}/80\%ZrO_2$-20%TiO$_2$ obtained with US, using the 40% of the maximum power, with US on for 0.1 s (pulse length) and off for 0.9 s, using a water/ZTNP + TTIP molar ratio equal to 30. The molar ratio between precursors and the sulfating agent $((NH_4)_2SO_4)$ and between the precursors and HNO$_3$ was kept constant at 0.15 and 0.21, respectively (Ardizzone et al., 2004, 2009). Water/precursors molar ratio was varied as indicated on Table 6.3.

In the case of traditional sol–gel method, the starting solution was prepared by mixing 1.23 ml of TTIP and 5.52 ml of ZTNP with 25 ml of the solvent (i-PrOH) for 30 min in a bath thermostated at 298 K, stirring at 300 rpm through a mechanical stirrer. The aqueous solution containing the sulfating agent and HNO$_3$ was then added to the mixture at the rate of 0.25 ml/min. After finishing the addition, the gel was left aging for additional 90 min under stirring.

TABLE 6.3

Sulfated ZrO$_2$-TiO$_2$ Systems

	Sample	Composition	US Power (% max power)	Pulses (on/off)	H$_2$O: Precursors mol ratio	Synthesis Time	Sonication Time
1	SZ	SO_4^{2-}/ZrO_2	–	–	30	123'0"	0"
2	SZT		–	–		123'0"	0"
2a	SZT_773_6h		–	–		123'0"	0"
3	USZT_20_1_30		20	1		43'0"	43'0"
4	USZT_40_0.1_30		40	0.1/0.9		43'0"	4'18"
5	USZT_40_0.3_30			0.3/0.7		43'0"	12'54"
6	USZT_40_0.5_7.5	$SO_4^{2-}/80\%ZrO_2$-20%TiO$_2$		0.5/0.5	7.5	17'30"	8'45"
7	USZT_40_0.5_15				15	26'0"	13'0"
8	USZT_40_0.5_30				30	43'0"	21'30"
9	USZT_40_0.5_60				60	77'0"	38'30"
10	USZT_40_0.7_30			0.7/0.3	30	43'0"	30'6"
11	USZT_40_1_15			1	15	26'0"	26'0"
12	USZT_40_1_30				30	43'0"	43'0"

In the case of US synthesis, a 20-kHz horn sonicator was used. The tip of the horn was placed inside the sol–gel mixture in a 100-ml water-jacketed reactor. The power was varied at 20% and 40% of amplitude of the maximum nominal power (450 W). Also US pulses were adopted for some syntheses.

The total synthesis time reported in Table 6.1 indicates the time required to perform the whole sol–gel process, whereas the sonication time indicates the fraction of the total synthesis time while US were functioning. As an example, consider entry 4 in Table 6.1. The total synthesis time is 43′0″, which is the sum of the time required for the addition of the aqueous solution to one of the precursors at a rate of 0.25 ml/min and 10 min for further aging. Since the US was powered in pulse mode with on/off ratio 9:1, the actual sonication time is given by (1/10) · 43′0″, i.e., 4′18″.

The temperature was monitored during the course of the ultrasonic synthetic experiments. The temperature increased up to 313 K during the first few minutes of the reaction, then remained constant until the end of the experiments.

All the samples were calcined in static air atmosphere at 773 K for 3 h with a heating rate of 2.5 K/min. To demonstrate that the choice of specific calcination conditions is fundamental to avoid the loss of a too large fraction of surface sulfates groups, thereby affecting the catalytic activity, the sample SZT was also calcined at 773 K for 6 h, employing the same heating rate. This sample is reported as SZT_773_6h in entry 2a, Table 6.3.

ZTNP 70% in 1-PrOH, TTIP 98%, i-PrOH, HNO3 69.5 wt%, and $(NH_4)_2SO_4$ were used and are all Fluka products used without further purification.

6.2.4 CATALYSTS CHARACTERIZATION

Specific surface area and pore volume were assessed for all the samples by N_2 adsorption. The samples obtained with ultrasound-assisted synthesis were also characterized for what concerns pores sizes distribution.

N_2 adsorption–desorption isotherms were measured at 77 K using a Micromeritics Tristar 3000 system. All samples were degassed at 433 K overnight on a vacuum line. The standard multipoint Brunauer–Emmett–Teller (BET) method was utilized to calculate the specific surface area (SSA). The pore size distributions of the materials were derived from the adsorption branches of the isotherms based on the Barett–Joyner–Halenda (BJH) model (Gregg and Singh, 1982).

FTIR spectra were run at 4 cm^{-1} resolution with a Bruker IFS 113v spectrophotometer equipped with an MCT detector. The powder samples were investigated in the form of thin layer depositions (~10 mg cm^{-2}) on a pure Si wafer, starting from aqueous suspensions of the powders. After drying, all samples were activated in a controlled atmosphere at 623 K in a homemade IR quartz cell, equipped with KBr windows, connected to a conventional vacuum glass line capable of a residual pressure < 1 × 10^{-5} torr, which allowed to perform strictly in situ adsorption/desorption experiments of molecular probes. The study of surface acidity was performed using as probe either 2,6-dimethylpyridine (2,6-DMP; Lu) or carbon monoxide (CO). The standard IR experiment of 2,6-DMP adsorption/desorption on the various samples was carried out as follows: (i) admission in the IR cell of an excess dose of 2,6-DMP vapor (~2 torr), and equilibration at beam temperature (hereafter BT), i.e., the temperature reached by samples in the IR beam, for 5 min; (ii) evacuation of the base excess at BT for 15 min; (iii) desorption of the strongly bonded 2,6-DMP fraction was eventually carried out at 423 K for 15 min.

In case of CO adsorption, approximately 100 torr of carbon monoxide was exposed to the samples at ambient temperature in order to determine whether differences existed in Lewis acidity of the surface of the samples of interest. Desorption was performed in stages allowing 50, 25, 10, 5, 2, 1, and 0 torr of carbon monoxide to remain in contact with the sample.

The quantification of the Brønsted acidity was carried out through ion exchange with NaCl saturated solution, leaving it in contact with the catalysts for 30 h (López et al., 2007). The evaluation of H^+ concentration was carried out using a pH meter. The number of H^+ milliequivalents released per gram of catalyst was then calculated.

The crystalline nature of the samples was investigated by X-ray diffraction (XRD) using a PW3050/60 X'Pert PRO MPD diffractometer from PANalytical working Bragg-Brentano, using as a source the high-power ceramic tube PW3373/10 LFF with a Cu anode equipped with Ni filter to attenuate Kβ. Scattered photons have been collected by a RTMS (Real Time Multiple Strip) X'celerator detector.

SEM-EDX (energy-dispersive X-ray spectroscopy) was used to determine the amount of Zr, Ti, and S in the samples.

TEM images were recorded using a JEOL 3010-UHR instrument (acceleration potential: 300 kV; LaB6 filament). Samples were "dry" dispersed on lacey carbon Cu grids. Philips XL-30CP with RBS detector of back scattered electrons was used for SEM-EDX analyses.

Sulfates amount of both SZT and SZT_773_6h samples was determined by ion chromatography (Metrohm model 883 Basic IC Plus) so as to assess the loss of sulfate groups generated by the different calcination procedures. The solutions injected in the ion chromatograph and containing the sulfates were obtained suspending the catalysts in a NaOH solution 0.1 M at ambient temperature and then filtering the liquid through a 0.45-μm PTFE filter, as already reported elsewhere (Sarzanini et al., 1995).

6.2.5 ACTIVITY TESTS

A list of the experimental conditions adopted in the activity tests are listed in Table 6.4 for all the catalysts.

The selected feedstock for all the activity tests was a commercial rapeseed oil (initial acidity 0.1 wt%), acidified with oleic acid up to 5 or 7.5 wt%, in order to obtain a high initial acidity. The stoichiometric alcohol/acid ratio for the esterification reaction is 1:1, but it is advisable to use a higher amount of alcohol to shift the reaction toward the desired products. For this reason, the added amount of methanol was calculated taking into account the oil mass to treat and not the reaction stoichiometry. Consequently, the methanol/oil weight ratio was 16:100, corresponding to about 4.5:1 MeOH/oil molar ratio. All the catalytic tests were performed with the catalyst suspended in the reaction medium, i.e., in slurry reactors using 25-ml magnetically stirred vials or 500 ml reaction vessel as esterification reactors. The

TABLE 6.4
FFA Esterification Experiments

Catalyst	Reactor	Temp. (K)	Initial Acidity (%wt)	Recycle of Use
From SZ1 to SZ5 (Table 6.1)	Vial	336	5.0	
ZS (Table 6.1)	Vial	336		
	Reaction vessel			yes
	Reaction vessel with methanol recycle	363		yes
From STTO_0 to STTO_20 (Table 6.2)	Vial	336		
STTO_20 (Table 6.2)	Reaction vessel	336		yes
SZ, SZT, USZT (Table 6.3)	Vial	336	7.5	

reactors were thermostated before starting the activity tests using an oil bath at 336 ± 2 K. The mechanical stirring was maintained at 300 rpm. All the activity tests were carried out for 6 h.

Oil acidity was determined through titration analyses with KOH 0.1 M in ethanol. Since oil and KOH are not soluble, each sample was dissolved before titration in a mixture of diethylether/ethanol 9:1 per volume in order to obtain a homogeneous mixture. Phenolphthalein was used as an indicator. The acidity percentage at reaction time t, a_t, was calculated as in Equation 6.1 (Russbueldt et al., 2009; Pasias et al., 2006; Boffito et al., 2012, 2013a,b):

$$a_t = \frac{M \times MW \times C}{W} \times 100, \tag{6.1}$$

where V is the volume of KOH solution employed for titration (ml), MW is the molecular weight of oleic acid (282.46 mg mmol^{-1}), C is the concentration of KOH (mmol ml^{-1}), and W is the weight of the analyzed sample (mg).

FFA conversions, i.e., the percentage of oleic acid converted in oleic methylester were determined as in Equation 6.2.

$$\text{FFA conversion } (\%) = \frac{a_{t=0} - a_t}{a_{t=0}} \times 100, \tag{6.2}$$

where $a_{t=0}$ is the acidity of oil at zero time and the acidity of the oil at reaction time t.

All the products used for catalytic tests were Fluka products of high purity.

As already evidenced in previous studies of the authors (Bianchi et al., 2010; Pirola et al., 2010), the experiment carried out in absence of the catalyst showed no FFA conversion.

6.3 RESULTS AND DISCUSSION

6.3.1 CHARACTERIZATION OF CONVENTIONALLY SYNTHESIZED CATALYSTS

The results of the characterization of sulfated Zr systems, i.e., BET surface area (SSA_{BET}), pore volume (V_p), and acidities (meq H^+/g) are given in Table 6.1. It can be seen that ZS is characterized by the highest acidity and lowest surface area. The coexistence of these two characteristics has already proven to be beneficial to the FFA esterification reaction activity (Boffito et al., 2012, 2013a). In this case, in fact, the active sites on the catalyst's surface can be easily reached by sterically hindered molecules such as FFA.

Differently, the other sulfated Zr-based catalysts have significantly lower acidities and higher SSA than ZS. This might result in a low activity in the FFA esterification.

The samples SZ2a and SZb, i.e., the ones obtained by impregnation of $Zr(OH)_4$ with sulfates exhibit the lowest acidity. A hypothesis might be related to the poor intercalation of sulfate groups in the ZrO_2 structure. Since both these samples are characterized by very low and similar acidities, despite being calcined at different temperatures, a role of the calcination temperature in determining acidic properties cannot be stated in this case. The quite high acidity and low surface area of sample SZ3 might potentially render this sample a better candidate for the FFA esterification than the other catalysts.

It has been reported (Lam et al., 2009) that the addition of sulfate groups causes part of the catalyst surface area to collapse. Nevertheless, the impregnated sulfate groups increase the pore width and pore volume of the catalysts. This phenomenon is important in biodiesel synthesis using solid acid catalysts, as catalysts with larger pores will minimize diffusion limitations especially for molecules having long alkyl chains. Primarily, catalysts with mesopores (10–50 nm) are preferred. On the other hand, the addition of a support such as SiO_2 or Al_2O_3 is reported to increase the BET surface area but retain its mesopores properties. This is the case of samples SZ4 and SZ5.

Table 6.2 shows the BET surface area (SSA_{BET}), pore volume (V_p), and acidities (meq H^+/g) for the different sulfated SnO_2-TiO_2 systems. The acidity of the samples, the acidity of the samples, which was found to be a key issue in the FFA esterification activity, increases along the TiO_2 content. There are no data in literature on the acidic properties of sulfated SnO_2-TiO_2 systems, but the increase of acidity with the TiO_2 content might be ascribable to the charge imbalance resulting from the heteroatoms linkage for the generation of acid centers, as already described by Kataoka and Dumesic in the case of the sulfated mixed Zr-Ti oxides (Kataoka and Dumesic, 1988). The content of TiO_2 seems not to affect the specific surface area in this case. Consequently, the number of acid active sites per unit of area increases with the content of TiO_2, as indicated in Table 6.2. This might result in a higher catalytic activity for a sample with the highest TiO_2 content, i.e., STTO_20.

In case of CO adsorption, approximately 100 torr of carbon monoxide was exposed to the samples at ambient temperature in order to determine whether differences existed in Lewis acidity of the surface of the samples of interest. Desorption was performed in stages allowing 50, 25, 10, 5, 2, 1, and 0 torr of carbon monoxide to remain in contact with the sample.

No spectra will be reported for the sake of brevity, as all SnO_2-TiO_2 samples exhibit almost the same trend, both in the background profile and in the differential spectral features obtained for CO adsorption. In summary, it can be reported that

(i) Surface sulfate groups exhibit the spectral features typical of covalent species, as reported in the literature so far (Sarzanini et al., 1995), and, as a function of the increasing TiO_2 content, the presence of polynuclear sulfate groups is evident;

(ii) Lewis acidity, as tested by CO adsorption at room temperature, is present in very little amount, and can be ascribed, on the basis of both experimental evidences and literature data, to the presence of coordinatively unsaturated (cus) Ti^{4+} species present at the surface of the examined samples (Cerrato et al., 2009). No specific trend is evident as a function of the TiO_2 content.

6.3.2 CHARACTERIZATION OF THE CATALYSTS SYNTHESIZED WITH ULTRASOUND

In Table 6.5, Brønsted acidity (measured through ion exchange), specific surface area, and porosity features of all the sulfated ZrO_2-TiO_2 samples listed in Table 6.3 are reported.

Note that the simple addition of TiO_2 to the sulfated ZrO_2 increases the acidity of the catalyst more than 2.5 times (compare entries 1 and 2 in Table 6.5). It is reported

TABLE 6.5

Acid Capacities (Ion Exchange), Specific Surface Areas (BET), and Porosity Features (Pore Volume, V_p; Average Diameter, D_p), Elemental Composition (EDX) of All Catalysts

Entry	Catalyst	Acid Capacity (meq H$^+$/g)	SSA (m^2 g^{-1})	V_p (cm^3 g^{-1})	Ave. BJH D_p (nm)	Zr/Ti Weight Ratio	S/(Zr+Ti) Atomic Ratio
1	SZ	0.30	107	0.20	6.0	100	0.090
2	SZT	0.79	152	0.19	5.0	79:21	0.085
2a	SZT_773_6h	0.21	131	0.20	5.0	n.d.	n.d.
3	USZT_20_1_30	0.92	41.7	0.12	12.5	80:20	0.095
4	USZT_40_0.1_30	1.03	47.9	0.11	9.5	81:19	0.067
5	USZT_40_0.3_30	1.99	232	0.27	4.5	81:19	0.11
6	USZT_40_0.5_7.5	1.70	210	0.20	5.0	78:22	0.086
7	USZT_40_0.5_15	2.02	220	0.20	5.0	80:20	0.13
8	USZT_40_0.5_30	2.17	153	0.20	5.0	78:22	0.12
9	USZT_40_0.5_60	0.36	28.1	0.10	10	79:21	0.092
10	USZT_40_0.7_30	1.86	151	0.16	5.0	78:22	0.11
11	USZT_40_1_15	3.06	211	0.09	7.0	80:20	0.15
12	USZT_40_1_30	1.56	44.1	0.09	7.0	80:20	0.17

in the literature that the addition of TiO_2 to ZrO_2 can increase the surface concentration of –OH groups (Neppolian et al., 2007; Das et al., 2002). This behavior is generally observed for mixed oxides, and it is probably due to the charge imbalance resulting from the heteroatoms linkage for the generation of acid centers, as already observed my Dumesic (Kataota and Dumesic, 1988). The addition of TiO_2 to ZrO_2 is also reported to decrease the particle size and increase the surface area (Fu et al., 1996) as observed for the samples synthesized in the current study.

It has been observed that with the use of US during the synthesis a further increase in acidity occurred, as well as in the surface area (compare entries 1 and 2 with entries from 3 to 12 in Table 6.5).

The improvement in the properties of the catalysts is probably due to the effects generated by acoustic cavitation. Acoustic cavitation is the growth of bubble nuclei followed by the implosive collapse of bubbles in solution as a consequence of the applied sound field. This collapse generates transient hotspots with local temperatures and pressures of several thousand kelvin and hundreds of atmospheres, respectively (Sehgal et al., 1979). Very high–speed jets (up to 100 m/s) are also formed. As documented by Suslick and Doktycz (1990), in the presence of an extended surface, such as the surface of a catalyst, the formation of the bubbles occurs at the liquid–solid interface and, as a consequence of their implosion, the high speed jets are directed toward the surface. The use of sonication in the synthesis of catalysts can therefore result in the enhancement of the distribution of the active phase on the support (Pirola et al., 2010), increasing the rates of intercalation by a variety of species into a range of chalcogenides solids, improving the nucleation production rate (i.e., sol–gel reaction production rate) and the production of surface defects and deformations with the formation of brittle powders (Suslick and Doktycz, 1990).

At equal water/precursors molar ratios, the acidity of the catalysts increases further when pulse-on time of higher than 0.3 are adopted (compare entries 5, 8, and 10 with entry 12 in Table 6.2). For these catalysts, an increase in the SSA, as well as in the acidity, is observed if compared with the sample obtained with continuous US.

An attempt was made to explain the effect of various pulse lengths and on/off ratios in terms of two times, $\tau 1$ and $\tau 2$ characteristics of each cavitation system (Henglein et al., 1989). $\tau 1$ is the time required to produce and then grow chemically active gas bubbles, $\tau 2$ is the time taken by all the gas bubbles generated during the previous US pulse to dissolve away. If $\tau 1$ is shorter than the US pulse length, the system activates. If $\tau 2$ is longer than the interval time from one US pulse to the following one, gas bubbles still present in the system can generate bubble nuclei able to be grown within the following US pulse. Hence, the active bubbles population is high at short $\tau 1$ with long $\tau 2$. This condition is more easily achieved when long pulse lengths and short intervals between one US pulse to the other are adopted. During the US pulse bubbles also coalesce leading to a "degassing effect." Moreover, at long US pulses with short intervals, including continuous sonication, degassing may result in a lower number of active bubbles: it can be stated that there is an optimal on/off ratio for US pulse times for which maximum cavitation efficiency (maximum number of active bubbles) can be achieved.

Considering the effect of different water/precursors ratios, a general observation can be made: the SSA decreases when higher water amounts are used (compare entries from 6 to 9 and 11 to 12 in Table 6.2). This may be ascribed to a decrease of US power per unit volume inside the sol–gel reactor. Increasing and decreasing the water amount can be seen in fact as increasing and decreasing, respectively, the reactor size (Neppolian et al., 2007). When the ultrasonic power per unit volume is increased as a consequence of the diminishment of the reactor size, higher energy is supplied to the system. That is the reason why a lower water amount might lead to a higher surface damage and, as a consequence, to the formation of extra surface area under the effect of US. The effect of the increase of water amount on the SSA is particularly evident for the sample USZT_40_0.5_60, i.e., the one obtained with the highest H_2O/precursors ratio (entry 9 in Table 6.5). For this catalyst, the SSA is in fact significantly lower than the ones of the samples obtained with lower H_2O/ ZTNP + TTIP ratios. From the BJH adsorption data (data not provided for the sake of brevity), it is evident that all the samples are characterized by similar pore size distribution. In particular, the distribution resulted to be rather narrow and mainly located in the lower region of the mesopores. In Table 6.5 the average pore diameter and the volume of the pores are reported for all the catalysts. No significant differences in the pores dimensions were observed among the samples.

For what concerns about how the water/precursors ratio affects the catalysts acidity, it appears that increasing it up to a certain amount brings an increase in the H^+ concentration (compare entries from 6 to 9 and 11 to 12 in Table 6.5). In fact, the rate of the hydrolysis, i.e., the rate of the sol–gel process, is enhanced when using higher water amounts. Moreover, the higher the water amount in the sol–gel medium, the higher the probability that H_2O molecules can be chemically bounded producing extra $-OH^+$ Brønsted acidity. Nevertheless, increasing the water/precursors ratio over a certain amount (30 for pulsed and 15 for continuous US, entries 8 and 11 in Table 6.5, respectively) seems to have a negative effect on the acidity concentration. In fact, the risk of the extraction of acid groups by the excess of water increases as well. In addition, a role might have been also played by the different power densities, which led to a more or less effective surface damage, as already explained for the differences in the specific surface areas.

For the same reason, the increase of the water amount produces a lowering of the overall Brønsted acidity and not just the decrease of the SSA for the sample USZT_40_0.5_60 (entry 9 in Table 6.2). As a consequence, these results may be considered due to the same phenomenon.

The XRD analysis highlights the total absence of crystalline structure in all materials. Curve 1 in Figure 6.1 (SZT sample) is typical of an amorphous system. For the sake of brevity, we reported only this diffractogram, as all the others show a similar pattern. The XRD pattern of SZT_773_6h is also reported in Figure 6.1, curve 2. In this case, the material exhibits a net crystalline structure, revealing a series of reflections consistent with the presence of both ZrO_2 and TiO_2 in the tetragonal and anatase phases, respectively. This result highlights the effects of the different calcination times (3 vs. 6 h) at the same temperature (773 K): it is evident that these conditions deeply affect the nature of the above materials.

HR-TEM and SEM investigations of all samples further confirm what XRD analysis indicated. HR-TEM images (reported in Figure 6.2) show the almost total amorphous nature of the particles in all cases: for the sake of brevity, we decided to report

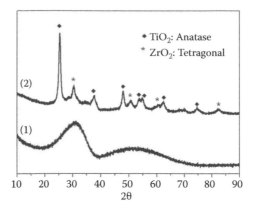

FIGURE 6.1 XRD patterns of SZT (1) and SZT_773_6h (2). (From Boffito, D.C., Crocellà, V., Pirola, C., Neppolian, B., Cerrato, G., Ashokkumar, M., Bianchi, C.L., Ultrasonic enhancement of the acidity, surface area, and free fatty acids esterification catalytic activity of sulphated ZrO_2-TiO_2 systems, *J. Catal.*, 2012b, http://dx.doi.org/10.1016/j.jcat.2012.09.013.)

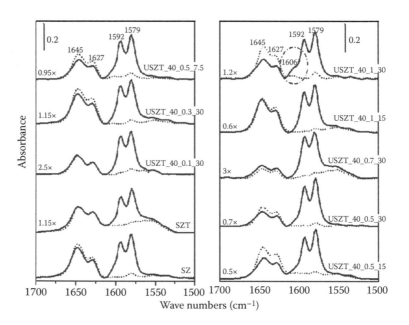

FIGURE 6.2 Differential absorbance spectra, normalized against the background spectrum of the starting sample activated at 623 K, relative to the adsorption/desorption of 2,6-DMP on some samples, as obtained in the vibrational region of 2,6-DMP 8a and 8b modes after contact at BT with the vapor pressure of 2,6-DMP (~2 torr) [solid curves], and after BT evacuation of the 2,6-DMP excess [broken curves]. (Boffito, D.C., Crocellà, V., Pirola, C., Neppolian, B., Cerrato, G., Ashokkumar, M., Bianchi, C.L., Ultrasonic enhancement of the acidity, surface area, and free fatty acids esterification catalytic activity of sulphated ZrO_2-TiO_2 systems, *J. Catal.*, 2012b, http://dx.doi.org/10.1016/j.jcat.2012.09.013.)

the features of SZT as this is typical of all systems (see the left hand image). On the contrary, crystalline particles with irregular shape (average diameter in the 4- to 7-nm range and exhibiting net fringe patterns) are easily observable in the case of the SZT_773_6h material (see the right hand image). SEM images are not reported here as the useful information about the morphology of the samples are already given in Figure 6.2 for the HR-TEM survey.

The sulfate concentrations evaluated by ion chromatography for the samples SZT and SZT_773_6h were 1.4 and 1.1 mg SO_4^{2-}/mg (ZrO_2-TiO_2), respectively. This indicates that the sample calcined for a long time (SZT_773_6h) significantly loses part of its surface sulfates groups, thus resulting in SO_4^{2-} content that is 20% less than SZT.

After all these preliminary characterization measurements, it can be stated that the choice of the calcination conditions reported in the catalysts preparation section (3 h at 773 K) allows to obtain materials that still possess a good amount of surface sulfate groups. In fact, although the increase of the calcination time and/or the employment of a higher calcination temperature would permit to generate crystalline materials, the subsequent consistent loss of surface sulfates might produce a drastic decrease of the catalytic activity.

The FTIR spectra, in the 4000- to 800-cm^{-1} range, were collected for all samples after activation at 623 K in O_2 atmosphere. They showed the typical spectral features of sulfate-modified zirconia. These systems have been widely studied in recent years by means of FTIR spectroscopy (Morterra et al., 1994, 1996, 1998, 2001, 2002), and therefore, these spectra have not been presented.

IR spectroscopic investigation of 2,6-DMP (Lu) adsorption/desorption is a useful analytical tool in surface chemistry because: (i) the strong base can interact, in a molecular form, with both weak and strong Lewis acid site [15,45]. The presence of methyl groups in the two α-positions of the heteroaromatic ring renders labile and reversible at BT, or just above that, the species adsorbed on weaker Lewis sites, whereas the species adsorbed on strong Lewis sites are much less labile and can thus be resolved; (ii) the interaction with acid protonic sites (Brønsted acid centers) yields lutidinium ions (LuH⁺), whose spectral features can be easily recognized and differentiated from those of all other 2,6-DMP adsorbed species [15,45].

Figure 6.2 reports, for some of the systems of interest, the main analytical spectral features in the region of the 8a–8b ring vibrational modes of adsorbed (solid lines) and desorbed (broken lines) 2,6-DMP at BT. When a large dose of 2,6-DMP is adsorbed, a band doublet at $\upsilon > 1615$ cm^{-1} (namely, ~1645 and ~1627 cm^{-1}) appears for all systems. This spectral feature clearly reveals the presence of abundant and strongly held LuH⁺ species [15,44], indicative of the presence of surface Brønsted acid centers.

For a better comparison among spectra of the different systems, the different spectral sets have been ordinate-magnified, as indicated on the curves.

As for the signals observed in the range below 1615 cm^{-1}, on the basis of literature data [15,44], they can be ascribed as follows:

(i) A strong band at 1580 cm^{-1} is due to the 8b mode of all 2,6-DMP species adsorbed in a molecular form (liquid-like physisorbed species and H-bonded species).

(ii) A second strong component located at 1592 cm^{-1} is ascribed to the partner 8a mode of 2,6-DMP H-bonded to surface OH groups.
(iii) A third higher-υ band, present as a weak shoulder at 1606 cm^{-1} only in the case of the USZT_40_1_30 sample, can be ascribed to 2,6-DMP interacting with medium-strong Lewis acid sites [coordinatively unsaturated (cus) Zr^{4+} or Ti^{4+} cations]. The presence of this component in one single sample indicates that only specific synthesis conditions (i.e., using 40% of the US maximum power, continuous US, and a water/precursors molar ratio equal to 30) are able to generate at the surface of the material the above mentioned sites.

When excess base is removed by vacuum at BT (see dotted lines in Figure 6.4), the bands due to protonated 2,6-DMP either remain virtually unchanged or slightly increase (meaning that the Brønsted-bound species are strongly adsorbed). On the contrary, the overall intensity of the envelope at υ < 1610 cm^{-1} decreases drastically, as expected of H-bonded and/or physisorbed species. Within this envelope, only in the case of the USZT_40_1_30 sample, part of the 8a band of Lewis-coordinated 2,6_DMP remains at 1608 cm^{-1}, together with a residual fraction of the 8b band at 1580 cm^{-1}, indicating the presence of a stronger fraction of Lewis acid sites. After evacuation at 423 K (these spectra are not shown in order to not overload the figure), only the signals at υ > 1610 cm^{-1} remain with virtually unchanged intensity (confirming that Brønsted-bound species are very strongly held), whereas no spectral components are present at lower frequencies yet, meaning that all the other species are desorbed during this mild treatment.

The results of the elemental analysis obtained by the SEM-EDX analysis are displayed in Table 6.2. It can be observed that the ratio between Zr and Ti corresponds to that used in the synthesis mixtures. Differently, the amount of sulfur, indicated as S/(Zr + Ti) atomic ratio in Table 6.2, varies for the different samples and, in some cases, it is far from the adopted 0.15. Indeed, note that some samples, i.e., USZT_40_0.5_15 (entry 7), USZT_40_1_15 (entry 11) and USZT_40_1_30 (entry 12) exhibit a higher sulfur on the surface, mainly ascribable to the different synthesis conditions adopted.

6.3.3 ACTIVITY TESTS OF THE TRADITIONALLY PREPARED CATALYSTS

In Figure 6.3 a typical trend of the FFA esterification reaction is represented as an example for the sample of zirconium sulfate (entry 6 in Table 6.1). The FFA conversion increases with the time until the achievement of the plateau of conversion, according the reaction equilibrium (Boffito et al., 2012; Bianchi et al., 2011; Pirola et al., 2011).

ZS allows to achieve very high FFA conversions within 6 h of reaction. This is ascribable to its high H$^+$ exchange capacity together with low SSA, for the reasons already highlighted in the section concerning catalysts characterization. A role in catalytic activity of ZS may be also played by Zr^{4+} Lewis acid sites, as evidenced in literature (Morterra et al., 1994, 1996, 1998, 2011, 2002). For this reason, ZS is widely studied as a catalyst for various reactions, among which the FFA

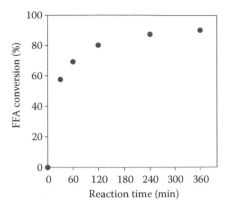

FIGURE 6.3   FFA conversion vs. reaction time for zirconium sulfates, 363 K.

esterification. Due to the high activity of ZS, it was decided to recycle it a few times in the conditions usually adopted for the FFA esterification, i.e., 363 K, and atmospheric pressure. Nevertheless, the major drawback of ZS lies in its fast deactivation due its partial solubility in water. Consequently, recycles of catalyst use at 373 K with continuous evaporation of methanol and water yielded by FFA esterification were also performed. In this case, fresh methanol was continuously added into the reactor. The results of these experiments are displayed in Figure 6.4. Note that the catalyst deactivation is just slightly delayed operating at a temperature of 363 K instead of 336 K. Karl Fischer analyses performed on both evaporated methanol and the deacidified oil after 6 h of reaction, detected the presence of water just in the evaporated methanol but not in the oil, indicating that water removal from the system was actually achieved in the experiments at 336 K. Nevertheless, leaching of active sulfate groups from ZS might occur in the presence of a liquid phase regardless of its polar nature. The presence of methanol might also have played a role in the loss of the sulfate groups from the catalyst.

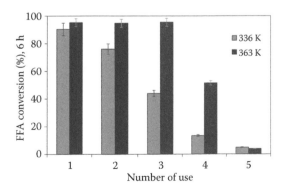

FIGURE 6.4   Recycle of use of catalyst ZS. Reaction conditions reported in Table 6.2.

Possible alternatives to zirconium sulfate are represented by the sulfated zirconias (Chen et al., 2007; Lotero et al., 2005) or sulfated supported catalysts (Jiménez-Morales et al., 2011; Juan et al., 2008).

The FFA esterification results of the sulfated ZrO_2 systems studied in this work are presented in Figure 6.5.

The results displayed in Figure 6.5 show that Zr-based sulfated systems do not provide a satisfactory performance in the FFA esterification, probably due to their low acid sites concentration related to their high SSA. Even if catalysts such as SZ3 and SZ4 exhibit higher surface area if compared with other catalysts, it is essential that this acidity is located mainly on the catalyst surface to be effectively reached by the FFA molecules. The authors have recently studied sulfonic ion exchange resins that are characterized by lower acidities than catalysts described in this chapter, but that perform better in the FFA esterification due to the surface localization of the acid sites (Boffito et al., 2012, 2013b; Bianchi et al., 2011, 2010; Pirola et al. 2010, 2011).

In Figure 6.6a, the results of the FFA esterification tests on the sulfated Ti and Sn mixed oxides are shown. Other conditions being equal, these catalysts perform better than the sulfated Zr-based systems just described. This is more likely due to the higher acidity along with a low surface area. With increasing the TiO_2 content, the acidity increases as well. This might be ascribable to the charge imbalance resulting from the heteroatoms linkage for the generation of acid centers, as already explained in the case of the mixed Zr-Ti oxides in the previous paragraph (Kataota and Dumesic, 1988).

As a consequence, the activity in the FFA esterification increases with the TiO_2 content along with the acidity of the samples. For the sake of clarity, in Figure 6.6b, the FFA esterification conversions are represented as a function of the number of active sites per unit of surface area of the samples: the FFA conversion increases with the number of meq H^+/m^2.

The best performing catalyst, i.e., the one with the highest TiO_2 content, was recycled a few times to verify its stability in the FFA esterification. Results are shown in

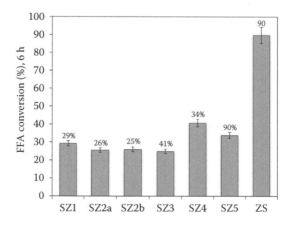

FIGURE 6.5 FFA esterification results of the Zr-based sulfated systems. Reaction conditions reported in Table 6.2.

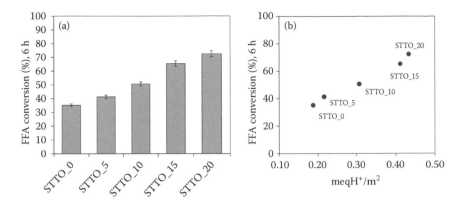

FIGURE 6.6 FFA esterification results of the sulfated SnO_2-TiO_2 systems for (a) different catalysts and (b) for different catalysts in function of the concentration of acid sites. (Boffito, D.C., Crocellà, V., Pirola, C., Neppolian, B., Cerrato, G., Ashokkumar, M., Bianchi, C.L., Ultrasonic enhancement of the acidity, surface area, and free fatty acids esterification catalytic activity of sulphated ZrO_2-TiO_2 systems, *J. Catal.*, 2012b, http://dx.doi.org/10.1016/j.jcat.2012.09.013.)

Figure 6.6a, where it can be noted that the activity of the STTO_20 drastically drops already after the first recycle. This might be due to the leaching of sulfate groups during the first use and their consequent loss during discharge operations. To confirm this hypothesis, the Sheldon test (Sheldon et al., 1998) was also carried out on sample STTO_20. The Sheldon test is a very simple way of checking if a catalyst is a real heterogeneous catalyst or if its activity is due to the active groups released into the reaction medium as a consequence of leaching. With this purpose, STTO_20 was removed from the reactor after 60 min of reaction. Results are represented in Figure 6.6b. As it can be seen, further FFA conversion was observed even after catalyst removal, meaning that the catalysis of FFA esterification is due in large part to the active species leached into the solution (Figure 6.7).

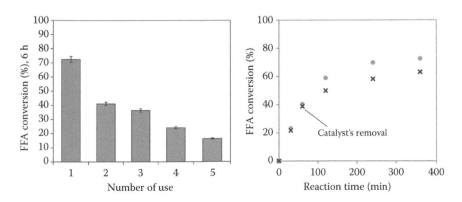

FIGURE 6.7 FFA esterification results of (a) the recycles of use of catalyst STTO_20 and (b) Sheldon test of catalyst STTO_20. Conditions reported in Table 6.3.

6.3.4 ACTIVITY TESTS OF THE CATALYSTS PREPARED WITH ULTRASOUND-ASSISTED SYNTHESIS

The results of the catalytic tests are shown in Figures 6.8a and 8b. In Figure 6.8a, the FFA conversions (%) obtained at the end of the reaction (6 h) are reported for the samples synthesized using the same H_2O/precursors ratio, whereas in Figure 6.8b, the conversions of the catalysts obtained using different H_2O/precursors ratio are shown. In Figure 6.8a, it can be observed that both the addition of TiO_2 to the SO_4^{2-}/ZrO_2 system and the use of US in specific conditions during the synthesis are able to improve the catalytic performances. As can be observed in Figure 6.8a, the addition of TiO_2 alone is able to improve the activity of the catalysts in the FFA esterification. The addition of TiO_2, in fact, as already highlighted in the discussion concerning the results of the characterization of the catalysts, is able to increase the Brønsted acidity and, as a consequence, the catalytic activity (compare entries 1 and 2 in Table 6.2).

Moreover, the SZT sample calcined for a long time (SZT_773_6h) exhibits almost no catalytic activity (results not shown for the sake of brevity). This catalytic behavior might be ascribable to the loss of part of the sulfates that occurred during the calcinations step. This hypothesis is also confirmed by the sulfates concentration evaluated by ion chromatography and whose results are reported in Table 6.3. The SZT_773_6h sample is characterized by a lower sulfate concentration and, as a consequence, by a lower acid capacity (see Table 6.2) than the "plain" SZT sample (calcined for only 3 h). Considering the samples obtained with the US pulses with on/

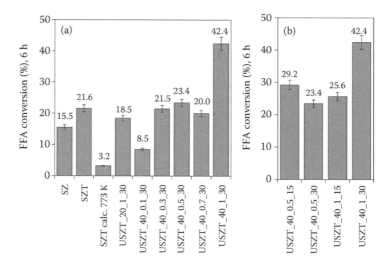

FIGURE 6.8   FFA esterification results of the sulfated Zr-Ti systems obtained using (a) the same and (b) different amounts of water during the US-assisted sol–gel synthesis. (Boffito, D.C., Crocellà, V., Pirola, C., Neppolian, B., Cerrato, G., Ashokkumar, M., Bianchi, C.L., Ultrasonic enhancement of the acidity, surface area, and free fatty acids esterification catalytic activity of sulphated ZrO_2-TiO_2 systems, *J. Catal.*, 2012b, http://dx.doi.org/10.1016/j.jcat.2012.09.013.)

off ratio from 0.3/0.7 on, the conversion does not increase much more if compared to the one achieved with the sample obtained via traditional sol–gel synthesis. Their conversions are in fact comparable (see samples USZ_40_0.3_30, USZ_40_0.5_30, USZ_40_0.7_30, and SZT in Figure 6.8a). The similarity in the catalytic performance of these catalysts may be ascribable to the fact that they are characterized by comparable values of SSA (entries 2, 5, 8, 10 in Table 6.2), and in the case of the catalysts obtained with pulses, also by comparable acidities (entries 5, 8, 10 in Table 6.2). A high SSA may in fact be disadvantageous for the catalysis of the reaction here studied. FFA are highly sterically hindered molecules, which might not be able to penetrate inside the pores as narrow as those of the catalysts here synthesized. The importance of the location of the active sites on the outer surface of the catalyst has already been evidenced by the authors previously in this chapter. Therefore, it is possible to assume in the present case that the active sites located on the catalyst's outer surface are the most responsible for the reaction to occur. For high SSA, it is more likely that the most part of the active sites is located inside the catalyst particles and, therefore, is not exploited in the catalytic process. As a consequence, for these samples, the high surface area results in a low catalytic performance.

The best catalytic performance is reached by the sample USZT_40_1_30, i.e., the one obtained using continuous US at higher power. This catalyst results in fact in a doubled catalytic activity with respect to the samples prepared either with the traditional synthesis or with the use of pulsed US. Despite the lower acidity of this catalyst compared with that of the samples obtained with the US pulses, it is characterized by a rather low surface area (entry 12 in Table 6.2) that can be associated with a localization of the active sites mainly on its outer surface. This feature results in many active sites immediately available for an efficient catalysis. As previously evidenced by the FTIR measurements, it is also important to highlight, that only in the case of the USZT_40_1_30 sample, a not negligible number of medium-strong Lewis acid sites is present at the surface, together with a high number of strong Brønsted acid centers. The good catalytic performances of this catalyst show that the presence of both these active sites is a necessary condition in order to obtain reasonably high values of conversion.

The low catalytic activity of the sample USZT_20_1_30 may be ascribable to the US power, probably not sufficient to activate the sol–gel reaction under the effect of the cavitation. In the case of the sample USZT_40_0.1_30, the unsatisfactory catalytic activity might be due to the too low on/off pulses ratio. Too short pulse times may in fact not be sufficient to initiate the acoustic cavitation in the medium, as previously explained and as evidenced in literature [21,32].

In Figure 6.8b, a comparison among the catalytic activity of the samples obtained with continuous US and with the use of 0.5 on/off pulses are displayed for different water/precursors ratios.

As can be noted, there is a remarkable difference between the samples USZT_40_1_15 and USZT_40_1_30, i.e., obtained using continuous US. On the contrary, the same difference is not evident in the case of the corresponding catalysts obtained with pulsed US, i.e., USZT_40_0.5_15 and USZT_40_0.5_30. This catalytic behavior can be explained considering that USZT_40_1_15 and USZT_40_1_30 possess distinct surface and morphological properties, resulting in different catalytic performances, whereas the samples, USZT_40_0.5_15 and USZT_40_0.5_30 are characterized by similar values of acidities and SSA, showing very close values of conversion. Noted that the SSA of USZT_40_1_15 is considerably higher than the SSA of the catalyst obtained in the same conditions but doubling the amount of water (USZT_40_1_30). This latter catalyst, as reported above, is found to be the best catalyst in the FFA esterification reaction.

In Figure 6.9, the FFA conversions as a function of the concentration of the acid sites normalized to the surface area are reported for the most significant samples. It can be observed that the best catalyst, i.e., USZT_40_1_30, possesses the highest ratio of acid sites over surface area, whereas the sample SZT_773_6h exhibits almost no conversion and shows the lowest content of acid sites/m2. Samples SZT, USZT_40_0.3_30, USZT_40_0.5_30, and USZT_40_0.7_30 exhibit very similar FFA conversion as well as similar concentrations of meqH+/m². The sample SZ is characterized by a lower meqH+/m² ratio and shows a lower catalytic activity than the above cited samples, with the exception of SZT_773_6h. This result confirms the importance of the localization of a high number of active acid sites on the outer surface of the catalyst for the reaction here studied. In the present work a high number of active acid sites on the outer surface of the catalyst was achieved adopting continuous ultrasound in the sol–gel synthesis of the sulfated ZrO_2-TiO_2 systems.

FFA conversion results may appear discouraging from a first approach, in particular if compared with the other results reported in literature for similar catalysts (Das et al., 2002). Nevertheless, it has to be taken into account that these catalysts are normally tested for the esterification of shorter chain carboxylic acids (Ardizzone et al., 2004, 2009; Juan et al., 2007), and that also in the case of the FFA esterification, they are usually experimented in the pure substrate, i.e., as pure fatty acid (Das et al.,

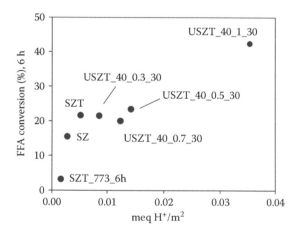

FIGURE 6.9   FFA esterification results of the sulfated Zr-Ti systems in function of the concentration of the active sites. (Boffito, D.C., Crocellà, V., Pirola, C., Neppolian, B., Cerrato, G., Ashokkumar, M., Bianchi, C.L., Ultrasonic enhancement of the acidity, surface area, and free fatty acids esterification catalytic activity of sulphated ZrO_2-TiO_2 systems, *J. Catal.*, 2012b, http://dx.doi.org/10.1016/j.jcat.2012.09.013.)

2002), and not directly in the acid oil. Taking into account these considerations, the FFA conversions obtained in this work appear encouraging.

Moreover, the outcome of this study is particularly important for what concerns the possibility of tuning the properties of acid catalysts based on mixed oxides by means of US (continuous or pulsed). This might be interesting also in view of the application of sulfated zirconias in other kinds of reactions, such as isomerizations, alkylations, and esterifications of other kinds of carboxylic acids.

6.4   CONCLUSIONS

Among all the tested sulfated Zr-based catalysts, commercial zirconium sulfate shows the highest activity in the FFA esterification, due to its high acidity. The main drawbacks of the use of $Zr(SO_4)_2$ still remain its partial solubility in water yielded by the reaction and the leaching of the active groups in methanol. Water removal from the system in the course of the reaction delayed catalyst deactivation but for insufficient times. Possible alternatives to $Zr(SO_4)_2$ are given by sulfated zirconia or supported sulfated materials. Nevertheless, these catalysts resulted in unsatisfactory catalytic performances due to the partial localization of the active sites inside the pores of the catalysts. This region cannot in fact be easily reached by sterically hindered molecules such as FFA.

Sulfated TiO_2-SnO_2 systems exhibit higher acidity than the sulfated ZrO_2 systems and their acidity increases with the TiO_2 content. Consequently, also their activity in the FFA esterification is higher.

Further results presented in this chapter have demonstrated how the surface acidity and specific surface area of sulfated zirconia can be increased by both adding TiO_2 and using ultrasound (US) in precise experimental conditions to assist the sol–gel synthesis of the catalysts. The beneficial effects of the use of US in the sol–gel synthesis of the SO_4^{2-}/ZrO_2-TiO_2 systems are ascribable to the occurrence of acoustic cavitation, which causes faster hydrolysis (sol–gel reaction) rates and surface damage. This effect is particularly evident for the catalysts obtained with the use of pulses longer than 0.3 s. For these catalysts, the acidity significantly increases along with the specific surface area if compared with the ones obtained with shorter pulses. The more efficient among the catalyst synthesized with US resulted to be the one obtained with continuous US and higher powers. For this sample, in fact, the lower surface area allows the location of the active acid sites mainly on the outer surface of the catalysts. In this way, the FFA molecules can undergo a faster catalytic transformation. Moreover, it has been demonstrated that this catalyst is the only one that possesses both Lewis and Brønsted acidity. Therefore, its good catalytic performances can be ascribed to the presence of both these active sites, together with a low value of SSA.

ACKNOWLEDGMENTS

The authors gratefully acknowledge Minstero delle Politiche Agricole, Alimentari e Forestali for the financial support (Project SUSBIOFUEL, D.M. 27800/7303/09) and the DEEWR (Department of Education, Employment and Workplace) of the

Australian Government for the Endeavour Research Fellowship grant. The authors are also thankful to Valentina Crocellà (Università degli Studi di Torino) for having carried out the FTIR measurements on Zr-Ti systems whose results are reported in this chapter.

REFERENCES

Ardizzone, S., Bianchi, C.L., Cappelletti, G., Porta, F. 2004. Liquid-phase catalytic activity of sulfated zirconia from sol–gel precursors: the role of the surface features. *J. Catal.* 227:470–478.

Ardizzone, S., Bianchi, C.L., Cappelletti, G., Annunziata, R., Cerrato, G., Morterra, C., Scardi, P. 2009. Liquid phase reactions catalyzed by Fe- and Mn-sulphated ZrO_2. *Appl. Catal. A: Gen.* 360:137–144.

Arzamendi, G., Arguinarena, E., Campo, I., Zabala, S., Gandia, L.M. 2008. Alkaline and alkaline-earth metals compounds as catalysts for the methanolysis of sunflower oil. *Catal. Today* 133:305–331.

Bianchi, C.L., Boffito, D.C., Pirola, C., Ragaini, V. 2010. Low temperature de-acidification process of animal fat as a pre-step to biodiesel production. *Catal. Lett.* 134:179–183.

Bianchi, C.L., Pirola, C., Boffito, D.C., Di Fronzo, A., Carvoli, G., Barnabè, D., Rispoli, A., Bucchi, R. 2011. Non edible oils: raw materials for sustainable biodiesel. In *Biodiesel Feedstocks and Processing Technologies*, ed. M. Stoytcheva and G. Montero, pp. 3–22. Intech, Rijeka, Croatia.

Boffito, D.C., Crocellà, V., Pirola, C., Neppolian, B., Cerrato, G., Ashokkumar, M., Bianchi, C.L. 2013a. Ultrasonic enhancement of the acidity, surface area and free fatty acids esterification catalytic activity of sulphated ZrO_2-TiO_2 systems, *J. Catal.*, 297:17–26.

Boffito, D.C., Pirola, C., Galli, F., Di Michele, A., Bianchi, C.L. 2013b. Free fatty acids esterification of waste cooking oil and its mixtures with rapeseed oil and diesel, *Fuel*, 108:612–619.

Boffito, D.C., Pirola, C., Bianchi, C.L. 2012. Heterogeneous catalysis for free fatty acids esterification reaction as a first step towards biodiesel production. *Chem. Today* 30:14–18.

Borges, M.E., Díaz, L. 2012. Recent developments on heterogeneous catalysts for biodiesel production by oil esterification and tranesterification reactions: a review. *Renew. Sustain. Energy Rev.*, 16:2839–2849.

Cerrato, G., Magnacca, G., Morterra, C., Montero, J., Anderson, J.A. 2009. Modification to the surface properties of titania by addition of India. *J. Phys. Chem. C* 113(47): 20401–20410.

Chen, X.R., Ju, Y.H., Mou, C.Y. 2007. Direct synthesis of mesoporous sulfated silica–zirconia catalysts with high catalytic activity for biodiesel via esterification. *J. Phys. Chem.* 111:18731–18737.

Das, D., Mishra, H.K., Pradhan, N.C., Dalai, A.K., Parida, K.M. 2002. Studies on structural properties, surface acidity and benzene isopropylation activity of sulphated ZrO_2-TiO_2 mixed oxide catalysts. *Microp. Mesop. Mat.* 80:327–336.

Dijs, I.J., Geus, J.W., Jenneskens, L.W. 2003. Effect of size and extent of sulfation of bulk and silica-supported ZrO2 on catalytic activity in gas- and liquid-phase reactions. *J. Phys. Chem. B* 107:13403–13413.

Ganesan, D., Rajendaran, A., Thangavelu, V. 2009. An overview on the recent advances in the transesterification of vegetable oils for biodiesel production using chemicals and biocatalysts. *Rev. Environ. Sci. Biotechnol.* 8:367–394.

Gregg, S.J. and Sing, K.S.W. 1982. *Adsorption, Surface Area and Porosity*, Academic Press, London.

Henglein, A., Ulrich, R., Lilie, J. 1989. Luminescence and chemical action by pulsed ultrasound. *J. Am. Chem. Soc.* 111:1974–1979.

Jacobson, K., Gopinath, R., Meher, L., Charan, D., Ajay, K. 2008. Solid acid catalyzed biodiesel production from waste cooking oil. *Appl. Catal., B: Environ.* 85(1–2):86–91.

Jiménez-Morales, I., Santamaría-González, J., Maireles-Torres, P., Jiménez-López, A. 2011. Calcined zirconium sulfate supported on MCM-41 silica as acid catalyst for ethanolysis of sunflower oil. *Appl. Catal. B: Environ.* 103:91–98.

Jitputti, J., Kitiyanan, B., Rangsunvigit, P., Bunyakiat, K., Attanatho, L., Jenvanitpanjakul, P. 2006. Transesterification of crude palm kernel oil and crude coconut oil by different solid catalysts. *Chem. Eng. J.* 116(1):61–66.

Juan, J.C., Jiang, Y., Meng, X., Cao, W., Yarmo, M.A., Zhang, J. 2007. Supported zirconium sulfate on carbon nanotubes as water-tolerant solid acid catalyst. *Mat. Res. Bullettin* 42:1278–1285.

Juan, J.C., Zhang, J., Yarmo, M.A. 2008. Study of catalysts comprising zirconium sulfate supported on a mesoporous molecular sieve HMS for esterification of fatty acids under solvent-free condition. *Appl. Catal. A: Gen.* 347(2):133–141.

Kataoka, T., Dumesic, J.A. 1988. Acidity of unsupported and silica-supported vanadia, molybdena, and titania as studied by pyridine adsorption. *J. Catal.* 112:66–79.

Lam, M.K., Lee, K.T., Mohamed, A.R. 2009. Sulfated tin oxide as solid superacid catalyst for transesterification of waste cooking oil: an optimization study. *Appl. Catal. B: Environ.* 93:134–139.

López, D.E., Goodwin, Jr., J.G., Bruce, D.A. 2007. Transesterification of triacetin with methanol on Nafion acid resins. *J. Catal.* 245:381–391.

Lotero, E., Goodwin, J.G., Bruce, D.A., Suwannakarn, K., Liu, Y., Lopez, D.E. 2006. The catalysis of biodiesel synthesis. *Catalysis* 19:41–83.

Lotero, E., Liu, Y., Lopez, D.E., Suwannakaran, K., Bruce, D.A., Goodwin, Jr., J.G. 2005. Synthesis of biodiesel via acid catalysis. *Ind. Eng. Chem. Res.* 44:5353–5363.

Ma, F., Hanna, M.A. 1999. Biodiesel production: a review. *Bioresour. Technol.* 70:1–15.

Marchetti, J.M., Errazu, A.F. 2008. Comparison of different heterogeneous catalysts and different alcohols for the esterification reaction of oleic acid. *Fuel* 87:3477–3480.

Matsuhashi, H., Hino, M., Arata, K. 1990. Solid catalyst treated with anion: XIX. Synthesis of the solid superacid catalyst of tin oxide treated with sulfate ion. *Appl. Catal.* 59(2):205–212.

Morterra, C., Cerrato, G., Ardizzone, S., Bianchi, C.L., Signoretto, M., Pinna, F. 2002. Surface features and catalytic activity of sulfated zirconia catalysts from hydrothermal precursors. *Phys. Chem. Chem. Phys.* 4:3136–3145.

Morterra, C., Cerrato, G., Di Ciero, S. 1998. IR study of the low temperature adsorption of CO on tetragonal zirconia and sulfated tetragonal zirconia. *Appl. Surf. Sci.* 126:107–128.

Morterra, C., Cerrato, G., Meligrana, G. 2001. 2,6-Dimethylpyridine as an analytical tool to test the surface acidic properties of oxidic systems. *Langmuir* 17:7053–7060.

Morterra, C., Cerrato, G., Pinna, F., Signoretto, M. 1994. Brønsted acidity of a superacid sulfate-doped ZrO_2 system. *J. Phys. Chem.* 98:12373–12381.

Morterra, C., Cerrato, G., Signoretto, M. 1996. On the role of the calcination step in the preparation of superacid sulfated zirconia catalysts. *Catal. Lett.* 41:101–109.

Neppolian, B., Wang, Q., Jung, H., Choi, H. 2009. Ultrasonic-assisted sol–gel method of preparation of TiO_2 nano-particles: characterization, properties and 4-chlorophenol removal application. *Ultrason. Sonochem.* 15:649–658.

Neppolian, B., Wang, Q., Yamashita, H., Choi, H. 2007. Synthesis and characterization of ZrO_2-TiO_2 binary oxide semiconductor nanoparticles: application and interparticle electron transfer process. *Appl. Catal. A* 333:264–271.

Okoronkwo, M.U., Galadima, A., Leke, L. 2012. Advances in biodiesel synthesis: from past to present. *Elixir Appl. Chem* 43:6924–6945.

Pasias, S., Barakos, N., Alexopoulos, C. 2006. Heterogeneously catalyzed esterification of FFAs in vegetable oils. *Chem. Eng. Technol.* 29:1365–1371.

Perego, C., Ricci, M. 2012. Diesel fuel from biomass. *Catal. Sci. Technol.* 1:1776–1786.

Pirola, C., Bianchi, C.L., Boffito, D.C., Carvoli, G., Ragaini, V. 2010. Vegetable oil deacidification by Amberlyst: study of catalyst lifetime and a suitable reactor configuration. *Ind. Eng. Chem. Res.* 49:4601–4606.

Pirola, C., Bianchi, C.L., Di Michele, A., Diodati, P., Boffito, D., Ragaini, V. 2010. Ultrasound and microwave assisted synthesis of high loading Fe-supported Fischer–Tropsch catalysts. *Ultrason. Sonochem.* 17:610–616.

Pirola, C., Boffito, D.C., Carvoli, G., Di Fronzo, A., Ragaini, V., Bianchi, C.L. 2011. Soybean oil deacidification as a first step towards biodiesel production. In *Recent Trends for Enhancing the Diversity and Quality of Soybean Products*, ed. D. Krezhova, pp. 321–344, Intech, Rijeka, Croatia.

Russbueldt, B.M.E., Hoelderich, W.F. 2009. New sulfonic acid ion-exchange resins for the preesterification of different oils and fats with high content of free fatty acids. *Appl. Catal. A* 362:47–57.

Santori, G., Di Nicola, G., Moglie, M., Polonara, F. 2012. A review analyzing the industrial biodiesel production practice starting from vegetable oil refining. *Appl. Energy* 92:109–132.

Sarzanini, C., Sacchero, G., Pinna, F., Cerrato, G., Morterra, C. 1995. Amount and nature of sulfates at the surface of sulfate-doped zirconia catalysts. *J. Mater. Chem.* 5:353–360.

Sun, Y., Ma, S., Du, Y., Yuan, L., Wang, S., Yang, J., Deng, F., Xiao, F.S. 2005. Solvent-free preparation of nanosized sulfated zirconia with Brønsted acidic sites from a simple calcination. *J. Phys. Chem. B* 109:2567–2572.

Suslick, K.S., Doktycz, S.J. 1990. The effects of ultrasound on solids. In *Advances in Sonochemistry*, ed. T.J. Mason, pp. 197–230, JAI Press, New York.

Veljković, V.B., Avramović, J.M., Olivera, S.S. 2012. Biodiesel production by ultrasound-assisted transesterification: state of the art and the perspectives. *Ren. Sust. Energy Rev.* 16:1193–1209.

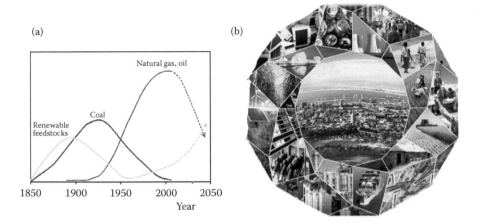

(a)

Natural gas, oil

Renewable feedstocks

Coal

1850   1900   1950   2000   2050

Year

(b)

FIGURE 2.1 (a) Raw materials basis of chemical industry in historical perspective. (Taken with permission from F. W. Lichtenthaler, S. Peters, *Comptes Rendus Chimie* 2004, 7, 65–90.) (b) A view on sustainable materials for a sustainable future. (Taken with permission from M. L. Green, L. Espinal, E. Traversa, E. J. Amis. *MRS Bulletin* 2012, 37, 303–309.)

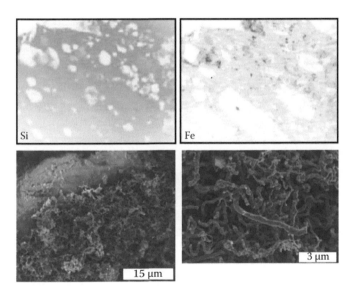

Si

Fe

15 μm

3 μm

FIGURE 2.2 SEM images showing the elemental distribution of Si and Fe in lava stone granulate (top) and CNFs grown on lava (bottom). (D. S. Su, *ChemSusChem* 2009, 2, 1009–1020. Copyright Wiley-VCH Verlag GmbH & Co. KGaA. Reproduced with permission.)

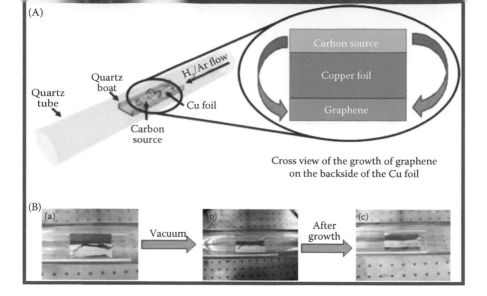

FIGURE 2.8 (A) Diagram of the experimental apparatus for the growth of graphene from food, insects, or waste in a tube furnace. On the left, the Cu foil with the carbon source contained in a quartz boat is placed at the hot zone of a tube furnace. The growth is performed at 1050°C under low pressure with a H₂/Ar gas flow. On the right is a cross view that represents the formation of pristine graphene on the backside of the Cu substrate. (B) Growth of graphene from a cockroach leg. (a) One roach leg on top of the Cu foil. (b) Roach leg under vacuum. (c) Residue from the roach leg after annealing at 1050°C for 15 min. The pristine graphene grew on the bottom side of the Cu film (not shown). (Reprinted with permission from G. Ruan, Z. Sun, Z. Peng, J. M. Tour, *ACS Nano* 2011, 5, 7601–7607. Copyright 2011 American Chemical Society.)

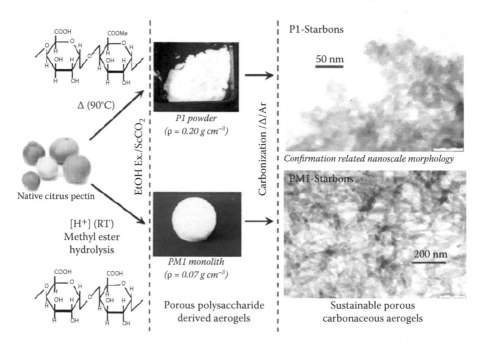

FIGURE 2.11 Representation of routes to porous polysaccharide-derived materials from pectin, and the corresponding TEM images of pectin-derived Starbon-type materials. (R. J. White, V. L. Budarin, J. H. Clark, *Chemistry—A European Journal* 2010, 16, 1326–1335. Copyright Wiley-VCH Verlag GmbH & Co. KGaA. Reproduced with permission.)

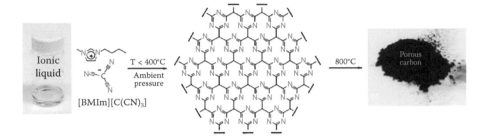

FIGURE 2.12 Reaction scheme of the trimerization of a nitrile-containing anion, leading to the formation of an extended framework. (J. S. Lee, X. Wang, H. Luo, S. Dai, *Advanced Materials* 2010, 22, 1004–1007. Copyright Wiley-VCH Verlag GmbH & Co. KGaA. Reproduced with permission.)

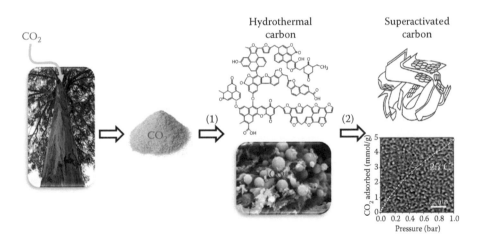

FIGURE 2.20 Schematic illustration of the overall synthesis procedure for HTC-based activated carbons and their application in CO_2 capture: (1) hydrothermal carbonization at 230°C–250°C (2 h), and (2) chemical activation with KOH.

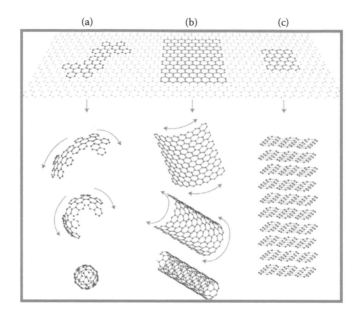

FIGURE 3.4 Schematic formation of carbon nanomaterials: (a) fullerene (zero-dimensional), (b) carbon nanotubes (one-dimensional), and (c) graphene (two-dimensional). (From A. K. Geim and K. S. Novoselov, *Nature Materials*, 2007, 6, 183–191.)

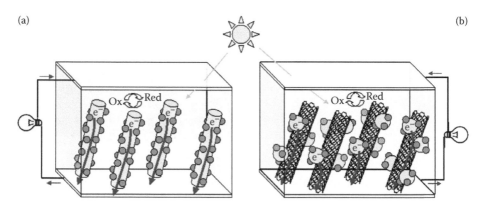

FIGURE 3.6 Directing the flow of photogenerated electrons across nanostructured semiconductor films: (a) nanotube/nanowires modified with light-absorbing dye molecules and (b) nanotubes as support architecture for anchoring dye modified semiconductor nanoparticles. (From P. Brown, K. Takechi and P. V. Kamat, *Journal of Physical Chemistry C*, 2008, 112, 4776–4782.)

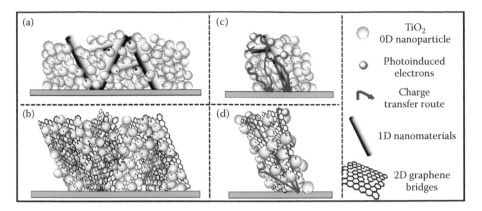

FIGURE 3.7 Differences between (a, c) 1D and (b, d) 2D nanomaterial composite electrodes. The transfer barrier is larger and the recombination is much easier to happen. (From N. L. Yang, J. Zhai, D. Wang, Y. S. Chen and L. Jiang, *ACS Nano*, 2010, 4, 887–894.)

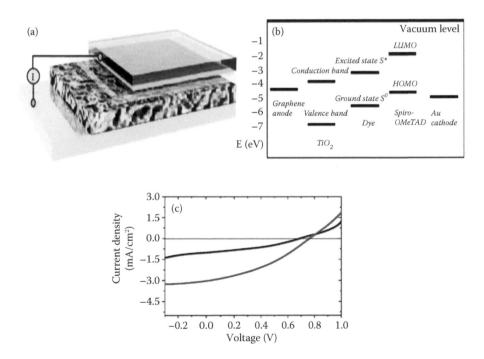

FIGURE 3.9 Illustration and performance of solar cell based on graphene electrodes. (a) Illustration of DSSC using graphene film as electrode, the four layers from bottom to top are Au, dye-sensitized heterojunction, compact TiO₂, and graphene film. (b) The energy-level diagram of graphene/TiO₂/dye/spiro-OMeTAD/Au device. (c) *I–V* curve of graphene-based cell (black) and the FTO-based cell (red), illuminated under AM solar light (1 sun). (From X. Wang, L. J. Zhi and K. Mullen, *Nano Letters*, 2008, 8, 323–327.)

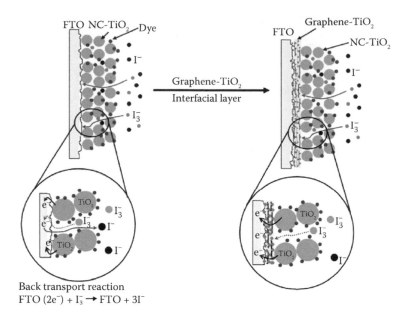

FIGURE 3.10 Schematic representation and mechanism of applied graphene–TiO$_2$ interfacial layer to prevent back-transport reaction of electrons. (From S. R. Kim, M. K. Parvez and M. Chhowalla, *Chemical Physics Letters*, 2009, 483, 124–127.)

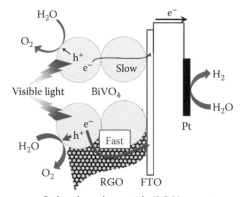

Reduced graphene oxide (RGO) promotes electron migration from BiVO$_4$ to FTO.

FIGURE 3.11 Schematic representation and mechanism of applied graphene–BiVO$_4$ photolysis cell. (From Y. H. Ng, A. Iwase, A. Kudo and R. Amal, *Journal of Physical Chemistry Letters*, 2010, 1, 2607–2612.)

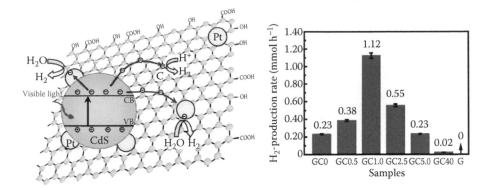

FIGURE 3.12 Schematic illustration of the charge separation and transfer in the graphene–CdS system under visible light. (From Q. Li, B. D. Guo, J. G. Yu, J. R. Ran, B. H. Zhang, H. J. Yan and J. R. Gong, *Journal of the American Chemical Society*, 2011, 133, 10878–10884.)

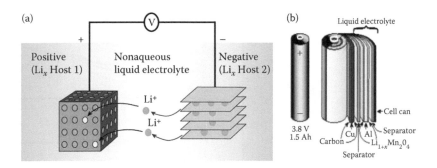

FIGURE 3.13 Schematic operation mechanism of lithium ion battery (a) and a setup of a cylindrical battery (b). (From J. M. Tarascon and M. Armand, *Nature*, 2001, 414, 359–367.)

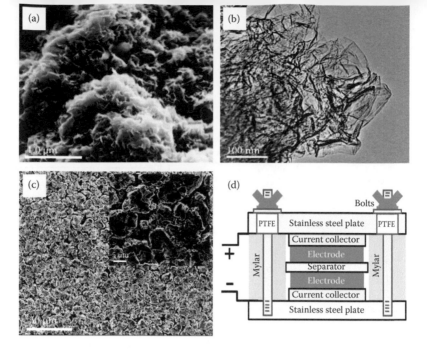

FIGURE 3.18 (a) SEM image of CMG particle surface, (b) TEM image of individual graphene sheets extended from CMG particle surface, (c) low and high (inset) magnification SEM images of CMG particle electrode surface, and (d) a schematic of test cell assembly. (From M. D. Stoller, S. J. Park, Y. W. Zhu, J. H. An and R. S. Ruoff, *Nano Letters*, 2008, 8, 3498–3502.)

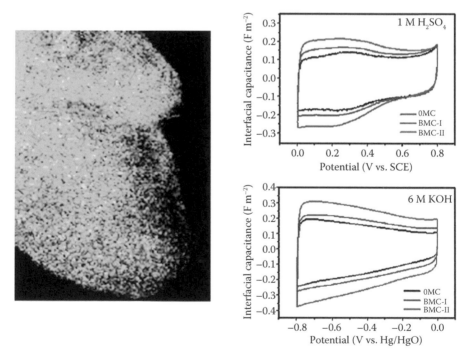

FIGURE 3.19 EELS mapping images of BMC-II sample: boron. Cyclic voltammetry plots of OMC, BMC-I, and BMC-II samples in 1 M H_2SO_4 electrolyte and 6 M KOH electrolyte recorded at 10 mV s[-1]. (From D. W. Wang, F. Li, Z. G. Chen, G. Q. Lu and H. M. Cheng, *Chemistry of Materials*, 2008, 20, 7195–7200.)

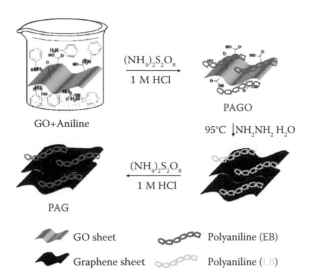

FIGURE 3.21 Illustration of the process for preparation of graphene–PANi composites. (From K. Zhang, L. L. Zhang, X. S. Zhao and J. S. Wu, *Chemistry of Materials*, 2010, 22, 1392–1401.)

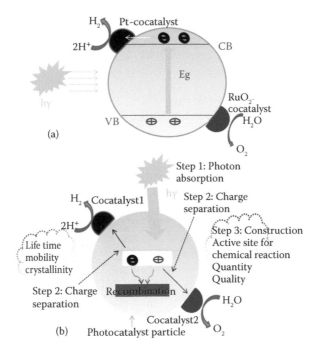

FIGURE 4.1 (a) Schematic diagram for overall water-splitting reaction on a solid photocatalyst. (b) Main process of electron transport during water-splitting reaction. (Reproduced by permission of The Bentham Science Publisher. Nuraje, N., R. Asmatulu, and S. Kudaibergenov. 2012. Metal oxide-based functional materials for solar energy conversion: a review. *Current Inorganic Chemistry* 2 (2):124–146.)

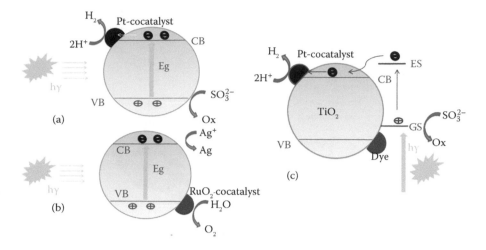

FIGURE 4.2 (a) Half reaction for hydrogen production using photocatalyst. (b) Half reaction for oxygen evolution using photocatalyst. (c) Dye-sensitized metal oxide particulate system for hydrogen evolution. (Reproduced by permission of The Bentham Science Publisher. Nuraje, N., R. Asmatulu, and S. Kudaibergenov. 2012. Metal oxide-based functional materials for solar energy conversion: a review. *Current Inorganic Chemistry* 2 (2):124–146.)

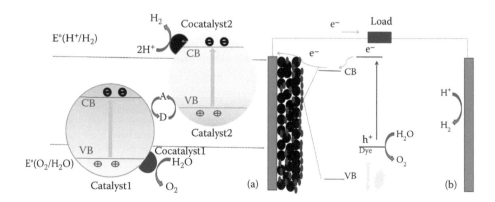

FIGURE 4.3 (a) Schematic diagram of dual-bed configuration for overall water-splitting reaction on two different solid photocatalysts, and (b) dye-sensitized photoelectrochemical cell for overall water splitting. (Reproduced by permission of The Bentham Science Publisher. Nuraje, N., R. Asmatulu, and S. Kudaibergenov. 2012. Metal oxide-based functional materials for solar energy conversion: a review. *Current Inorganic Chemistry* 2 (2):124–146.)

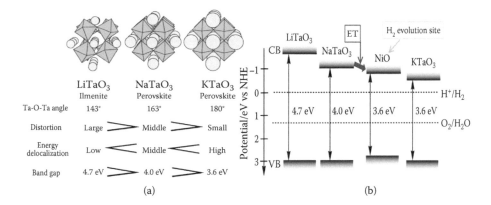

FIGURE 4.4 (a) Crystal and energy structures of alkali tantalate and (b) band structures of alkali tantalates with NiO co-catalyst. (From Kato, H., and A. Kudo. 2001. Water splitting into H_2 and O_2 on alkali tantalate photocatalysts $ATaO_3$ (A = Li, Na, and K). *J. Phys. Chem. B* 105 (19):4285–4292. doi: 10.1021/jp004386b. Reprinted with permission of AAAS.)

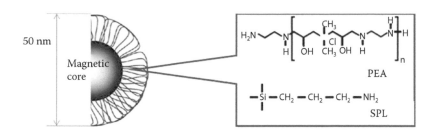

FIGURE 7.2 Magnetic nanoparticles covered with polyethylene amine (PEA) and silicon polymer layer (SPL) used for lipase immobilization. (From M. Tudorache et al., *Applied Catalysis A: General*, 437–438, 90–95, 2012.)

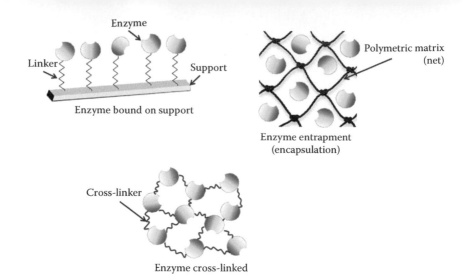

FIGURE 7.3 Enzyme immobilization on the preparation of heterogeneous biocatalyst.

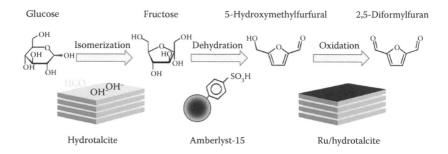

FIGURE 8.10 Reaction sequence to obtain DFF from glucose using different nano-structured catalysts. (Reprinted with permission from A. Takagaki, M. Takahashi, S. Nishimura, K. Ebitani, *ACS Catal.* 2011, 1, 1562–1565. Copyright 2011 American Chemical Society.)

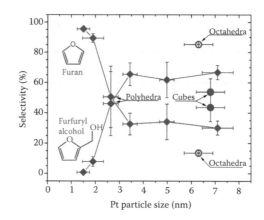

FIGURE 8.12 Steady-state product selectivities to furan (blue) and furfuryl alcohol (red) versus TEM-projected Pt particle sizes. (Reprinted with permission from V. V. Pushkarev, N. Musselwhite, K. An, S. Alayoglu, G. A. Somorjai, *Nano Lett.* 2012, 12, 5196–5201. Copyright 2012 American Chemical Society.)

(a)

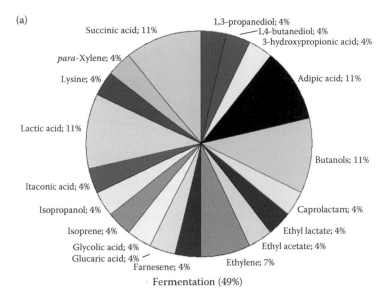

Fermentation (49%)

(b)

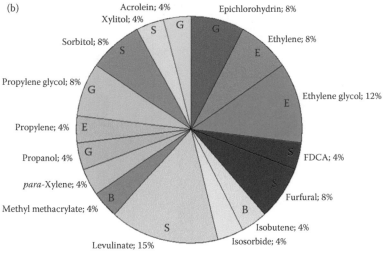

Catalytic (46%)

FIGURE 9.1 Announced commercial processes and research projects up to 2016, as announced June 2012. (a) Fermentation and (b) chemocatalytic. Feedstocks used: S, sugar; E, ethanol; B, butanol; G, glycerol.

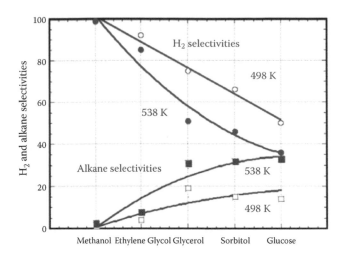

FIGURE 9.11 Aqueous phase reforming of 1 wt% methanol, ethylene glycol, glycerol, sorbitol, and glucose. (From R.D. Cortright, 2012. *Aqueous-Phase Reforming Process*, 2nd Annual Wisconsin Bioenergy Summit.)

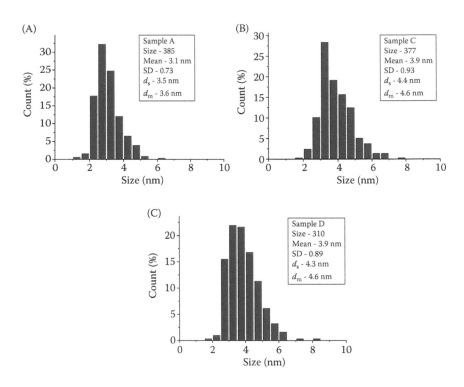

FIGURE 10.6 Au particle size distribution of catalysts A, B, and C.

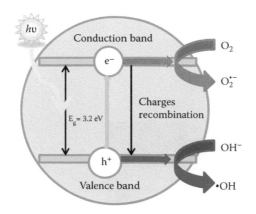

Photoexcitation: $\quad\quad\quad\quad\quad\quad$ $TiO_2 + h\upsilon \rightarrow e_{CB}^- + h_{VB}^+$
Charge carrier trapping: $\quad\quad\quad\quad$ $e_{CB}^- \rightarrow e_{TR}^-$
Charge carrier trapping: $\quad\quad\quad\quad$ $h_{VB}^+ \rightarrow h_{TR}^+$
Electron-hole recombination: $\quad$ $e_{TR}^- + h_{VB}^+(h_{TR}^+) \rightarrow e_{CB}^- + heat$

FIGURE 11.1 Photocatalytic process over nano-TiO_2.

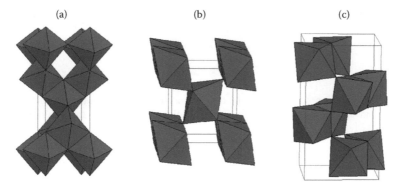

FIGURE 11.2 Crystalline structure of anatase (a), rutile (b), and brookite (c). (Adapted from http://ruby.colorado.edu/~smyth/min/tio2.html.)

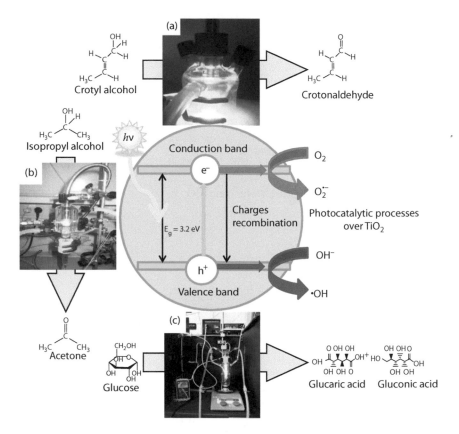

FIGURE 11.3 Representative scheme of a variety of selective catalytic photooxidation of alcohols and sugars to aldehydes/ketones/carboxylic acids in gas (b) and liquid phase (a,c).

7 Nanoheterogeneous Design of Biocatalysts for Biomass Valorization

Madalina Tudorache, Simona Coman, and Vasile I. Parvulescu

CONTENTS

7.1 INTRODUCTION

Biocatalysis (enzymatic and whole cell catalysis) has been extensively applied in white biotechnology for the production of fuels and fine chemicals in recent years.[1] Furthermore, biocatalysis is currently on the policy agendas because it is considered one of the most promising routes for a bio-based economy. Biocatalysis plays a key role in the development of biorefinery for conversion of biomass into a wide array of fuels, chemicals, and materials.[2–4] Expanding from traditional applications, biocatalysis is nowadays providing the initial steps toward the production of chemicals commodities[5,6] for sustainable energy supply and environment of long-term care throughout the world. Consequently, the importance of biocatalysis as a biotechnological tool on biomass transformation is increasing every day as well.[7,8]

Chemical production from biomass can be followed on two different directions, which are the biomass biodegradation and further the biotransformation of the biomass derivatives to final value-added products (Scheme 7.1). The first refers to the

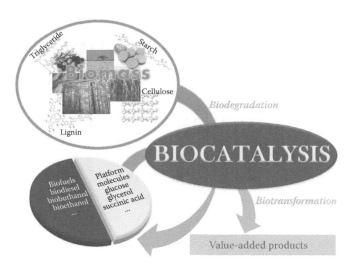

SCHEME 7.1   Biocatalysis on biomass applications.

enzymatic treatments of cellulose, starch, lignin, chitin, protein, and oils, directly components of the biological sources, whereas the second category includes the biocatalytic conversion of simple molecules derived from biomass components, e.g., free sugars, organic acids, and alcohols.[6] For both directions, biocatalysis provides substantial advantages compared with chemical catalysis, such as high efficiency, high degree of selectivity (e.g., regio-, chemo-, and enantio-), "green" reaction conditions (i.e., mild reaction conditions) and thus low energy consumption. Besides less waste amounts and production of toxic wastes, biocatalytic processes led to high atom economy reactions. Nowadays, these are the main characteristic for the biotransformations.[9] Moreover, the heterogeneous design of the biocatalyst (e.g., immobilized enzyme) avoids the product contamination and provides the possibility to recover the biocatalyst and its reuse in successive reaction cycles, with often improving the operational stability and shelf-life.[10] Consequently, under these conditions, a low amount of biomaterial (enzyme or whole cell) is requested for the biocatalyst preparation, thus leading to a low cost of the technology.

The combination of the biocatalysts advantages with those of the nanochemistry and nanomaterials may provide high achievements in the future. Such a goal can be realized either by chemical agglomeration of enzymes as nanoparticles or by synthesis of nanocomposites incorporating enzymes.

7.2   IMMOBILIZED ENZYMES AS BIOCATALYSTS

Enzymes are used as biocatalysts due to their excellent functional properties (e.g., activity, selectivity, and specificity). However, in most of the cases, naturally extracted enzymes are not providing optimal conditions for industrial applications due to differences between the environment of the living cell and practical settings of the industrial approach.[11] Enzymes have been modified during evolution in order to optimize the catalytic behavior and to control the selectivity of complex catalytic chains of the biotechnological processes. Thus, it appears more comfortable to suit the biocatalyst to one of the achievement ways via engineering of existing enzymes leading to biocatalysts compatible with the targeted industrial process.[12]

Beside the engineering methods, the improvement of the biocatalytic properties of the enzymes could be performed via immobilization techniques, which in addition are less invasive versus the protein (enzyme) structure.[13,14] Thus, it appears that the important traits (e.g., thermostability, activity, selectivity, and tolerance toward organic solvents) of biocatalysts can be optimized and the resulting heterogeneous biocatalyst will be reasonably stable under conditions that may be far from the enzyme physiological environment.[15,16] Such biocatalysts are much more suitable for the chemical industry, where mild and cost-effective conditions are requested to synthesize complex and useful compounds.[17] In addition, there are many examples demonstrating that biocatalysis can also enable reactions that are difficult to be realized with classical chemical catalysts. A famous example is the production of acrylamide using a nitrilase enzyme,[6] which can catalyze the nitrile hydrolysis in the presence of an ester or an amide. This reaction is almost impossible to be carried out using traditional chemical routes.

7.3 NANOSUPPORTS FOR HETEROGENEOUS BIOCATALYSTS

The selection of the support used for the attachment of the biomolecules represents an important requirement for the enzyme immobilization. The support matrix should provide a biocompatible and inert environment for enzymes, without any interference with the native structure of the protein, which thereby could compromise its biological activity.[18] Consequently, a number of processes for the enzyme immobilization in silica tubes, on self-assembled monolayers, polymer matrices, Langmuir–Blodgett films, galleries of R-zirconium phosphates, polystyrene latex, gold particles, etc. have already been developed.[18–24] In addition, it was suggested that the shape and, importantly, the size of the immobilization surface of the support could influence the catalytic capacity of the immobilized biomolecules. Therefore, in the last time, different nanomaterials have been extensively employed for the "heterogeneization" of the biocatalysts (e.g., nanorods, nanotubes, nanowires, nanorings, etc.[25–30]) (Figure 7.1). Their behavior proved this approach as being indeed powerful in manipulating protein–protein and protein–environment interactions. Furthermore, this approach can even generate unique biotransformations.[6]

The use of nanoscale materials for enzyme immobilization enables us to reduce the effects of the diffusion limitations and maximize the functional surface area to increase the enzyme loading.[31,32] Also, the catalytic activity of the enzyme could be improved by the physical characteristics of the nano-supports such as enhanced diffusion and particle mobility.[33] Thermal stability, increased surface areas, and irradiation resistance represent other advantages, which make the corresponding nanoheterogeneous biocatalysts as potential candidates for industrial applications.[25] In addition, the use of the nano-support materials points to an interesting transition region between heterogeneous and homogeneous catalysis. In fact, this is a direct consequence of the support mobility governed by the particle size and solution

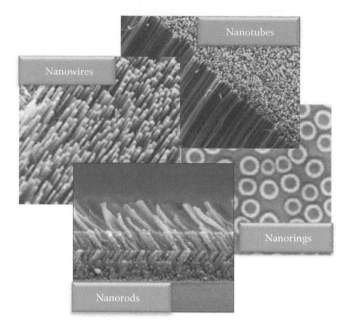

FIGURE 7.1   Nanomaterials used as supports for biocatalyst preparation.

viscosity, with impact on the biocatalyst activity.[33] Based on these properties, efforts have been made in exploring the effect of the texture and intrinsic characteristics of the nanomaterials on the biocatalyst structure and behavior. In this scope, nanotubes, nanocapsules, nanofibers, nanopaticles, nanocomposites, and nonoporous matrices have already reported.[6]

In the recent years, functionalized magnetic particles (i.e., Fe_3O_4 nanospheres coated with polymeric layer(s)) (Figure 7.2) have shown attractive support properties for lipase immobilization with the final goal to prepare biocatalysts for biomass transformations.[34] Magnetic nanoparticles were chosen as supports for enzyme immobilization due to their fast and facile separation from the reaction medium that can be effectively controlled by applying a magnetic field.[35,36] Besides the easy way of manipulation, the catalytic properties of the enzyme–magnetic particle composites

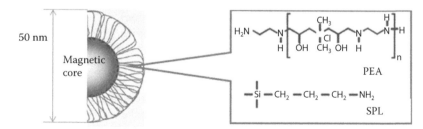

FIGURE 7.2   (See color insert.) Magnetic nanoparticles covered with polyethylene amine (PEA) and silicon polymer layer (SPL) used for lipase immobilization. (From M. Tudorache et al., *Applied Catalysis A: General*, 437–438, 90–95, 2012.)

were often improved compared with the free enzyme.[37–39] Minimized diffusion limitation, improved reaction kinetics, larger exposed surface areas for immobilization, and avoiding the nonchemical separation of the biocatalyst from the reaction mixture are only a few of the specific advantages generated by the valuable features of the nanomagnetic supports.[40] As a specific application, the lipase immobilization strategy on these materials appeared as a promising alternative for the synthesis of performing and sustainable biocatalysts.[34,39,41–43]

An innovative way to adapt the enzyme biocatalyst to organic media was to emulsify an aqueous enzyme solution using hydrophobic SiO_2 nanoparticles (Pickering emulsion).[44] Different enzymes (e.g., lipase A and B from *Candida antarctica*, benzaldehyde lyase from *Pseudomonas fluorescens*) showed significantly enhanced catalytic performance after immobilization using this route for hydrolysis of various oils.

Surface modification of nanoparticles with enzyme molecules provided unique bio-functionality to the heterogeneous biocatalytic structure.[45] An eloquent example is the lipase enzyme, which is a water-soluble protein with catalytic activity for triacylglycerol hydrolysis in the aqueous phase and transesterification in microaqueous media. Lipase is characterized by an interfacial activity that can be easily achieved if suitable solid supports able to fulfill this requirement are used for the enzyme immobilization. The catalytic behavior of the immobilized enzyme is strongly influenced by the support properties (i.e., nature of the support surface, and the composition and conformation of the support).[34] In the specific case of lipase, the catalytic behavior is dictated by two conformations in equilibrium, i.e., the open-active and closed-inactive conformation, respectively. Modulation of the support properties may lead to a shift of the equilibrium to one of these enzyme forms, thus improving or even diminishing the enzyme catalytic capacity.[46] The deposition of lipase on a hydrophobic surface favors the shift of the equilibrium between the two phases in the favor of the "open-active" conformation. In addition, the hydrophobic interactions between the support and enzyme may help a better docking of the lipase on the magnetic particle surface.[47] Also, an "aerated" structure of the polymeric layer promotes an easy exchange of water molecules between beads coating matrix and external medium.[48] Consequently, the "water" microenvironment created around the biocomposite ("necessary water" to afford the lipase activity) assures a better catalytic behavior of the lipase enzyme in the transesterification process.[49] The catalytic activity and the stability at fixed temperatures, pH, and organic solvent can be improved using different alternatives such as enzymes adsorbed on cyclodextrin-based carbonate nanosponge,[50] enzymes covalently bound on magnetic nanoparticles covered with a polyethylene amine layer,[51,52] or enzymes encapsulated in polycaprolactone (PCL) nanofibers.[53] The examples in this direction are continuously published in the literature today.

7.4   IMMOBILIZATION APPROACH

There are various immobilization methods that can be applied for the attachment of the enzyme to a solid surface. Basically, the heterogeneous structure of a biocatalyst can be obtained if the enzyme is bound to a support (carrier), entrapped (encapsulated),[54] and cross-linked[55] (Figure 7.3). Also, the enzyme binding to the surface of a solid support

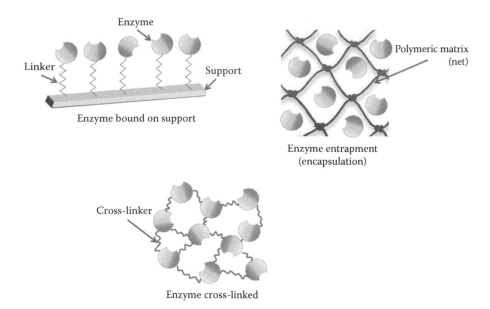

FIGURE 7.3 **(See color insert.)** Enzyme immobilization on the preparation of heterogeneous biocatalyst.

can be physically achieved, but in this case, the interactions between enzyme and support are too weak for keeping enzyme fixed to the support. Ionic and covalent bindings are generally strong interactions avoiding the enzyme leaching from the surface. The covalent binding is often used for the biocatalyst preparations since it offers the advantage of almost unlimited pH and ionic strength values that is valuable for industrial applications. However, this kind of immobilization has also the disadvantage of irreversible deactivation of the enzyme catalytic capacity in the case of improper orientations of the protein molecule on the support surface. The better efficiency of the covalent attachment comparing to physical immobilization was demonstrated for the composite of cellulase enzyme and mesoporous silica nanoparticles support.[56] The investigation of the effect of the loading amount demonstrated that the catalytic activity and stability of the immobilized cellulase were very high (80% glucose yield).[56]

An example of the covalent binding of the enzyme onto a solid support is a lipase biocatalyst used for glycerol conversion to glycerol carbonate (GlyC) based on an excess of dimethyl carbonate (DMC).[51] Lipase enzyme from the *Aspergillus niger* source was covalently immobilized onto a functionalized magnetic nanoparticle surface using two different immobilization approaches, i.e., 1-ethyl-3-(3-dimethylaminopropylcarbodiimide) (EDC) and glutaraldehyde (GA) (Scheme 7.2). The catalytic behavior of the resulting biocomposites was evaluated in a solvent-free system designed for GlyC synthesis (e.g., 48.6% glycerol conversion and 85% selectivity in GlyC).[57] To perform an efficient coupling enzyme–support, several compulsory conditions were emphasized. Thus, the incipient lipase molecules must be in a neutral form (i.e., the negative and positive enzyme charge must be equal), which means that the pH of the immobilization solution has to be close to the isoelectric point (IP) of

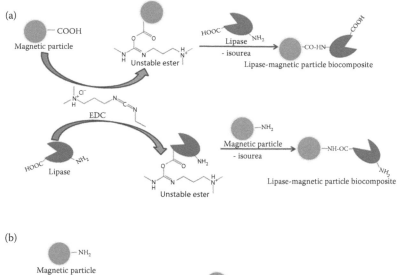

SCHEME 7.2 Covalent immobilization of the enzyme on the solid support using (a) EDC and (b) GA immobilization route.

the enzyme (IP 4.0–4.5).[58] Such a designed biocatalyst exhibited a higher catalytic efficiency (3.52×10^5 h^{-1} TOF), longer stability, and higher recyclability (90 operation hours) compared to the free lipase (i.e., 1.16×10^5 h^{-1} TOF and 16 operation hours).[51,57]

Entrapment (encapsulation) of the enzymes involves their inclusion in a polymeric matrix (e.g., organic polymer, silica sol–gel, hollow fibre, and microcapsule).[59] To prevent the enzyme leakage, the enzyme could be supplementarily retained in the support cage via a covalent attachment. However, in this case, the entrapment method is likely to be confused with the support binding. To avoid this confusion, the tendency is to call "support binding" the case, where the enzyme is attached on the prefabricated support, while the "entrapment" is used for the cases where the support is built up during the immobilization approach.[55]

Nanoparticle biocatalysts were prepared by encapsulation of α-chymotrypsin cross-linked with polyacrylamide hydrogel leading to reversed micelles structures.[60] The biocatalyst with 84 nm size had relatively low activity due to the mass transfer resistance through the hydrogel layer surrounding the enzyme.[60] Higher activity and stability was achieved when a single-enzyme molecule (α-chymotrypsin) was separately encapsulated using a porous organic/inorganic network that was less than a few nanometers thick.[61] However, the entrapment of a lipase within the matrix of a polymer was reported to lead to much more stable configurations than physically adsorbed lipase,[62] allowing to maintain the lipase activity and stability.[63] Other applications have also been developed for trapping lipases in encapsulation form.[64,65] As an example, lipase from *Pseudomonas cepacia* was investigated in immobilized form within a sol–gel support for the transesterification of soybean oil with methanol and ethanol.[66] The gel-entrapped lipase was prepared by polycondensation of hydrolyzed tetramethoxysilane and *iso*-butyltrimethoxysilane. The activity of this biocatalyst corresponded to 67 and 65 mol% methyl and ethyl esters in only 1 h of reaction. The immobilized lipase also proved to be stable and lost little activity when it was subjected to repeated cycles.[66]

Enzyme cross-linked as aggregates or crystals were considered other approaches for the heterogeneization of the biocatalysts. These methodologies are typically called cross-linked enzyme aggregation or crystallization (e.g., CLEA or CLEC). CLEA is still a relatively new, effective, and less explored methodology.[67–69] It involves the enzyme precipitation from an aqueous solution using salts or organic solvents followed then by the cross-linking of the enzyme aggregates with a bifunctional reagent such as glutaraldehyde.[68] CLEA is applicable to a wide variety of enzymes (e.g., lipases, oxidoreductases, lyases, etc.) generating stable, recyclable catalysts, sometimes with an even higher catalytic activity than the free one.[69] The methodology brings together the enzyme purification and its immobilization in one step. The CLEA approach can also be applied for a cocktail of the enzymes leading to a multienzyme biocatalyst that can be used in multisteps or cascade synthesis.[70] CLEA is a less structure developed carrier usually with sizes in the macroscale range. For this reason, it is not proper to include these catalysts among nanoheterogeneous biocatalysts. However, a slight modification of the CLEA approach to the so-called CLEMPA (crosslinked enzyme onto magnetic particles surface aggregates) may correct this mistake.

CLEMPA biocatalyst preparation involves a sequence of two physicochemical steps starting with the enzyme precipitation and followed by the cross-linking aggregation of the enzyme around the magnetic particles (Scheme 7.3). For this purpose, lipase enzyme and magnetic nanoparticles were dispersed in an aqueous phase

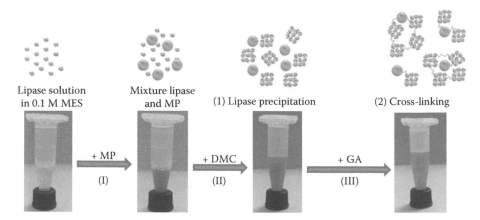

SCHEME 7.3 CLEMPA preparation using cross-linked immobilization approach. MP, magnetic nanoparticles; DMC, dimethyl carbonate; GA, glutaraldehyde.

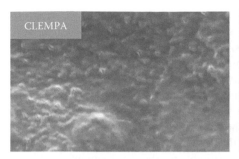
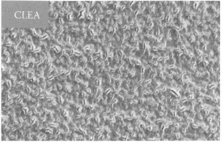

FIGURE 7.4   CLEMPA and CLEA structure of lipase biocatalyst.

(buffer solution). Then, the lipase was separated from the mixture adding a precipitation agent leading to enzyme aggregates. Consecutively, the cross-linker was added allowing the formation of the clusters of the enzyme aggregates and magnetic nanoparticles (Scheme 7.3). The structure of CLEMPA is different than the CLEA's (Figure 7.4).

CLEMPA biocatalyst was investigated in the glycerol carbonate synthesis using different glycerol sources separated from various methodologies.[71,72] Glycerol was converted to glycerol carbonate based on glycerol carbonylation with DMC in excess, in order to ensure the solvent-free reaction conditions (e.g., glycerol/DMC = 1:10 molar ratio). The process was catalyzed by the CLEMPA biocatalyst (5% CLEMPA) at 60°C for 6-h incubation time. The performance of the biocatalytic system for pure glycerol corresponded to 61% glycerol conversion, 55% glycerol yield, and 90% selectivity in glycerol carbonate.[71] The CLEMPA approach offered the opportunity to recover the biocatalyst from the reaction mixture using a magnetic field and its reuse in further reaction cycles. Considering this advantage, the stability and recyclability of the CLEMPA biocatalyst was investigated in 20 successive reaction cycles without any notice of significant losses of the catalytic capacity.[71] The CLEMPA biocatalyst was also used for the conversion of "crude" glycerol taken directly from the biodiesel process of sunflower oil without any pretreatment step. The prepared biocatalyst catalyzed with similar performance the carbonylation of the "crude" and pure glycerols (e.g., 60% glycerol conversion, 54.7% glycerol carbonate yield and 90% selectivity in glycerol carbonate).[71]

7.5   BIOFUELS DERIVED FROM THE BIOCATALYTIC CONVERSION OF BIOMASS USING NANOHETEROGENEOUS DESIGNED BIOCATALYSTS

7.5.1   BIODIESEL

Today's biodiesel sources are mostly found in vegetables and seed oils.[73] In addition, the algae with high lipid content could be also used as a biodiesel facility. The biodiesel process of the triglyceride transesterification with a short chain alcohol (e.g., methanol and ethanol) could be developed in the presence of the biocatalyst. The process is successfully catalyzed by the lipase enzymes.

Nanobiocatalysts with an innovative design for the biodiesel process were reported by the Kim group.[74] The biocatalyst support was a mesostructured onion-like silica (Meso-Onion-S) with a 200- to 300-nm size and exhibited highly curved mesopores of 10 nm diameter in a multishell structure. The nanoscale enzyme reactor was prepared via the lipase enzyme adsorption and cross-linking into highly curved mesopores of Meso-Onion-S-type, thus protecting the enzyme against denaturation and leakage. The biocatalyst activity was 23 times higher than those of the free lipase that provided an additional example of the importance of the lipase stabilization in the form of an immobilized enzyme in the presence of organic solvents.[74] Lozano et al. prepared a heterogeneous biocatalyst suitable for the development of a biodiesel process under supercritical CO_2 conditions.[75] Nanostructured supports, based on 1-decyl-2-methyimidazolium cation covalently attached to a polystyrene divinylbenzene porous matrix, were used as carrier to immobilize *Candida antarctica* lipase B. Such a designed nanobiocatalyst favored the biodiesel production with maximum yields of 95% and operational stability of 45 cycles (85% biodiesel yield after the last reaction cycles at 45°C and 18 MPa).[75]

Xie et al. reported the preparation of a lipase biocatalyst for the transesterification of the vegetable oil based on enzyme covalently attached onto magnetic functionalized nanoparticles using the EDC immobilization approach.[76] The immobilized lipase exhibited better resistance to temperature and pH inactivation compared to the free enzyme. Using these nanocatalysts, the biocatalytic conversion of soybean oil reached around 90% after three reaction cycles, while the catalytic activity was conserved.[76] Lipase biocatalysts deposited on magnetic particles were also developed by Chang and coworkers.[77] In this specific case, the biocatalyst support had a core–shell design with a silica shell coating the Fe_3O_4 core. Lipase produced from an isolated strain *Burkholderia* sp. C20 was immobilized on these magnetic nanoparticles to catalyze the biodiesel synthesis, after a pretreatment of the magnetic support with dimethyl octadecyl [3-(trimethoxysilyl) propyl] ammonium chloride. The immobilized lipase catalyzed the transesterification of over 90% olive oil with methanol to produce fatty acid methyl esters within 30 h in batch operation conditions where 11 wt% immobilized lipase were employed. The biocatalyst has been reused for 10 cycles without any significant loss in its catalytic activity.[77]

Heterogeneous nanobiocatalysts were also successfully applied for the conversion of microalgal biomass into biodiesel fuels.[78] Lipase from a strain of *Burkholderia* sp. C20 was immobilized on hybrid nanomaterials (Fe_3O_4–SiO_2) grafted to a long chain alkyl group as support. The obtained biocatalyst was used to convert oil produced by an isolated microalgal strain *Chlorella vulgaris* ESP-31 after it was pretreated by sonication to destroy the cell walls.[78]

7.5.2 BIOETHANOL

Bioethanol is the largest volume of biofuels produced in refineries nowadays. The most popular approach to produce the bioethanol is yeast fermentation of sugar from starch (cornstarch) or sugar crops. Also, the biochemical conversion can be applied to depolymerize more abundant "cellulose" biomass sources such as grasses, trees, and most of the agricultural residues. The sugar substrate from the bioethanol production is successfully unlocked from the cellulosic matrix using enzymatic processes.[79] β-Glucosidase immobilized on superparamagnetic nanoparticles was reported for the enzymatic hydrolysis of various biomass sources such as wheat straw, eucalyptus globules, and pulp with good conversions of 76.1%, 83.6%, and 75.6%, respectively.[80] The immobilization of enzymes provided an opportunity to reduce their costs in the bioethanol production process.[80] The enzyme amyloglucosidase physically adsorbed or covalently immobilized onto single-walled and multiwalled carbon nanotubes can also catalyze the cellulose hydrolysis.[81] Carbon nanotubes (CNTs) are ideal support systems. Obviously they show no diffusion limitations displayed by the porous texture and have the advantage of an easy further functionalization of the surface (e.g., by the $-NH_2$ terminal groups). Using these supports, higher activities were reported for enzymes physically adsorbed onto CNTs compared with enzymes covalently bound on the CNTs surface. Like in other examples, the possibility of inducing magnetic properties into CNTs offered the opportunity of an easy separation of the catalysts from the original solution.[81]

7.6 BIOMASS DERIVATIVES AS ADDED-VALUE PRODUCTS AS A RESULT OF USING NANOHETEROGENEOUS DESIGNED BIOCATALYSTS (GLYCEROL CONVERSION TO VALUE-ADDED PRODUCTS)

The valorization of glycerol as a biomass-based derivative resulted as waste of the biodiesel process (via the transesterification of various oils such as mustard, rapeseed, canola, corn, soybean, sunflower, or waste cooking oil) is of a great importance. As already mentioned, it represents the main co-product of the biodiesel process (10% w/w) regardless of the biomass source.[82] This glycerol offers varied opportunities for chemical reactions due to its structure, properties, bioavailability, and renewability. There are already many applications converting glycerol, as low-cost feedstock, to value-added products in various sectors as fuels, chemicals, automotive, pharmaceuticals, detergents, or the building industry.[83] Propylene glycol, 1,3-propanediol, syngas, acrolein, epychlorohydrin, or glycerol carbonate are only a few examples of value-added products obtained using chemical/biochemical glycerol transformation.[84,85]

1,3-Dihydroxyacetone is a value-added chemical derived by glycerol oxidation. Industrially, oxidation of glycerol is performed based on a glycerol dehydrogenase (GDH) biocatalyst.[86] Silica-coated magnetic Fe_3O_4 nanoparticles were activated with an amino-silane reagent for covalent immobilization of GDH via a glutaraldehyde linkage. Using this system, the catalytic capacity of the enzyme was conserved after immobilization, while the resistance to temperature and pH variation was improved. Furthermore, an excellent reusability was observed for immobilized GDH such that 10 cycles of reuse with only 9% loss of enzyme activity.[86] More complex enzymatic systems involving xylose reductase together with GDH allowed the cofactor regeneration simultaneously with the 1,3-dihydroxyacetone production.[87] Both enzymes were immobilized on polymeric nanoparticles improving the productivity and stability of the native enzymes, i.e., total turnover number of 82 leading to the yield of 160 g/g immobilized GDH for 1,3-dihydroxyacetone production.[87]

Glycerol carbonate (4-hydroxymethyl-1,3-dioxolan-2-one) (GlyC) is another value-added product, which can be synthesized from renewable glycerol. GlyC is a relative new compound in the chemical industry with a great potential in a large synthesis area (e.g., "green" solvent due to its ideal physicochemical properties such as high stability, low toxicity, low evaporation rate, and low flammability,[88] solvent in cosmetics, personal care items, and medicine,[89] valuable intermediate for the production of resins and plastics, in the pharmaceutical and cosmetics industry,[90] the main electrolyte ingredient of lithium-based batteries[91] and precursor for the production of coatings, adhesives and lubricants via polymerization or via reaction with izocyanates/acrylates[92]).

Glycerol was converted to GlyC using an advanced biocatalytic process involving the glycerol carbonylation with dimethyl carbonate assisted by a heterogeneous enzyme nanobiocatalyst (lipase covalently attached on magnetic nanoparticles surface).[51] Glycerol used as raw material for GlyC synthesis was produced based on a conventional transesterification process[93] applied to both crude and residual sunflower oil. Comparable performances were obtained for glycerol from residual oil and standard glycerol (e.g., glycerol conversion of 36% and 45% with the selectivity in GlyC of 90% and 92%, respectively), while the use of glycerol from residual oil led to conversion of only 27% glycerol with a selectivity in GlyC of 95%.[51]

7.7 CONCLUSIONS

The promising potential of biocatalysis in the chemical industry was already demonstrated by an increasing number of industrial applications. Due to its environmental compatibility and chemical and structural selectivity, biocatalysis allowed reactions that are difficult to be carried out with conventional catalysis. Recent applications of biocatalysis in the biomass valorization open new opportunities due to the high economic advantages and high selectivities. Under these conditions, the use of nano-heterogeneous designed biocatalysts is still very challenging. The combination of easily recoverable magnetic nanoparticles showed the advantage not only of a better separation but also of a better activation of enzymes that led in these conditions to higher activities.

REFERENCES

1. A. L. Demain, *Industrial Biotechnology*, 2007, **3**, 269–283.
2. H. Ohara, *Applied Microbiology and Biotechnology*, 2003, **62**, 474–477.
3. B. Kamm and M. Kamm, *Applied Microbiology and Biotechnology*, 2004, **64**, 137–145.
4. C. Dumon, L. Song, S. Bozonnet, R. Fauré and M. J. O'Donohue, *Process Biochemistry*, 2012, **47**, 346–357.
5. M. G. Adsul, M. S. Singhvi, S. A. Gaikaiwari and D. V. Gokhale, *Bioresource Technology*, 2011, **102**, 4304–4312.
6. P. Wang, in *Bioprocessing for Value-Added Products from Renewable Resources*, ed. Y. Shang-Tian, Elsevier, Amsterdam, 2007, pp. 325–350.
7. S. L. Neidleman, *Current Opinion in Biotechnology*, 1991, **2**, 390–394.
8. W. Saibi, S. Abdeljalil, K. Masmoudi and A. Gargouri, *Biochemical and Biophysical Research Communications*, 2012, **426**, 289–293.
9. P. Grunwald, *Biocatalysis: Biochemical Fundamentals and Applications*. I. C. Press, London, 2009.

10. R. A. Sheldon, *Advanced Synthesis & Catalysis*, 2007, **349**, 1289–1307.
11. M. Wang, T. Si and H. Zhao, *Bioresource Technology*, 2012, **115**, 117–125.
12. M. Goldsmith and D. S. Tawfik, *Current Opinion in Structural Biology*, 2012, **22**, 406–412.
13. D. A. Cowan and R. Fernandez-Lafuente, *Enzyme and Microbial Technology*, 2011, **49**, 326–346.
14. A. Illanes, in *Comprehensive Biotechnology* (second edition), ed. M.-Y. Murray, Academic Press, Burlington, 2011, pp. 25–39.
15. K. Hernandez and R. Fernandez-Lafuente, *Enzyme and Microbial Technology*, 2011, **48**, 107–122.
16. P. V. Iyer and L. Ananthanarayan, *Process Biochemistry*, 2008, **43**, 1019–1032.
17. C. Jäckel and D. Hilvert, *Current Opinion in Biotechnology*, 2010, **21**, 753–759.
18. D. T. Mitchell, S. B. Lee, L. Trofin, N. Li, T. K. Nevanen, H. Söderlund and C. R. Martin, *Journal of the American Chemical Society*, 2002, **124**, 11864–11865.
19. S. W. Hung, J. K. Hwang, F. Tseng, J. M. Chang, C. C. Chen and C. C. Chieng, *Nanotechnology*, 2006, **17**, S8-S13.
20. V. I. Troitsky, T. S. Berzina, L. Pastorino, E. Bernasconi and C. Nicolin, *Nanotechnology*, 2003, **14**, 597–602.
21. F. Caruso and H. Möhwald, *Journal of the American Chemical Society*, 1999, **121**, 6039–6046.
22. C. H. Chan, C. C. Chen, C. K. Huang, W. H. Weng, H. S. Wei, H. Chen, H. T. Lin, H. S. Chang, W. Y. Chen, W. H. Chang and T. M. Hsu, *Nanotechnology*, 2005, **16**, 1440–1444.
23. W. Z. Lee, G. W. Shu, J. S. Wang, J. L. Shen, C. A. Lin, W. H. Chang, R. C. Ruaan, W. C. Chou, C. H. Lu and Y. C. Lee, *Nanotechnology*, 2005, **16**, 1517–1521.
24. C. V. Kumar and G. L. McLendon, *Chemistry of Materials*, 1997, **9**, 863–870.
25. S. A. Ansari and Q. Husain, *Biotechnology Advances*, 2012, **30**, 512–523.
26. K. Min, J. Kim, K. Park and Y. J. Yoo, *Journal of Molecular Catalysis B: Enzymatic*, 2012, **83**, 87–93.
27. M. Pérez-Cabero, A. B. Hungría, J. M. Morales, M. Tortajada, D. Ramón, A. Moragues, J. E. Haskouri, A. Beltrán, D. Beltrán and P. Amorós, *Journal of Nanoparticle Research*, 2012, **14**.
28. L. Yang, S. Shan, R. Loukrakpam, V. Petkov, Y. Ren, B. N. Wanjala, M. H. Engelhard, J. Luo, J. Yin, Y. Chen and C. J. Zhong, *Journal of the American Chemical Society*, 2012, **134**, 15048–15060.
29. H. H. P. Yiu and M. A. Keane, *Journal of Chemical Technology and Biotechnology*, 2012, **87**, 583–594.
30. J. Zhu and G. Sun, *Reactive and Functional Polymers*, 2012, **72**, 839–845.
31. F. Caruso and C. Schüler, *Langmuir*, 2000, **16**, 9595–9603.
32. M. B. F. Martins, S. I. D. Simões, M. E. M. Cruz and R. Caspar, *Journal of Materials Science: Materials in Medicine*, 1996, **7**, 413–414.
33. H. Jia, G. Zhu and P. Wang, *Biotechnology and Bioengineering*, 2003, **84**, 406–414.
34. M. M. M. Elnashar, *Journal of Biomaterials and Nanobiotechnology*, 2010, **1**, 61–77.
35. S. Huang, M. Liao and D. Chen, *Biotechnology Progress*, 2003, **19**, 1095–1100.
36. W. Xie and J. Wang, *Biomass and Bioenergy*, 2012, **36**, 373–380.
37. D. S. Jiang, S. Y. Long, J. Huang, H. Y. Xiao and J. Y. Zhou, *Biochemical Engineering Journal*, 2005, **25**, 15–23.
38. M. M. Zheng, L. Dong, Y. Lu, P. M. Guo, Q. C. Deng, W. L. Li, Y. Q. Feng and F. H. Huang, *Journal of Molecular Catalysis B: Enzymatic*, 2012, **74**, 16–23.
39. C. H. Kuo, Y. C. Liu, C. M. J. Chang, J. H. Chen, C. Chang and C. J. Shieh, *Carbohydrate Polymers*, 2012, **87**, 2538–2545.
40. I. Magario, X. Ma, A. Neumann, C. Syldatk and R. Hausmann, *J. Biotechnol.*, 2008, **134**, 72–78.

41. Y. Yong, Y.-X. Bai, Y.-F. Li, L. Lin, Y.-J. Cui and C.-G. Xia, *Process Biochemistry*, 2008, **43**, 1179–1185.
42. J. L. Malcos and W. O. Hancock, *Applied Microbiology and Biotechnology*, 2011, **90**, 1–10.
43. X. Liu, L. Lei, Y. Li, H. Zhu, Y. Cui and H. Hu, *Biochemical Engineering Journal*, 2011, **56**, 142–149.
44. C. Wu, S. Bai, M. B. Ansorge-Schumacher and D. Wang, *Advanced Materials*, 2011, **23**, 5694–5699.
45. Z. Cheng, A. Al Zaki, J. Z. Hui, V. R. Muzykantov and A. Tsourkas, *Science*, 2012, **338**, 903–910.
46. R. Fernandez-Lafuente, *Journal of Molecular Catalysis B: Enzymatic*, 2010, **62**, 197–212.
47. D. H. Zhang, L. X. Yuwen, Y. L. Xie, W. Li and X. B. Li, *Colloids and Surfaces B: Biointerfaces*, 2012, **89**, 73–78.
48. www.chemicell.com.
49. K.-E. Jaeger and T. Eggert, *Current Opinion in Biotechnology*, 2002, **13**, 390–397.
50. B. Boscolo, F. Trotta and E. Ghibaudi, *Journal of Molecular Catalysis B: Enzymatic*, 2010, **62**, 155–161.
51. M. Tudorache, L. Protesescu, A. Negoi and V. I. Parvulescu, *Applied Catalysis A: General*, 2012, **437–438**, 90–95.
52. Y. Cui, Y. Li, Y. Yang, X. Liu, L. Lei, L. Zhou and F. Pan, *Journal of Biotechnology*, 2010, **150**, 171–174.
53. J. Song, D. Kahveci, M. Chen, Z. Guo, E. Xie, X. Xu, F. Besenbacher and M. Dong, *Langmuir*, 2012, **28**, 6157–6162.
54. X.-Y. Yang, G. Tian, N. Jiang and B.-L. Su, *Energy & Environmental Science*, 2012, **5**, 5540–5563.
55. R. A. Sheldon, *Advanced Synthesis and Catalysis*, 2007, **349**, 1289–1307.
56. R. H.-Y. Chang, J. Jang and K. C. W. Wu, *Green Chemistry*, 2011, **13**, 2844–2850.
57. M. Tudorache, L. Protesescu, S. Coman and V. I. Parvulescu, *Green Chemistry*, 2012, **14**, 478–482.
58. R. Sharmaa, Y. Chistib and U. Chand Banerjee, *Biotechnology Advances*, 2001, **19**, 627–662.
59. C.-H. Lee, T.-S. Lin and C.-Y. Mou, *Nano Today*, 2009, **4**, 165–179.
60. Y. L. Khmelnitsky, I. N. Neverova, A. V. Gedrovich, V. A. Polyakov, A. V. Levashov and K. Martinek, *European Journal of Biochemistry*, 1992, **210**, 751–757.
61. J. Kim and J. W. Grate, *Nano Letters*, 2003, **3**, 1219–1222.
62. W. Hartmeier, *Trends in Biotechnology*, 1985, **3**, 149–153.
63. J. F. Kennedy and E. H. M. Melo, *Chemical Engineering Progress*, 1990, **86**, 81–89.
64. M. T. Reetz, *Advances Materials*, 1997, **9**, 943–954.
65. D. Avnir, S. Braun, O. Lev and M. Ottolenghi, *Chemistry of Materials*, 1994, **6**, 1605–1614.
66. H. Noureddini, X. Gao and R. S. Philkana, *Bioresource Technology*, 2005, **96**, 769–777.
67. P. Lopez-Serrano, L. Cao, F. van Rantwijk and R. A. Sheldon, *Biotechnology Letters*, 2002, **24**, 1379–1383.
68. L. Cao, L. M. Van Langen, F. Van Rantwijk and R. A. Sheldon, *Journal of Molecular Catalysis B: Enzymatic*, 2001, **11**, 665–670.
69. R. A. Sheldon, R. Schoevaart and L. M. Van Langen, *Biocatalysis and Biotransformation*, 2005, **23**, 141–147.
70. R. A. Sheldon, *Applied Microbiology and Biotechnology*, 2011, **92**, 467–477.
71. M. Tudorache, A. Nae, S. Coman and V. I. Parvulescu, *RSC Advances*, 2013, **3**, 4052–4058.

72. M. Tudorache, A. Negoi, B. Tudora and V. I. Parvulescu, *Applied Catalysis B: Environmental*, 2013, 10.1016/j.apcatb.2013.02.049.
73. P. T. Vasudevan and B. Fu, *Waste and Biomass Valorization*, 2010, **1**, 47–63.
74. S. H. Jun, J. Lee, B. C. Kim, J. E. Lee, J. Joo, H. Park, J. H. Lee, S. M. Lee, D. Lee, S. Kim, Y. M. Koo, C. H. Shin, S. W. Kim, T. Hyeon and J. Kim, *Chemistry of Materials*, 2012, **24**, 924–929.
75. P. Lozano, E. García-Verdugo, J. M. Bernal, D. F. Izquierdo, M. I. Burguete, G. Sánchez-Gómez and S. V. Luis, *ChemSusChem*, 2012, **5**, 790–798.
76. W. Xie and N. Ma, *Biomass and Bioenergy*, 2010, **34**, 890–896.
77. D. T. Tran, C. L. Chen and J. S. Chang, *Journal of Biotechnology*, 2012, **158**, 112–119.
78. D.-T. Tran, C.-L. Chen and J.-S. Chang, *Bioresource Technology*, 2013, **135**, 213–221.
79. J. G. Elkins, B. Raman and M. Keller, *Current Opinion in Biotechnology*, 2010, **21**, 657–662.
80. R. Valenzuela, J. F. Castro, C. Parra, J. Baeza, N. Durán and J. Freer, *Clean Technology*, 2011, **3**, 119–122.
81. J. T. Cang-Rong and G. Pastorin, *Nanotechnology*, 2009, **20**, 255102–255121.
82. E. Santacesaria, G. M. Vicente, M. Di Serio and R. Tesser, *Catalysis Today*, 2012, **195**, 2–13.
83. C.-H. Zhou, J. N. Beltramini, Y.-X. Fan and G. Q. Lu, *Chemical Society Reviews*, 2008, **37**, 527–549.
84. M. Pagliaro, R. Ciriminna, H. Kimura, M. Rossi and C. Della Pina, *Angewandte Chemie International Edition*, 2007, **46**, 4434–4440.
85. S. S. Yazdani and R. Gonzalez, *Current Opinion in Biotechnology*, 2007, **18**, 213–219.
86. M. Zheng and S. Zhang, *Biocatalysis and Biotransformation*, 2011, **29**, 278–287.
87. Y. Zhang, F. Gao, S. P. Zhang, Z. G. Su, G. H. Ma and P. Wang, *Bioresource Technology*, 2011, **102**, 1837–1843.
88. J. Huntsman Corporation, *Glycerine Carbonate*, 2009, http://www.huntsman.com/performance products/Media/JEFFSOL%C2%AE GlycerineCarbonate.pdf., 2009.
89. D. Herault, A. Eggers, A. Strube and J. Reinhard, *DE101108855A1*, 2002.
90. J. Rousseau, C. Rousseau, B. Lynikaite, A. Sakcus, C. de Leon, P. Rollin and A. Tatibouet, *Tetrahedron*, 2009, **65**, 8571–8581.
91. A. S. Kovvali and K. K. Sirkar, *Industrial and Engineering Chemistry Research*, 2002, **41**, 2287–2295.
92. G. Rokicki, P. Rakoczy, P. Parzuchowski and M. Sobiecki, *Green Chemistry*, 2005, **7**, 529–539.
93. M. Verziu, B. Cojocaru, J. Hu, R. Richards, C. Ciuculescu, P. Filip and V. I. Parvulescu, *Green Chemistry*, 2008, **10**, 373–381.

Section III

Production of High–Added-Value Chemicals from Biomass Using Nanomaterials

8 Nanostructured Solid Catalysts in the Conversion of Cellulose and Cellulose-Derived Platform Chemicals

*Marcus Rose, Peter J. C. Hausoul,
and Regina Palkovits*

CONTENTS

8.1 INTRODUCTION

Lignocellulose is one of the most abundant renewable raw materials for the future substitution of fossil resources [1]. In contrast to herbal biomass such as oils or carbohydrates such as starch and various sugars, the utilization of lignocellulose as an energy carrier and chemical feedstock poses no competition to the production of edibles. Thus, it can be considered that the second generation of biogenic raw materials is more sustainable in production and utilization. Lignocellulosic plant materials can be obtained from agriculture and forestry in the form of by-products such as straw, wood chips, and crop residues, as well as fast growing grasses. Even municipal solid waste can be utilized. Additionally, no valuable soil has to be used for the cultivation of monocultures of certain plants, and no plant-protecting agents or fertilizers have to be applied, rendering it a rather "green" resource.

The annual worldwide growth of plant material via photosynthesis is estimated to be 170 billion tons with a fraction of approximately 95% of lignocellulose [2]. Thus, the production of significant amounts of basic chemicals from this feedstock seems feasible, especially in terms of tailor-made platform chemicals. In contrast, thermochemical routes for biomass conversion to known platform chemicals require enormous amounts of energy [3]. These processes include the biomass to liquid (BtL) process, which is based on gasification of biomass to synthesis gas and subsequent Fischer–Tropsch synthesis to the respective chemicals. Alternatively, biomass can be converted to so-called biocrude requiring also a high energy input. This oil can be further converted in existing processes of the mineral oil industry. However, thermochemical routes are energy intense. Thus, the direct conversion of biogenic feedstocks to valuable intermediates is the preferable route.

One of the main challenges connected to the utilization of lignocellulose is the high degree of functionalization resulting in highly polarized molecules and biopolymers. In contrast, established chemical processes are based on mineral oil and secondary products with a low degree of functionalization. Thus, catalytic transformations are typically carried out in the gas phase while the utilization of biomass as feedstock depends on liquid phase processes with significantly different requirements.

Lignocellulose is the major component of plant material. It consists of three main compounds: cellulose, hemicellulose, and lignin. Cellulose representing 40%–70% of a plant is a linear polymer of glucose monomer units connected via β-1,4-glycosidic bonds; 10%–40% are accounted by hemicellulose, which is a similar biopolymer. A difference from cellulose is that it consists of various pentoses and hexoses as building blocks with branched polymer chains. The third compound lignin is also a biopolymer based on aromatic building blocks. It accounts for 10%–30% of the plant. By a high degree of branched and crosslinked chains, it provides stability to the plant skeleton. Since cellulose poses the main fraction of nonedible plant material and it is chemically well-defined in contrast to hemicellulose and lignin, it is considered the most promising feedstock for a chemocatalytic conversion into various platform chemicals. Thus, this chapter gives an overview on recent advances in the chemocatalytic conversion of cellulose. In particular, it focuses on the development and investigation of novel nanostructured solid catalysts. In the first part different classes of solid acidic catalysts are discussed for the hydrolytic depolymerization of cellulose to its basic building block glucose as well as to other platform chemicals by further acid-catalyzed dehydration such as 5-hydroxymethylfurfural (5-HMF) and levulinic acid. Subsequently, catalytic concepts utilizing solid bifunctional catalysts combining acid and hydrogenation sites for the hydrolytic hydrogenation of cellulose to sugar alcohols are presented. In the second part, recent developments in the conversion of cellulose-derived furfural-based platform chemicals by reductive and oxidative reaction paths utilizing solid and nanostructured catalysts are presented.

8.2 CELLULOSE TO PLATFORM CHEMICALS

8.2.1 CHALLENGES IN CELLULOSE PROCESSING

The main challenge for the transformation of cellulose into platform chemicals is defined by its appearance in combination with the other biopolymers hemicellulose and lignin as well as its structural characteristics. Lignocellulose is a natural high-performance composite material. It is insoluble in common solvents and possesses a high stability against chemical disintegration. Cellulose itself occurs as a semicrystalline polymer due to the high functionalization degree with hydroxyl groups resulting in a high degree of crosslinking of the polymeric chains by hydrogen bonds.

For further chemical transformation of its monomer unit glucose, the cellulose chains have to be depolymerized first by acid- or enzyme-catalyzed reactions. Although the latter ones can be carried out with a high selectivity they are not suitable for large-scale applications due to rather low space-time yields, their high substrate selectivity respective substrate limitation as well as the high cost of the enzymes. The significantly faster acid-catalyzed depolymerization utilizing mineral acids is known for nearly a century and was even commercialized, e.g., within the Bergius (Rheinau) or the Scholler (Tornesch) process [4]. Therein, ethanol was produced by wood saccharification with subsequent fermentation of the glucose. Today, a large-scale cellulose conversion under process conditions similar to those is not aspired to due to the large amounts of acids that are required and that have to be neutralized afterward resulting in stoichiometric amounts of salts as by-products. Also, the high corrosion potential poses a challenge with regard to the respective corrosion resistant process technology. Additionally, the selectivity to glucose is lower by the formation of by-products due to side-reactions catalyzed by the strong acids.

However, recent research focuses on the utilization of solid acids for hydrolysis reactions as well as bifunctional catalysts with solid acid-supported metal species for a combined hydrolytic hydrogenation or oxidation reaction in heterogeneously catalyzed processes enabling a simple catalyst recovery and of course, environmentally favorable recycling. This is most important as more than half of the investment costs in the chemical industry are used for separation processes including recycling and regeneration of the catalysts [5]. Therefore, solid catalysts are highly preferred in industrial processes in comparison to molecular catalysts despite their higher selectivity. The challenge in the utilization of solid catalysts in the conversion of cellulose is the accessibility of the catalytic active centers since cellulose is a solid as well as the catalyst. Therefore, viable pretreatment methods are crucial for an efficient interaction of the catalyst and the solid substrate and thus, should be discussed briefly in the following.

The crystalline cellulose shows very low to no solubility at all in common polar solvents such as water due to the highly hydrogen-bonded supramolecular structure of the cellulose chains. Thus, one main target of pretreatment methods is to decrease the degree of crystallinity and obtain amorphous cellulose in which most of the hydrogen bonds are destroyed [6]. Additionally, such pretreatment methods result in an increased reactivity of cellulose for further catalytic transformation. Unfortunately, there seems to be a lack of systematic studies investigating the connection between crystallinity and reactivity of cellulose.

In the simplest case, the decrease of crystallinity is achieved by mechanical grinding or fiber explosion methods involving steam, carbon dioxide, or even ammonia as well as by different variations of the organosolv process [6–7]. Even cellulose pretreatment in supercritical water is considered and investigated in detail [8]. Various studies are available dealing with cellulose dissolution in ionic liquids and the catalytic conversion therein, which proved highly effective [9]. However, for large-scale processes, the use of ionic liquids can only be efficient with a quantitative solvent recovery, which is not possible for IL by classical separation technology. An efficient approach is the utilization of organic electrolyte solutions, i.e., small amounts of ionic liquids in organic solvents, for a rapid dissolution of cellulose [10]. Besides dissolving cellulose or just decreasing the degree of crystallinity in the solid state, also a partial decomposition of the cellulose chains to cellooligomers seems favorable due to an increase in reactivity compared with the mechanically treated chains. Recently, Meine et al. proposed a pretreatment process combining the impregnation of cellulose with catalytic amounts of mineral or organic acids followed by a mechanical milling process [11]. Completely water-soluble oligosaccharides were obtained within 2 h. A subsequent hydrolysis at 130°C for only 1 h resulted in over 90% conversion into the monomeric building blocks.

Despite all different pretreatment methods that have been proposed and investigated, catalyst development in literature seems to be carried out relatively independently from these results. In recent years, various findings have been published dealing with the utilization of solid catalysts in the conversion of different forms of cellulose and cellulose-derived platform chemicals. In the following chapters, we give an overview of recent examples of the utilization of nanostructured catalysts for selected reactions.

8.2.2 CELLULOSE HYDROLYSIS BY SOLID ACIDS

The acid-catalyzed hydrolysis of cellulose into glucose is mandatory for a further utilization as a versatile biogenic feedstock (Scheme 8.1) [4]. Solution-based processes as mentioned above with molecular mineral (H_2SO_4, HCl, HF) or organic acids (oxalic, maleic, and fumaric acid) can be considered proven technologies and have been summarized in various reviews [4,12]. For the development of efficient processes for future biorefinery schemes the utilization of solid catalysts is desirable. In recent years, various solid acidic materials have been proposed for the hydrolysis of solid cellulose. Their potential has been evaluated on the laboratory-scale often utilizing cellulose with differing degrees of crystallinity and general properties from different laboratory-chemical providers. This additionally has to be kept in mind

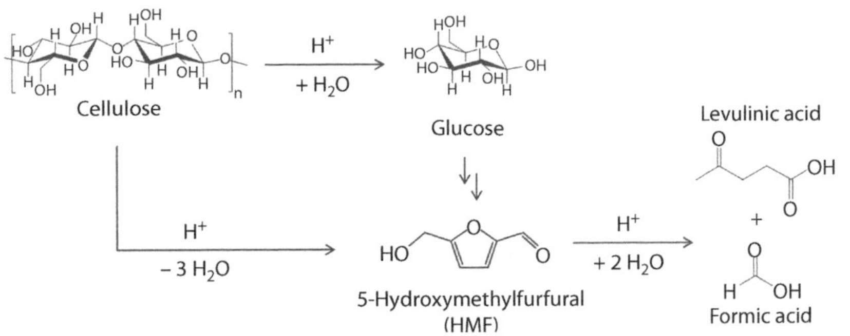

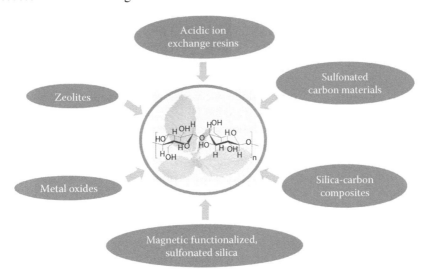

SCHEME 8.1   Acid-catalyzed transformation of cellulose into versatile platform chemicals.

when directly comparing hydrolysis results of different catalyst materials under more or less similar conditions in the various published studies. One of the main challenges as mentioned above is the accessibility of the catalytic active sites upon use of a solid substrate and a solid support. Often, high catalyst/substrate ratios are applied to obtain a satisfactory cellulose conversion. Often a ratio of 1:1 is exceeded rendering the whole process highly inefficient in terms of utilization of the catalyst capacity. Another limitation regarding the pure hydrolysis of cellulose to glucose is the limited thermal stability of glucose. Thus, temperatures below 150°C are often used for this reaction.

Solid acid materials that have been investigated so far in the hydrolysis of cellulose are of various types ranging from classical zeolites and metal oxides over all kinds of acid-functionalized carbon materials and acidic ion exchange resins up to various composite materials (Scheme 8.2). Selected examples are presented and discussed in the following.

SCHEME 8.2   Solid acidic catalysts that have been applied in the heterogeneously catalyzed hydrolytic cellulose depolymerization.

Zeolites are the main catalyst species utilized in large scale processes of the petroleum refining industry when acidic catalysts are required such as in the fluid catalytic cracking (FCC) process or isomerization and alkylation reactions. Zeolites are alumosilicates with a highly crystalline and mainly microporous structure [13]. They can act as Brønsted as well as Lewis acids. In conventional catalytic processes in the industry, zeolites are used in gas phase processes at temperatures of several hundreds of degree celsius. In contrast, the hydrolytic decomposition of cellulose has to be carried out in the liquid state at moderate temperatures and in polar and maybe even protic solvents such as water.

The hydrolysis of disaccharides such as maltose, sucrose, or cellobiose as well as of the water-soluble polysaccharide starch applying zeolites has been reported in the literature for several years [14]. Reaction temperatures between 80°C–150°C in aqueous solution resulted in high conversions and yields of >90% of the respective sugar monomers. One of the main limitations of zeolites is definitely their mainly microporous structure. For example, Dhepe et al. showed that HZSM-5 is not active in the conversion of sucrose let alone starch due to its small pore diameter of 0.56 nm and thus, the inaccessibility for the substrates [14d].

More recent investigations dealing with the hydrolysis of cellulose dissolved in ionic liquids gave deviating results. Rinaldi et al. investigated solid acids with different pore sizes for the partial depolymerization of cellulose into cellooligomers for further processing [15]. They found that acidic silica/alumina materials as well as zeolite Y and ZSM-5 with pore sizes of 4.5, 0.7, and 1.2 nm, respectively, are inaccessible for the dissolved cellulose chains. Additionally, the contribution of the very low external surface area is negligible. In contrast, other groups observed catalytic activity for small pore zeolites in ionic liquid solutions of cellulose. Zhang et al. applied HY-zeolites with a low Si/Al ratio and high specific surface areas in this reaction [16]. Applying microwave heating they were able to produce glucose with a yield of 37% within 8 min. Presumably, the energy input by the microwave irradiation is involved in the increased reaction rate and leaching of, e.g., aluminum species cannot be excluded. However, no detailed explanation is given. But also without microwave heating zeolites have been successfully applied in cellulose hydrolysis.

Recently, Cai et al. investigated the interaction of the zeolite and the ionic liquid solvent in the cellulose hydrolysis reaction [17]. By adding small amounts of water gradually over the whole reaction time at a temperature of 130°C highest glucose yields of 50% were obtained utilizing the HY-zeolite. Beside glucose as the main product, the overall yield of 97% contained the glucose dimer cellobiose as well as the dehydration product 5-hydroxymethylfurfural (HMF). It was shown that the ionic liquid 1-butyl-3-methylimidazolium chloride (BMIMCl) enters the pore system and blocks the pores. Due to the zeolite ion exchange with the IL-cation, the in situ released protons appear as the active species and thus, are responsible for the high catalytic activity. The authors claim that the zeolite framework is stable upon loading with the IL although cell parameters appeared to be enlarged due to a dilation effect. However, zeolite regeneration by calcination and recycling experiments have been conducted successfully. Despite the good catalysis results, a process in which an IL incorporated in the pores is burned off for catalyst reuse seems not favorable.

Beside other acidic solids, Lanzafame et al. applied the H-BEA and H-MOR zeolites for cellulose hydrolysis in aqueous environment [18]. They used a cellulose/catalyst ratio of 10:1 at reaction temperatures of 190°C in a static, not-stirred autoclave and obtained product mixtures containing glucose, HMF, and other water-soluble by-products. The microporous zeolites total cellulose conversions of 4%–6% were reported, while all other catalysts with larger pore systems did not exceed 20% total conversion. The authors state that definitely an efficient solid-solid interaction between the cellulose and the external surface of the catalyst particles is required. Additionally, the micropore structure limits the further conversion of glucose to oligomeric/polymeric by-products due to size-confining properties.

In contrast to microporous zeolites polymeric materials such as ion exchange resins gave promising results in recent investigations. Ion exchange resins are mainly macroreticular polymers, e.g., polystyrene/-divinylbenzene, exhibiting an open macroporous structure [19]. Their surface is covered with sulfonic acid groups that covalently bond to the aromatic building blocks of the resin frameworks. In literature, ion exchange resins of the Amberlyst® series are used very often [20]. The hydrolysis of cellulose, which is dissolved in the ionic liquid BMIMCl, seems to work very well with ion exchange resins. At a temperature of 100°C and an IL/cellulose/catalyst mass ratio of 100:5:1, the cellulose is depolymerized to cellooligomers and reduces sugars to a large extent after a few hours of reaction time [15]. Especially Amberlyst-15 and Amberlyst-35 with specific surface areas of 36 and 40 m^2 g^{-1}, respectively, show a very similar high activity. In contrast, Amberlyst-70 with a significantly lower surface area of <1 m^2 g^{-1} as well as the gel-type resin Nafion with only 0.02 m^2 g^{-1} showed a comparably poor performance in the hydrolysis reaction due to the significantly lower specific surface area and thus hindered accessibility of the pore system. The authors also observed an initial induction period for the production of reducing sugars of approximately 1 h. This effect was not observed when using dissolved *para*-toluenesulfonic acid as a homogeneous catalyst. In a subsequent study, the authors investigated this reaction in detail to obtain information about the mechanism as well as influencing and controlling reaction parameters [21]. They observed that the initial induction period is related to a continuous in situ generation and release of protons by ion exchange of the resin with the IL cation, thus controlling the initial reaction rate of the depolymerization reaction. The Amberlyst resin can be easily reactivated by washing with sulfuric acid and reused.

Few publications are available dealing with the cellulose hydrolysis with ion exchange resins in an aqueous environment. The main challenge connected to the use of sulfonated ion exchange resins in water is the hydrolysis of the sulfonic acid groups releasing sulfuric acid at elevated temperatures. Thus, one has to be careful by describing catalytic results as heterogeneously catalyzed while maybe the acidic species was leached out. Recently, Ahlkvist et al. reported the one pot production of levulinic acid and formic acid by direct conversion of pulp from the paper industry with the low surface area ion exchange resin Amberlyst-70 at temperatures of 150°C–200°C [22]. Amberlyst-70 is known to pose a higher thermal stability than the other Amberlyst resins [20]. The reaction was carried out in water under Ar/CO_2 pressure of 50 bar. Maximum yields of 53% formic acid and 57% levulinic acid have been reported after approximately 10 h. Recently, Benoit et al. reported the cellulose hydrolysis with Amberlyst-35 at 150°C in water for 3 h with different pretreatment methods [23]. While non-pretreated Avicel-cellulose yielded >1% of glucose, pretreatment by ball milling or with an IL yielded up to 14% glucose. A special pretreatment avoiding the use of chemicals or solvents by a non-thermal atmospheric plasma increased the subsequent glucose yield in the catalyzed reaction up to 22% with Amberlyst-35.

The cheapest acidic catalyst material providing a wide range of tuning parameters in terms of specific surface area, pore size, and functionalizability, as well as concentration, type, and strength of acidic groups are activated carbons (AC) or in more general terms carbonaceous materials [24]. They have been known for hundreds of years and are used in widespread applications ranging from the chemical industry up to everyday applications. Also catalysis carbon materials are widely used as a catalyst themselves or as catalyst supports. In terms of porosity, carbon materials are the type of classical porous materials, which can reach up to several thousand square meters per gram, with pore sizes ranging exclusively from micropores (<2 nm) up to macropores (>50 nm) and of course hierarchical pore systems of all kinds.

In recent years, various reports have been published investigating mainly sulfonated carbon materials in the hydrolytic cellulose depolymerization. Onda et al. sulfonated an activated carbon by treating it with sulfuric acid [25]. A subsequent hydrothermal treatment was carried out to prevent leaching of sulfate ions by hydrolysis of sulfonic acid groups during the catalytic reaction ($c_{sulfate}$<0.03 mmol·L^{-1}). The hydrothermally treated acid catalyst showed a specific surface area of 806 m^2 g^{-1} and a concentration of sulfonic acid groups of 1.63 mmol·g^{-1}. The conversion of cellulose was carried out in water at 150°C for 24 h. With a high ratio of catalyst/cellulose (50 mg/45 mg) a conversion of 43% was achieved with a selectivity to glucose of 95%. In comparison to other acidic solids such as Amberlyst resins, zeolites, or sulfated zirconia, a significantly higher conversion and the highest glucose selectivity was observed. Due to the excess ratio of the microporous catalyst it can be assumed that only acid groups on the external surface are responsible for the hydrolytic depolymerization. A TON of 5 was calculated with regard to the total amount of sulfonic acid groups. Additionally, the recyclability was tested. The catalyst has been reused for up to three cycles without change in activity.

In 2004, Hara et al. reported the production of sulfonated carbons with detailed investigations of the hydrolytic stability of the sulfonic acid groups [26]. They proposed that sulfonated carbon materials with a high concentration of sulfonic acid groups and a high stability against hydrolysis and leaching in an aqueous environment can be produced from aromatic precursors by an incomplete carbonization with a simultaneous sulfonation in sulfuric acid. Such a production process results in materials with a higher degree of graphene-like sp^2 hybridized carbon in comparison to a disordered sp^3 hybridized carbon. In recent years, the same group investigated different precursors for the production of stable acidic carbon materials especially for the hydrolysis of cellulose in an aqueous environment. They identified a cellulose-derived carbon-based solid acid (CCSA) as the best suited catalysts for this reaction [27]. The catalyst is produced by heating microcrystalline cellulose to 400°C–650°C under nitrogen atmosphere for 5 h for an incomplete carbonization. The precursor is then treated in oleum at 80°C for 10 h and washed until the filtrate is neutral. The acidic carbon materials are not porous and exhibit an external surface area below 5 m^2 g^{-1} with acid concentrations of 0.75–1.8 mmol·g^{-1} [27e]. Their catalytic performance was investigated using cellobiose and cellooligomers, as well as microcrystalline cellulose. Using a catalyst:cellobiose ratio of 1:1 at 100°C a glucose yield of 35% has been reported after 6 h [27e]. For the conversion of microcrystalline cellulose excess amounts of the catalyst (0.3 g with 25 mg of cellulose) have been used [27a]. After 3 h at 100°C only 4% glucose has been observed. The major reaction product was β-1,4-glucan, a water-soluble cellooligomer, in up to a 64% yield. In contrast to many other sulfonated carbon materials, the leaching of acid groups is negligible for the nonporous materials. They have been tested in up to 25 cycles without loss of activity. The structural feature of these catalysts, which is made responsible for the enhanced stability, are the graphene-like carbon sheets in the size of 1 nm, which are functionalized with sulfonic acid as well as with hydroxyl and carboxylic acid groups along the edges (Figure 8.1). The catalytic activity of these materials is attributed to a supporting effect of the phenolic hydroxyl and carboxylic acid groups. They are weak acid sites with pK$_a$ values of approximately 10 (OH) and 4.7 (COOH), whereas sulfonic acid groups pose a pK$_a$ of −2.8 [28]. It is known that OH and COOH groups should not contribute to the hydrolysis reaction due to their weak Brønsted acidity [29]. However, they are responsible for a decreased hydrophobicity of the surface and thus are responsible for a better interaction with the polar solvent and substrate. Additionally, they are regarded to serve as adsorption sites for glucan units, water-soluble cellooligomers, by the formation of hydrogen bonds in close proximity to the catalytically active sulfonic acid groups, thus enhancing the hydrolysis reaction once soluble oligomers are available [27a].

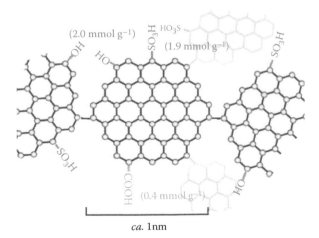

FIGURE 8.1 Idealized schematic representation of cellulose-derived carbon-based solid acids (CCSA) consisting of graphene-like sheets functionalized with sulfonic acid, carboxylic acid and hydroxyl groups along the edges. (Reprinted with permission from S. Suganuma, K. Nakajima, M. Kitano, D. Yamaguchi, H. Kato, S. Hayashi, M. Hara, *J. Am. Chem. Soc.*, 2008, 130, 12787–12793. Copyright 2008 American Chemical Society.)

Similar carbon catalysts derived from lignocellulosic biomass were applied in the hydrolysis of microcrystalline cellulose with an excess amount of the substrate (0.2 g cellulose and 0.1 g catalyst) and using microwave irradiation for heating [30]. After 1 h and a reaction temperature of 90°C up to 20% glucose and 8% water-soluble cello-oligomers have been obtained. The authors claim the microwave radiation responsible for two beneficial effects: (1) an additional destruction of the crystalline cellulose structure, which was confirmed by XRD, and (2) an enhanced contact of solid cellulose particles with the external surface area of the catalyst particles. Adsorption experiments of glucose and cellooligomers proved a selective adsorption of the latter species. The adsorbed amount was directly proportional to the concentration of phenolic hydroxyl groups. In contrast, Amberlyst-15 containing exclusively sulfonic acid groups showed no adsorption at all. From these results, the authors concluded that exclusively the hydroxyl groups are responsible for a preferred binding of the cellooligomers by strong hydrogen bonds to the oxygen atoms in the β-1,4-glycosidic bonds. Also, for these materials, the recyclability was investigated. The microwave heating seems not only to increase the hydrolysis rate of the cellulose depolymerization but also the hydrolytic desulfonation with leaching of sulfuric acid since a gradual decrease in activity was observed for each cycle.

Beside microporous activated carbons and nonporous graphene-like carbon materials also templated ordered mesoporous carbon materials have been investigated in the hydrolysis of cellulose. Especially the "carbons mesostructured by KAIST" (CMK) have gained an enormous interest in the last decade for a variety of applications due to the combination of an ordered mesopore structure with a high specific surface area solely made of carbon [31]. CMK-3 is a carbon replica of the mesoporous silica SBA-15 with a specific surface area of 1520 m^2 g^{-1}, a total pore volume of 1.3 cm^{3}g^{-1}, and a pore size of 4.5 nm that can be varied within a few nanometers by the pore size of the template [32]. For its production, the silica template is impregnated with a carbon precursor such as a sucrose solution and thermally carbonized. Subsequently, the silica matrix is removed by extraction with hydrofluoric acid or sodium hydroxide solution.

Pang et al. prepared a sulfonated CMK-3 by a pretreatment with nitric acid for oxidation, and thus hydrophilization of the inner surface, and a subsequent sulfonation in concentrated sulfuric acid at temperatures of 150°C–300°C for 24 h [33]. At 250°C, highest acid densities (–SO$_3$H) of up to 0.44 mmol·g^{-1} have been observed. The specific surface area decreased to 762 m^2 g^{-1}. The cellulose hydrolysis test reaction was conducted with a slight excess amount of catalyst (0.3 g with 0.27 g cellulose). At 150°C for 24 h total cellulose conversions of up to 94% have been reached with total glucose yields of 74% (the highest reported up to date with solid acids) exceeding the performance of all other tested sulfonated carbon materials such as cellulose- and resin-derived carbons, acetylene carbon black or multiwall carbon nanotubes. Recycling experiments showed that the CMK-3 catalyst is stable over at least four cycles without hydrolysis of the sulfonic acid groups.

Kobayashi et al. used CMK-3 as a support material for Ru nanoparticles [34]. They found that this hybrid material without acid functionalization functions as a water-tolerant and stable catalyst for the hydrolysis of cellulose to glucose and other soluble products. If the cellulose hydrolysis in water is carried out at a temperature of 230°C without a catalyst, a conversion of 24% into different water-soluble species (5% glucose) is observed. Pure CMK-3 (50 mg catalyst with 324 mg cellulose) increases the conversion to 54%, and the glucose yield to 21%. Ru/CMK-3 was prepared by impregnation of CMK-3 with dissolved $RuCl_3$ and subsequent drying and reduction under hydrogen atmosphere. A varying Ru loading of 2, 5, and 10 wt% gave increased cellulose conversions of 59%, 62%, and 68%, and glucose yields of 28%, 29%, and 34%, respectively. The catalyst can be reused without structural change. A Ru/Al_2O_3 catalyst was used for comparison. In this case, a rapid deactivation due to a structural transformation of the support has been observed. To clarify the role of the supported Ru and the support CMK-3, experiments of the hydrolysis of cellulose have been compared with cellobiose hydrolysis. It was found that CMK-3 preferably accelerates the hydrolysis of cellulose chains while it does not increase the hydrolysis of cellobiose to glucose. If Ru is involved in the reaction, an increased formation of glucose is observed, although its role in the acid catalyzed reaction was not clear.

Further studies on that system including a more detailed investigation of the catalyst clarified the role of the supported Ru species [35]. After the reduction of the Ru precursor a passivation step was conducted by treating the sample at ambient temperature in oxygen to yield RuO_2 on the surface. In an aqueous environment this species is hydrated to form $RuO_2 \cdot 2H_2O$. Particles with a size of 1.1 ± 0.2 nm inside the CMK-3 pore system have been identified. Two explanations are given why this Ru species gives an enhanced hydrolytic behavior (1) by desorption of water molecules from the hydrated species free Lewis acid sites are accessible or (2) the Ru species is involved in the heterolysis of water molecules releasing protons, thus acting as Brønsted acid–generating site ($[Ru(H_2O)_6]^{3+}$ has a low pK_a of 2.9 at 25°C).

Sulfonated silica–carbon hybrid materials for the hydrolysis of cellulose have been recently reported by de Vyver et al [36]. The intention was to combine the flexibility and functionalizability of an organic-based carbon material with a high mechanical and thermal stability of silica-based materials. For the synthesis of the catalyst, an evaporation-induced co-assembly method of three compounds has been applied to obtain a composite structured on the nanoscale. Tetraethyl orthosilicate (TEOS) and sucrose were used as a precursor for the silica and the carbon, respectively, whereas the Pluronic F127 triblock copolymer was used as a structure-directing agent. The solution of those compounds was evaporated and carbonized under inert gas at 400°C–550°C decomposing the F127 and carbonizing the sucrose to a hydrophobic carbon species within the silica matrix. The samples were sulfonated by treatment in sulfuric acid. The resulting acidic nanocomposite materials showed specific surface areas between 100–500 $m^2 g^{-1}$ and acid group densities of 0.15–0.57 mmol·g^{-1}. Their catalytic activity in the hydrolysis of cellulose was investigated at 150°C for 24 h. A total conversion of up to 61% with a total glucose yield of 50% has been achieved. A recycling test showed a significant activity drop in the second cycle remaining constant in the following. For higher silica/carbon ratios, higher TOFs with regard to the acid groups have also been observed. This has been interpreted with a favored adsorption of the substrate, respective soluble cellooligomer intermediates, on the hydrophilic silica surface enhancing the hydrolysis reaction on the sulfonated carbon sites.

When the solid substrate cellulose is not quantitatively transformed into soluble products, the separation of the solid catalyst poses a great challenge especially in reaction systems in which both the substrate and the catalyst are solids. To overcome this problem, recently, Lai et al. proposed the production of a magnetically function-alized solid acid catalyst [37]. They produced an ordered mesoporous silica (SBA-15) by co-condensation of TEOS and mercaptopropyl trimethoxysilane (MTPNS) with a structure directing agent and involving the direct oxidation of the mercapto group catalyzed by Fe_3O_4 nanoparticles and hydrogen peroxide. The resulting com-posite material is an ordered mesoporous silica matrix functionalized with propylene sulfonic acid groups (1.09 $mmol \cdot g^{-1}$) and incorporating magnetite nanoparticles of approximately 20 nm in diameter within the silica matrix. The magnetic particles enable a simple separation of the catalyst particles from the reaction mixture using magnets, which is an essential step in systems where the substrate as well as the catalyst is a solid. The composite catalyst was investigated in the hydrolysis of cel-lulose with a cellulose/catalyst ratio of 1:1. At 150°C microcrystalline cellulose was hydrolyzed to yield 26% glucose after 3 h and 12% glucose/42% levulinic acid after 12 h of reaction time. Starting from amorphous cellulose with a slight excess of the catalyst after 3 h, even 50% glucose have been obtained. A recycling of the catalyst material was proven without signs of deactivation in up to three cycles.

Taking into account the nature of the solid and often not soluble substrate cellu-lose it seems more beneficial to apply nanoparticulate catalysts with a high external surface area to increase the interface for an optimal utilization of the functional acid groups. Takagaki et al. combined the concept of acid functionalized nanoparticles with the magnetic functionalization for a simple catalyst recovery (Figure 8.2) [38]. They prepared co-ferrite ($CoFe_2O_4$) by a simple sol–gel approach. Subsequently, 3-sulfanylpropyltrimethoxysilane (SPTMS) was used as a silica precursor and con-densed in a basic aqueous solution to cover the ferrite nanoparticles. In the third step, the thiol groups were oxidized to sulfonic acid groups with hydrogen peroxide. The final composite material consisted of aggregated ferrite nanoparticles of 20–50 nm in diameter in the center of each particle covered with a silica layer of approximately

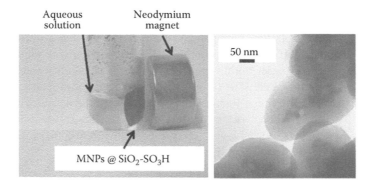

FIGURE 8.2 Magnetic separation of $CoFe_2O_4$-NP@SiO_2-SO_3H after sucrose hydro-lysis (left) and TEM micrographs of the catalyst particles (right). (From A. Takagaki, M. Nishimura, S. Nishimura, K. Ebitani, *Chem. Lett.* 2011, 40, 1195–1197.)

50–60 nm with sulfonic acid groups on the external surface (0.5 mmol·g^{-1}), rendering this catalyst a submicrometer material rather than a nanomaterial. However, its catalytic performance was investigated with a catalyst/cellulose ratio of 1:1 at 150°C for 3 h. A total yield of reducing sugars of 30% was achieved with a glucose yield of 7%. Compared with Amberlyst-15, a 9.5 times higher TON was achieved with regard to the sulfonic acid groups due to the increased interface and their improved accessibility. Recently, conventional inorganic materials with acidic properties have also been considered for the conversion of cellulose. Especially layered metal oxides can act as acid catalysts by providing the possibility of a cation exchange of metal ions by protons. Connected to the cation exchange is often the exfoliation and aggregation of the metal oxide layered structure, thus resulting in inorganic nanosheets with a high external surface area and Brønsted acid properties. This class of materials was recently reviewed by Takagaki et al. [39].

First results utilizing the layered transition metal oxide HNbMoO$_6$ in the hydrolytic conversion of polysaccharides have been reported by the same group [40]. With a specific surface area of 5 m^2 g^{-1} and an acid density of 1.9 mmol·g^{-1}, a high rate of glucose production from sucrose and cellobiose as well as significantly higher TOFs in comparison to other solid acid materials have been observed. At a temperature of 130°C and 12-h reaction time, microcrystalline cellulose has been hydrolyzed to yield 8.5% cellobiose and glucose. Also, other transition metal oxides with acidic behavior such as mesoporous Ta$_x$W$_{10-x}$ oxides and HTaWO$_6$ nanosheets have been considered and investigated in the hydrolysis of disaccharides such as sucrose and cellobiose [41]. These results are not outstanding in comparison to other solid acid materials. However, due to a great variety in structure and type of these solid acids, as well as tunable surface properties of the acidic nanosheets, great potential for further development is given.

In contrast to the above-presented concepts, Hick et al. proposed an economically and ecologically much more feasible process for the depolymerization of cellulose [42]. They used clay minerals such as kaolinite, a layered alumosilicate with acidic properties in solvent-free solid-solid mechanocatalytic reactions, i.e., mechanical grinding of cellulose with the solid catalyst. They could show that after 3 h of milling up to 84% of microcrystalline cellulose can be converted into water-soluble compounds such as cellooligomers, glucose, and its dehydration products levoglucosan, levoglucosenone, and HMF in a single solvent-free step. In this process, the crystalline cellulose particles are physically and chemically broken down, and thus, even whole lignocellulosic raw material can be utilized and soluble compounds from cellulose and hemicellulose can be extracted afterward. Since kaolinite is a naturally occurring clay mineral it is quiet cheap ($80/ton) [42]. Together with the fact that it can be simply reused and no by-products occur or other solvents are necessary, a process with high atom efficiency seems economically and ecologically feasible.

8.2.3 HYDROLYTIC HYDROGENATION AND HYDROGENOLYSIS OF CELLULOSE BY BIFUNCTIONAL SOLID CATALYSTS

Beside the pure acid-catalyzed hydrolytic depolymerization of cellulose, a combination with the hydrogenation or hydrogenolysis poses a promising concept. A significant amount of further valuable platform chemicals, especially in the form of sugar

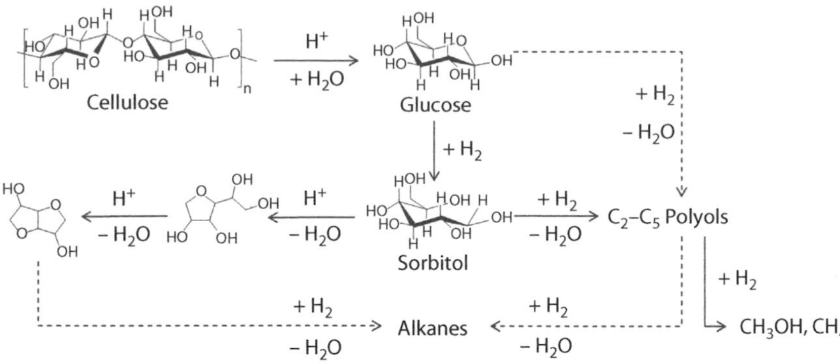

SCHEME 8.3 Possible pathways for the conversion of cellulose to platform chemicals by hydrolytic hydrogenation and hydrogenolysis using multifunctional catalysts.

alcohols, is accessible (Scheme 8.3). Especially reactions under reducing conditions involving hydrogen result in a partial defunctionalization of the highly oxygen-functionalized polysaccharides. Therefore, extensive investigations have been carried out in the last few years. Numerous review papers give a comprehensive overview on the different applied catalytic systems [12,43].

For one-pot processes involving the acid-catalyzed hydrolysis of polysaccharides as well as a further reduction of the intermediates to sugar alcohols the development of bifunctional catalyst materials is crucial. However, hydrogenolysis, i.e., the cleavage of C–O and C–C bonds with hydrogen catalyzed with respective catalyst materials is also a reaction path of great importance. Recently, an extensive and comprehensive review on the catalytic hydrogenolysis of carbohydrates has been published by Ruppert et al. [44]. Since this review also covers all kinds of supported metal-catalysts, this topic is not treated in too great detail within this book chapter. We rather give an overview on more specialized nanostructured catalyst materials and discuss specific investigations on structure–activity relationships. The one pot reaction combining a hydrolytic depolymerization and hydrogenation to C_4–C_6 sugar alcohols such as sorbitol, xylitol, and erythritol, but also C_1–C_3 alcohols, poses a promising approach. Especially the subsequent in situ transformation of glucose to sugar alcohols with a higher chemical and thermal stability prevents the decomposition of glucose to unwanted by-products such as furfural derivatives and humins. It also allows a higher reaction temperature and thus, a faster conversion of the polymeric substrate cellulose.

Few catalyst systems using non-noble metal catalysts have been reported. Ding et al. used a nickel phosphide catalyst that provided acid sites as well as hydrogenation sites (Ni_2P/AC) [45]. At 225°C and 60 bar of hydrogen pressure, sorbitol yields of up to 48% were achieved at 100% cellulose conversion. Also lower sugar alcohols were formed due to hydrogenolysis reactions. With regard to long-term use these catalysts lack stability resulting in leaching of phosphorous species under the relatively harsh synthesis conditions.

Ji et al. investigated the catalytic performance of nickel-promoted tungsten carbide supported on activated carbon ($Ni-W_2C/AC$) depending on different preparation methods [46]. With a post-impregnation in comparison to a co-impregnation method for the introduction of the Ni species, the post-impregnation method did not result in sintering of the supported nanoparticles, thus giving a significantly higher dispersion of the catalytic active species. However, a selective hydrolytic hydrogenation to C_6 sugar alcohols was not possible due to a very high activity as hydrogenation catalysts. Thus, they obtained 100% cellulose conversion with up to 73% yield of ethylene glycol at 245°C and 60 bar of hydrogen pressure.

In the field of noble metal catalysts, especially Ru nanoparticles supported on different materials such as carbon or oxides gained enormous interest due to a high activity and selectivity for the hydrogenation of intermediate sugar compounds [47]. In recent years, the development of bifunctional catalysts was also facilitated. The main aim is the combination of acid sites for the hydrolysis with catalyst species suitable for the hydrogenation of the sugar intermediates to sugar alcohols. Han et al. reported the utilization of porous sulfonated activated carbons as supports for Ru nanoparticles (Figure 8.3) [48]. They obtained a high dispersion of the nanoparticles with a small particle size distribution and a diameter of approximately 10 nm at a loading of 10 wt% via a conventional impregnation route. At 165°C and 50 bar hydrogen pressure, they converted up to 81% cellulose with 59% yield of sorbitol. The difference consisted mainly of C_5 and C_4 sugar alcohols produced by hydrogenolytic C–C bond cleavage. No change in the activity was obvious within five reaction cycles.

Using acidic zeolites as metal supports for the conversion of polysaccharides by hydrolytic hydrogenation was reported and even patented already in the 1980s [49]. Latest investigations in that field by Geboers et al. proved the excellent performance of those hybrid materials [50]. They used a cellulose/catalyst ratio of 2:1 with Ru supported on a H-USY zeolite in an aqueous system. Trace amounts of HCl (35 ppm) are added to facilitate the cellulose depolymerization to cellooligomers while the acidic zeolite is responsible for the further hydrolysis of the soluble intermediates to glucose. Within 24 h at 190°C and 50 bar of hydrogen, 100% conversion of cellulose was achieved with a total yield of hexitols (including sorbitan as a dehydrated species of sorbitol) of 86%.

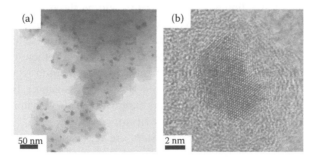

FIGURE 8.3 TEM (a) and HR-TEM (b) images of 10 wt% Ru nanoparticles on a sulfonated activated carbon support. (Reprinted from *Catal. Commun.*, 19, J. W. Han, H. Lee, Direct conversion of cellulose into sorbitol using dual-functionalized catalysts in neutral aqueous solution, 115–118, Copyright 2012, with permission from Elsevier.)

As discussed in the previous chapter, microporous acidic catalysts might not be the ideal choice in terms of nonsoluble cellulose as substrate. In fact, materials with a high external surface area due to small particle sizes are the preferred kind of materials in terms of increased interactions of the solid catalyst with the substrate. In recent years, many investigations have been published utilizing dissolved heteropoly acids (HPA) in combination with supported metals as hydrogenation catalysts [51]. However, by partial substitution of HPA protons by metal ions, low to nonsoluble species still providing Brønsted acidity can be obtained with a high stability in an aqueous environment [52]. Geboers et al. applied nonstoichiometric Cs salts of phosphor- and silicotungstic acid in combination with carbon-supported Ru in the hydrolytic hydrogenation of cellulose [53]. They showed that a postsynthetic hydrotreatment of the Cs-HPAs at 190°C increased activity as well as selectivity by increasing the hydrophobicity of the material. At 170°C and 50-bar hydrogen pressure, they achieved full conversions of cellulose with up to 90% yields of hexitols. The high selectivity to hexitols was ascribed to the suppression of the hydrogenolysis as a most important side-reaction by the Cs-HPAs. Unfortunately, the Cs-HPAs cannot be regarded as purely heterogeneous in their performance due to partial solubility under the applied reaction conditions. However, through a simple recrystallization procedure, the catalysts could be fully recovered and reused.

Liu et al. even went a step further [54]. They used $Cs_3PW_{12}O_{40}$ as support for Ru nanoparticles. The Cs-HPA was added to a colloidal solution of preformed Ru nanoparticles with a mean size of 1.6 nm. The nanoparticles were adsorbed on the surface of the support material. In all cases the Ru content was kept at 1.0 wt%. Under relatively mild conditions of 160°C and 20-bar hydrogen, after 24 h, a total yield of 47% sorbitol and minor amounts of cellobiose, glucose, and mannitol have been observed. Since $Cs_3PW_{12}O_{40}$ does not provide Brønsted acid sites, the catalytic activity in the hydrolytic depolymerization was attributed to the formation of protons by a heterolytic dissociation of hydrogen on the Ru-catalyst. Unfortunately, the catalyst activity decreases successively after each use. Additionally, no information is given on the stability and leaching behavior of the Ru nanoparticles under reaction conditions in the aqueous environment.

The approach of avoiding a solid acid and providing the Brønsted acidity by heterolytic dissociation of hydrogen on a metal catalyst has been known for several years [55] and has been used before in the conversion of cellulose to sugar alcohols. In recent years, especially alumina-supported platinum and ruthenium catalysts have been reported showing this behavior [43b,56]. However, often significantly, lower hexitol yields (<50%) have been observed due to hydrogenolysis side-reactions. Additionally, the often-used alumina lacks support of hydrothermal stability under the applied harsh conditions in aqueous environment.

Ruthenium nanoparticles and nanoclusters are reported to be efficient catalytic materials especially in hydrogenation reactions. Yan et al. showed how highly disperse Ru nanoclusters can be used in the efficient conversion of cellobiose (100%) with a selectivity to the hydrogenation product sorbitol of 100% [57]. The Ru nanoclusters exhibited a diameter of 2.4 ± 0.4 nm and were applied for this catalytic reaction in aqueous cellobiose solution at 120°C and 40 bar of hydrogen pressure.

A very elegant catalyst system was proposed by Zhu et al. using ionic liquid stabilized Ru nanoparticles in the direct conversion of cellulose to sorbitol [58]. The nanoparticles showed a narrow particle size distribution of approximately 4 nm. As a reaction medium, the ionic liquid BMIMCl was used. Additionally, an ionic liquid containing a boronic acid functional group was added as a binding agent. Its function is supposed to be a reversible covalent binding to hydroxyl groups of the cellulose chains, thus disabling hydrogen bond, facilitating dissolution as well as enhancing the contact with the Ru nanoparticles stabilized in the ionic liquid solution. After 5 h under 10 atm of hydrogen pressure and at a very low temperature of 80°C, the pure mixture of BMIMCl and the binding agent converted 95% of the cellulose to 87% glucose while under the same conditions but with the Ru nanoparticles 98% of the cellulose were converted to 89% of sorbitol as the wanted hydrogenation product.

In terms of nanostructured support materials, interesting results have been achieved using carbon nanotubes/nanofibers (CNT/CNF) as catalyst supports. Deng et al. utilized CNTs with 20–60 nm outer and 3–5 nm inner diameter, which they treated in nitric acid to remove amorphous carbon and Ni residues from the synthesis and to obtain hydroxyl and carboxylic acid groups for anchoring metal nanoparticles [59]. Ru nanoparticles with a mean size of 8.7–8.9 nm supported on these CNTs were prepared by a conventional impregnation route resulting in a composite material with a specific surface area of 142 $m^2 g^{-1}$. Using these composite catalysts to convert commercially available microcrystalline cellulose, 40% yield of hexitols, thereof 36% sorbitol, have been achieved after 24 h at 185°C and 50 bar of hydrogen pressure. The advantage of CNTs as carriers is the graphene-like character of the carbon surface exhibiting an increased stability compared with amorphous, disordered carbon materials as well as an improved functionalizability, e.g., by oxidative treatment. Similar results have been reported by Wang et al. [60]. They also used Ru nanoparticles supported on CNTs but at temperatures above 200°C. They observed a significantly increased formation of hydrogenolysis products such as C_2–C_5 sugar alcohols.

Another interesting catalyst concept for the production of sugar alcohols from cellulose was reported by de Vyver et al. in 2010 [61]. They synthesized Ni-containing CNFs (3 wt% Ni) by a catalytic vapor deposition process of methane catalyzed by Ni nanoparticles supported on γ-Al_2O_3 (Figure 8.4), where a lifting of the Ni particles from the original support on the tip of the CNFs occurs. The resulting fibers are of a diameter of 60 ± 40 nm with a mesopore structure due to agglomeration and a specific surface area of 76 $m^2 g^{-1}$. At 60 bar of hydrogen pressure and 210°C after 24 h 87% of microcrystalline cellulose have been converted with sorbitol and mannitol yields of 30% and 5%, respectively. Due to the elevated temperature hydrogenolysis by-products have also been observed. However, in comparison to other supported Ni catalysts, Ni/CNF shows quiet low amounts of these hydrogenolysis by-products and an unexpectedly higher yield of hexitols. The authors claim the higher selectivity as a result of the Ni nanoparticles being reshaped upon the growth of the CNFs (Figure 8.4d), thus being less active for undesired C–C and C–O bond cleavage. However, by such a significant influence of the particle shape, more detailed investigations are necessary to understand the effect of particle shapes in these reaction types to enable tailoring the particle size and shape for a highly selective conversion of cellulose.

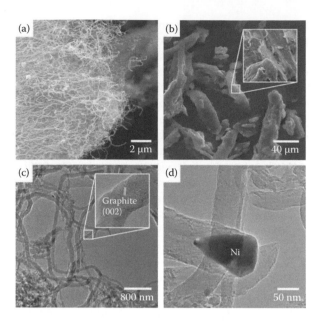

FIGURE 8.4 SEM images of carbon nanofibers (a) and microcrystalline cellulose (b) as well as TEM images of the carbon nanofibers consisting of graphene layers (c) and a pear-shaped Ni particle at the tip of a carbon nanofiber from its synthesis (d). (S. V. de Vyver, J. Geboers, M. Dusselier, H. Schepers, T. Vosch, L. Zhang, G. Van Tendeloo, P. A. Jacobs, B. F. Sels, Selective bifunctional catalytic conversion of cellulose oven reshaped Ni particles at the tip of carbon nanofibers. *ChemSusChem.* 2010, 3, 698–701. Copyright Wiley-VCH Verlag GmbH & Co. KGaA. Reproduced with permission.)

8.3   PLATFORM CHEMICALS TO VALUE-ADDED PRODUCTS

As discussed above, considerable challenges exist in the efficient transformation of cellulose toward suitable platform chemicals. With the ongoing research and advances in the field of cellulose depolymerization, it is expected that these difficulties will be overcome in the near future and allow the efficient products of platform chemicals such as glucose, xylose, and their dehydration products such as hydroxymethylfurfural (HMF) and furfural (FF). The development of processes that utilize these platform chemicals to produce biomass-derived substitutes and analogs for chemical intermediates currently obtained from petrochemical sources is of the utmost importance. It is particularly desirable to develop processes that lead to products that can be directly added to existing infrastructures. In this regard, HMF and FF offer several opportunities for the production of value-added products (Scheme 8.4) [62]. HMF and FF contain a carbonyl function, furan ring, and in the case of HMF, an alcohol group. Via catalytic oxidation and reduction reactions, these can be converted to a wide range of compounds that include mono- and diacids, dicarbonyls, esters, tetrahydrofurans, diols, and anhydrides. These products can in turn be applied as polymer intermediates [63], fine chemicals, pharmaceutical precursors, solvents, and fuel additives [64].

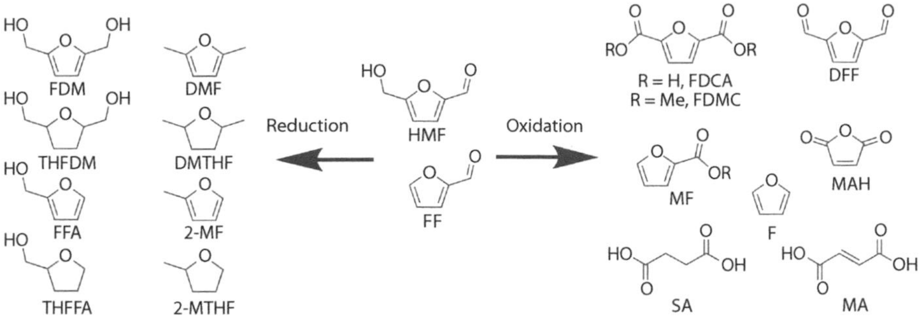

SCHEME 8.4 Value-added products obtainable via catalytic hydrogenation and oxidation reactions of furan and HMF.

In what follows, an overview concerning the recent developments of the preparation of value-added products from furan-based chemicals by application of nano-structured catalysts in oxidation and reduction reactions will be discussed.

8.3.1 Oxidative Transformation of Platform Chemicals by Solid Catalysts

A survey on the potential uses of biomass-based platform chemicals by the U.S. Department of Energy has identified 2,5-furandicarboxylic acid (FDCA) as one of the twelve chemicals that can be used as chemical building blocks in the future [65]. Obtainable by the full oxidation of HMF, FDCA constitutes a suitable substitute for terephthalic acid [63], which is used in the large-scale synthesis of poly(ethylene terephthalate) (PET) and other materials and is currently obtained from petrochemical resources. Besides this, other applications of FDCA include its use in the manufacture of polyureas [66] and polyester polyols [67]. The preparation of FDCA from HMF can be achieved using stoichiometric oxidants (e.g., $KMnO_4$) [68] or homogeneous catalysts (e.g., Co/Mn/Br) [69]. However, from a green chemistry perspective, the use of molecular oxygen in combination with recoverable catalysts is clearly preferred [1b,70]. As shown in Scheme 8.5, several different products can be obtained from the catalytic oxidation of HMF. Oxidation of the aldehyde function to carboxylic acid

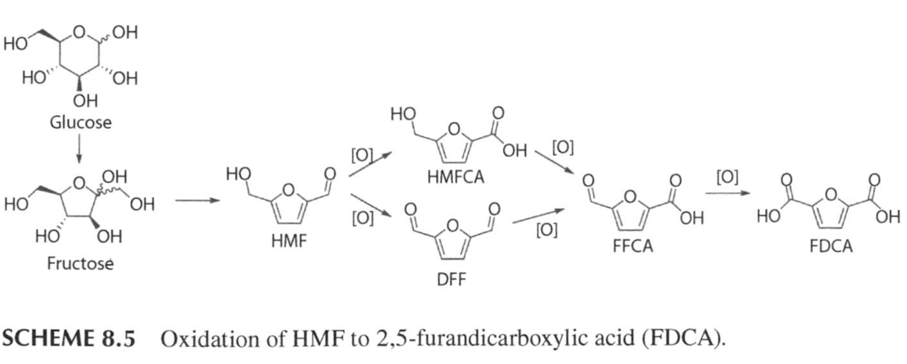

SCHEME 8.5 Oxidation of HMF to 2,5-furandicarboxylic acid (FDCA).

yields 5-hydroxymethylfuran-2-carboxylic acid (HMFCA), whereas oxidation of the alcohol results in 2,5-diformylfuran (DFF). Further oxidation of these products will initially result in 5-formylfuran-2-carboxylic acid (FFCA) and ultimately furan-2,5-di-carboxylic acid (FDCA). Because FDCA is considered a highly suitable alternative for terephthalic acid, the total oxidation of HMF to FDCA has been most widely studied.

8.3.1.1   Aqueous Oxidation of HMF

Early reports on the oxidation of HMF to FDCA describe the use of supported noble metal catalysts such as Pt/C, [71] Pt/Al$_2$O$_3$, [72] Pt/ZrO$_2$, [73] and Pt-Pb/C [74] under oxygen pressure in aqueous medium. The addition of bases such as NaOH, KOH, or Na$_2$CO$_3$ prior or during the reaction was found to be required to steer the reaction to completion, i.e., when omitted from the reaction the mixture becomes increasingly acidic and DFF and HMFCA are preferentially formed. More recent reports describe the aqueous, aerobic gold-catalyzed conversion of HMF to FDCA. The applied catalysts consist mainly of gold nanoparticles supported on various metal oxide supports. As with the platinum catalysts, the presence of stoichiometric amounts of base was required to reach high conversion and selectivity to FDCA. Riisager et al. reported the use of 1 wt% Au/TiO$_2$ (4–8 nm particles, BET: 49 m^2 g^{-1}) in H$_2$O with 20 eq NaOH at 30°C under 20–30 bar O$_2$ [75]. Analysis of the reaction with HPLC showed that HMF is quickly converted to HMFCA in yields up to 70%. As the reaction progresses, HMFCA is slowly converted to FDCA with selectivities up to 75%, leaving 25% HMFCA as the only by-product. DFF was not observed in these reactions, which shows that the oxidation of the carbonyl group to carboxylic acid is faster than oxidation of the hydroxyl group. The amount of added NaOH was varied, and the results showed that selectivity and conversion are significantly improved with 5 eq giving an optimal result. Interestingly, when no base was added, conversion only amounted to 13% giving 12% HMFCA. It was proposed that the precipitation and/or deposition of FDCA, which is largely insoluble in water, may lead to deactivation or inhibition of the catalyst. Reuse of the catalyst was attempted; however, activity was reduced 10%–15%, which could attribute to leaching. Corma et al. compared the influence of the support on conversion and selectivity [76]. Reactions were run at 65°C with 4 eq NaOH under 10 bar air. At a HMF/Au molar ratio of 150, 2.6 wt% Au/CeO$_2$ (3.5 nm particles, BET: 180 m^2 g^{-1}) prepared by co-precipitation and a commercial 1 wt% Au/TiO$_2$, both gave full conversion with >99% selectivity after 8 h. In contrast, Au/Fe$_2$O$_3$ (4.6 wt%) and Au/C (1.5 wt%, 3.5 nm average particle size) were much slower, respectively, resulted in 45% and 78% FDCA after 24 h with HMFCA as the only side-product. Comparison of the Au/TiO$_2$ and Au/CeO$_2$ catalysts at 130°C (HMF/Au = 640) showed that CeO$_2$ gives significantly fewer by-products (e.g., decarboxylation, ring opening) than TiO$_2$. A hot filtration test showed that the catalyst can be completely removed from the product stream. Metal leaching was found not to occur; however, reuse proved ineffective. It was proposed that poisoning of the catalyst leads to suppression of the aldehyde oxidation activity and side-product formation. Nevertheless when the reaction was carried out in a two-step protocol, switching between 25°C and 130°C, activity could be maintained for 4 cycles before the eventual loss of activity. Davis et al. compared the activity of supported Pt, Pd, and Au catalysts in the aqueous aerobic oxidation HMF [77]. Reactions were run at 22°C, with 2 eq NaOH under 6.9 bar O$_2$. Over Au/C and Au/TiO$_2$, HMFCA was formed in high yield, but did not continue to react substantially to FDCA, whereas over Pt/C and Pd/C FDCA was formed under identical conditions. For Pt/C, up to 80% FDCA was obtained after 6 h. It was also demonstrated that by increasing the O_2 pressure and base concentration FDCA could also be obtained for the supported Au catalysts.

In an attempt to avoid the use of stoichiometric base, Ebitani et al. reported the use of hydrotalcite materials as recoverable catalyst support and heterogeneous base [78]. A 1.9 wt% Au/HT catalyst was prepared by deposition precipitation using aqueous NH_3 followed by calcination at 200°C. TEM analysis of the catalyst revealed a narrow particle size distribution with an average particle size of 3.2 nm (Figure 8.5). The oxidation was performed at 95°C with a HMF/Au ratio of 40 and under O_2 flow at atmospheric pressure. After 7 h, near full conversion with more than 99% selectivity was obtained. Like in previous studies, the kinetic plot shows a rapid buildup of HMFCA, which is subsequently converted to FDCA. FFCA was also established as a reaction intermediate in this case. Hot filtration of the catalyst confirmed heterogeneity and ICP showed no leaching of gold into the product solution. The catalyst could be reused up to three times, after which activity dropped.

Cavania et al. compared the HMF oxidation activity of bimetallic Au-Cu/TiO₂ (1:1) catalysts with their monometallic counterparts Au/TiO₂ and Cu/TiO₂ [79]. Catalysts with 0.5–2 wt% metal(alloy) on TiO₂ (BET: 60–90 m² g⁻¹) were prepared by the deposition of PVP-stabilized nanoparticles. Average particle sizes of 6–8 nm were obtained from HAADF images and XEDS spectra of single Au–Cu particles confirmed formation of 1:1 alloys (Figure 8.6). Reactions were run for 2 h at 60°C with 4 eq NaOH, with an HMF/Au ratio of 100 under 10 bar O_2. In the absence of metal, full degradation of HMF to levulinic and formic acid occurred, and the Cu-loaded catalyst showed no oxidation activity. The Au based catalysts gave 10%–15% yield FDCA, independent of the metal loading (0.5–2 wt%). In contrast, the bimetallic Au–Cu catalysts gave 20%–35% yield, demonstrating the promoting effect of Cu on Au. Further optimization of the reaction conditions resulted in the selective production of FDCA with up to 90%–99% yield without signification formation of by-products. The Au–Cu catalysts showed excellent reusability as conversion and selectivity were maintained for 5 runs, whereas for the Au catalysts, selectivity dropped significantly after the second run.

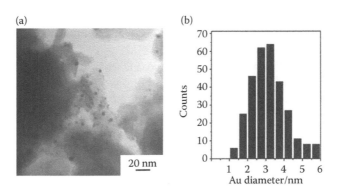

FIGURE 8.5 TEM and particle size distribution of Au/HT. (From N. K. Gupta, S. Nishimura, A. Takagaki, K. Ebitani, *Green Chem.* 2011, *13*, 824–827.)

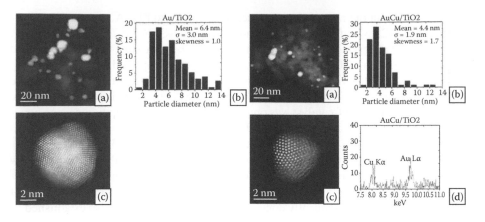

FIGURE 8.6 Low and high magnification HAADF images and particle size distribution of monometallic Au/TiO$_2$ (left) and bimetallic Au-Cu/TiO$_2$ catalysts (right, with XEDS spectrum). (From T. Pasini, M. Piccinini, M. Blosi, R. Bonelli, S. Albonetti, N. Dimitratos, J. A. Lopez-Sanchez, M. Sankar, Q. He, C. J. Kiely, G. J. Hutchings, F. Cavania, *Green Chem.* 2011, *13*, 2091–2099.)

In studying the role of heterogeneous bases in Au-catalyzed oxidation reactions, Davis et al. used Au/TiO$_2$ catalysts and added HT separately to the reaction mixture [80]. Using the same conditions as reported by Ebitani et al., similar conversions and selectivities to FDCA were found as for the Au/HT system. However, ICP analysis of the resulting product solution showed that an equimolar amount of Mg^{2+} was leached into the solution, providing strong evidence that the HT is consumed by the formed FDCA. Interestingly the amount of Al in solution was very low, which suggests that the HT is converted to an Al(OH)$_3$-based material. The same group also reported a detailed mechanistic study regarding the role of O$_2$ in the Au- and Pt-catalyzed oxidation reactions of HMF to FDCA [81]. Using ^{18}O-labeled O$_2$ and water, it could be established that the oxygen incorporated in the acid originates from solution rather than O$_2$. Based on this, it was proposed that O$_2$ acts as an electronic scavenger to regenerate the metal surface.

Besides platinum- and gold-based catalysts, ruthenium-based catalysts have also been investigated in the aerobic oxidation of HMF to FDCA. The group of Riisager immobilized Ru(OH)$_x$ on a variety of metal oxide supports with a wide range of surface areas (5–145 m^2 g^{-1}) [82]. Catalysts with 2.4 wt% Ru were prepared by NaOH induced precipitation of Ru(OH)$_x$ from RuCl$_2$. Oxidation reactions were carried out in water at 140°C under 1–40 bar O$_2$ pressure without added base. The concentration of HMF was kept low to ensure full dissolution of the formed FDCA. At low O$_2$ pressures, the Ru/TiO$_2$-catalyzed oxidation was very slow, and HMF was rather converted to formic acid and levulinic acid. Increasing the oxygen pressure to 2.5 and 20 bar improved selectivity to the oxidation products (i.e., HMFCA, DFF, and FDCA) and decreased the degradation of HMF. The influence of the catalyst support on selectivity was studied at 2.5 bar O$_2$. Basic materials such as MgO, MgO·La$_2$O$_3$, and HT gave the best selectivities and conversions to FDCA. However, pH measurements and ICP analysis of these product solutions revealed that the basic support dissolved to different extents during the reaction. In contrast, CeO_2, Fe_2O_3, $MgAlO_4$, and hydroxyapatite supported catalysts remained stable and gave acidic product solutions. This was further demonstrated in a follow-up study on Mg-based supports, which showed that $MgAlO_4$ constitutes a stable and reusable catalyst support in this reaction [83]. Although conversions were somewhat lower, the selectivity to oxidation products was very high and the degradation of HMF to formic acid was low. In case of Fe_2O_3 and hydroxyapatite, insoluble humins were also formed with selectivities up to 30%. Following this screening, CeO_2 was identified as the most promising support for the Ru-catalyzed oxidation of HMF. Reuse of the catalyst was possible for three times without significant loss of activity or selectivity.

Recently, Bhaumik et al. reported the first example of an Fe(III) catalyzed aerobic oxidation of HMF to FDCA[84] using porphyrin-based porous organic polymer supports [85]. In contrast to the noble metal-based systems discussed above, no base was added to the reactions. Reactions were carried out in water at 100°C under 1 atm O_2. A kinetic study of the reaction showed that for this system the oxidation of the alcohol function proceeds more readily as DFF and FFCA were identified as the main intermediate products of the reaction. At full conversion a FDCA selectivity of 75% was obtained with 9% DFF and 5% FFCA as the main by-products. These results constitute a clear improvement over the gold based catalysts as high selectivity and conversion are obtained under base free conditions using a non-noble metal.

8.3.1.2 In Situ Oxidation of HMF

When considering the potential process designs for the dehydration of carbohydrates to HMF and subsequent oxidation to FDCA, the use of aqueous systems for both reactions is highly preferable due to the benign nature of the solvent. Hence, most of the reports on the oxidation of HMF use water as solvent or cosolvent. If the preceding dehydration of fructose or glucose to HMF is also performed in water, these reactions can potentially be coupled together or carried out as a one-pot reaction. Vorlop et al. reported the direct dehydration and oxidation in a one-pot two-phase system [86]. A solid acid catalyst from Lewatit (SPC 108) was selected for the dehydration of fructose in water and a bimetallic PtBi/C (5%/5%) for the oxidation of HMF in MIBK. When performed separately, the dehydration of fructose at 80°C led to a maximum yield of 12% HMF and a final yield of 79% levulinic acid. Thus, if the in situ oxidation reaction is sufficiently fast, a maximum yield of 79% of FDCA could be attainable. The oxidation reactions that were performed at 80°C using O_2 from bubbling air at 1 atm showed that the presence of small amounts of water in MIBK is beneficial for the extent of oxidation, and a total yield of 75% oxidation products (FDCA 10%, FFCA, DFF) was obtained after 70 h. The coupled reactions were carried out in two different setups (Figure 8.7). In one case, a PTFE membrane (pore size 0.45 μm) was used to keep the catalysts separated but allowing solutes to pass through. The results of the membrane reactor demonstrated that the reactions can be successfully coupled as 25% FDCA was obtained with total selectivity of 32% toward oxidation products. In the other setup, no membrane was present and only water was used as solvent. To prohibit oxidation of fructose, the PtBi/C catalyst was encapsulated in MIBK swollen silicone beads with diameters

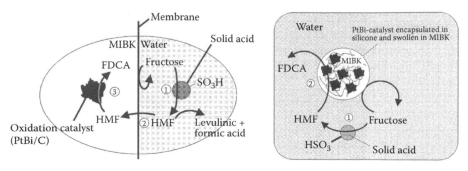

FIGURE 8.7   Schematic representation of the dehydration and oxidation processes occurring in the membrane reactor (left) and the batch reactor (right). (With kind permission from Springer Science+Business Media: *Top. Catal.,* A new approach for the production of 2, 5-furandicarboxylic acid by in situ oxidation of 5-hydroxymethylfurfural starting from fructose, 13, 2000, 237–242, M. Kröger, U. Prüße, K.-D. Vorlop.)

ranging from 63 μm to 2 mm. As such the oxidation catalyst remains submersed in a supported liquid phase (SLP) of MIBK, which selectively dissolves HMF but not fructose. Like in the case of the membrane reactor, FDCA was obtained in 25% yield with a maximum total selectivity of 50% to oxidation products. These data provide a strong proof of concept and show that the dehydration of fructose and in situ oxidation of HMF can potentially be realized. Ribeiro and Schuchardt also reported on the one-pot conversion of fructose to FDCA [87]. However, instead of using two separate catalysts, the authors proposed cobalt containing SiO_2 gels prepared by coprecipitation as a bifunctional catalyst for the dehydration and oxidation reactions. Comparison of the dehydration activities of SiO_2, Co(acac)$_3$ and Co/SiO$_2$ in water after 8 h at 88°C showed that Co/SiO$_2$ (46% HMF) exhibits equal activity as SiO_2 (47%), whereas the activity of Co(acac)$_3$ was much less (13%). Interestingly, the authors note that the selectivity to HMF was very high in these reactions as no levulinic acid or formic acid was formed. For the one-pot conversion of fructose to FDCA, the reaction was performed in water at 160°C under 20 bar synthetic air for 65 min. A maximum yield of 72% FDCA was reported for the Co/SiO$_2$, demonstrating the high feasibility of the approach.

Recently, Dumesic et al. also reported a study aimed at integrating the various catalytic steps commonly associated with the biomass refinery scheme using biomass-based solvents such as γ-valerolactone (GVL) and 2-methyltetrahydrofuran (MTHF) as novel reaction media [88]. HMF solutions obtained from these reactions were further converted to FDCA via Au/TiO$_2$ catalyzed oxidation. Reactions were carried out using optimized conditions for Au catalysis (NaOH, 20 bar O$_2$, 22°C) and resulted in overall yields of FDCA of 35%–38%, relative to glucose.

8.3.1.3 Nonaqueous Oxidation of HMF

Although performing follow-up reactions in an aqueous medium is highly preferred, many studies have shown that the dehydration of carbohydrates in water is generally low yielding due to side reactions such as levulinic acid and humin formation.

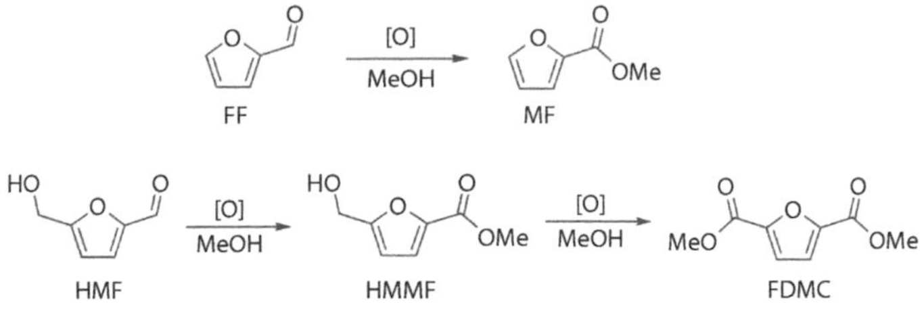

SCHEME 8.6 Oxidation of FF and HMF in methanol.

Therefore numerous studies focus on the employment of nonaqueous systems (e.g., ionic liquids, DMSO, alcohols, etc.) to improve selectivity. When the oxidations of FF or HMF are performed in alcohols, the corresponding mono- and diesters of the alcohol solvent are obtained (Scheme 8.6). Via a simple hydrolysis step, the corresponding acids can be obtained, thus allowing application for the same purposes as mentioned above. In case of furfural oxidation, the corresponding alkylfuroates are obtained, which are applied as flavors and fragrances but can also be used for fine chemical syntheses.

Christensen et al. reported the Au/TiO_2 catalyzed oxidation of FF and HMF in methanol [89]. In contrast to the aqueous oxidation where the formed acids drastically lower the pH of the reaction, oxidation in methanol yields methylesters, which do not influence pH, negating the need for stoichiometric bases. Still, because bases increase the oxidation performance of the catalyst a catalytic amount NaOMe base (8% w.r.t HMF) was added to the reaction. After 3 h at 130°C under 4 bar O_2, 98% dimethyl furan-2,5-carboxylate (FDMC) was obtained at full HMF conversion. As with the aqueous gold-catalyzed reactions, oxidation of the aldehyde group is considerably faster and (hydroxymethyl)methylfuroate (HMMF) was identified as the main intermediate product. In fact, when the oxidation was carried out for 3 h at 22°C under 1 bar O_2, HMMF could be selectively obtained in yields up to 95%. Under the same conditions, FF was fully converted to methylfuroate (MF) after 10–12 h.

Corma et al. studied the same reactions using 2.1 wt% Au/CeO_2 catalysts (3.5 nm Au particles, BET: 180 m^2 g^{-1}) in methanol without any added base [90]. Kinetic plots of the reaction run at 130°C, under 10 bar N_2 with O_2 bubbled through the solution, showed several intermediate reaction products. Besides the main reaction products HMMF and FDMC, minor amounts of methyl acetals were also detected. After 4–5 h, near full conversion of HMF is obtained with 99% selectivity to FDMC. Under the same conditions, FF was oxidized to MF with 91% selectivity after only 2 h. The influence of the support was studied by comparing the performance of Au catalysts supported on Fe_2O_3, TiO_2, CeO_2, and C. The results show that Au/TiO_2 and Au/CeO_2 are most selective for the oxidation reaction, with Au/CeO_2 giving higher conversions and increased selectivity in shorter reaction time than Au/TiO_2. Interestingly, the dispersion of the support was found to have a significant effect on the catalytic activity. When bulk CeO_2 was used instead of nanoparticulate CeO_2, oxidation proceeded much slower and 72 h were required to reach 85% selectivity to FDMC. Reuse of the catalyst was studied for five consecutive runs, with vacuum filtration, washing with methanol, and drying performed after each reaction. It was found that after the first use, the activity dropped considerably as 10 h were required to reach full conversion in the second run as opposed to 5 h for the first run. Nevertheless, after this initial drop in activity, the activity remained constant for three additional runs. Leaching of the metal into solution was found not to occur to a noticeable degree. Importantly, the authors also noted that the oxidation of methanol occurs to a significant extent, leading to the formation of side-products such as methylformate, formaldehyde dimethylacetal, and CO_2. For longer-chain alcohols, similar products were found as well as inhibition of the desired reaction. Consequently, longer reaction times were required to reach full conversion. The presence of water (10%–20%) was also found to decrease activity but did not lead to the formation of FDCA.

Boccuzi et al. also studied the base-free, aerobic gold-catalyzed oxidation of FF to MF using zirconia supported catalysts [91]. Catalysts containing 1.5 wt% Au were prepared by base-induced deposition–precipitation of $HAuCl_3$. After filtration and drying, the catalysts were calcined at different temperatures between 150°C and 600°C to study the influence of the pretreatment. Reactions were carried out at 120°C under 6 bar O_2. The results showed that activity increased with increasing calcination temperature up to 500°C. However, at higher temperatures (600°C–650°C) the activity dropped considerably. HRTEM data of these catalysts showed that with increasing calcination temperature the particle size distribution shifted to larger sizes due to enhanced sintering (Figure 8.8). Correlation of this data with the catalytic data suggested that particles larger than 3 nm are considerably less active. Reuse of the catalyst after washing and drying was attempted, but both activity and selectivity were significantly decreased. Activity and selectivity could, however, be restored by performing an additional oxidation step after the recovery procedure. Analysis of the spent catalyst showed that the particle distribution is stable to both catalysis and the recovery procedure. Additional TPO data of the spent catalyst indicated the presence of adsorbed species. Based on this, the authors proposed that fouling of the catalyst occurs and results in decreased activity and selectivity.

Abu-Omar et al. investigated the aerobic oxidation of HMF to FDCA with a homogeneous Co/Zn/Br-based catalyst and with heterogeneous Au-based catalysts in acetic acid [92]. After 3.5–4 h at 90°C under 1 atm O_2, the homogeneously catalyzed reaction selectively gave DFF in 96% yield without formation of FFCA or FDCA. In contrast, with 5 mol% trifluoroacetic acid (TFA) in the reaction, the product mixture contained DFF, FFCA, and FDCA in approximately equal amounts. This indicates a shift toward the formation of higher oxidized products. Optimization of the homogeneous system gave 60% of FDCA and 29% FFCA at 90% conversion. For comparison, the activity of 1.9 wt% Au/TiO_2 (285 m^2 g^{-1}) and 1.4 wt% Au/CeO_2 (72 m^2 g^{-1}) catalysts were tested in acetic acid for 3 h at 130°C under 10 bar O_2. It was found that also for the Au-based catalysts, the addition of TFA improved the conversion. However, contrary to the homogeneous system, the reaction primarily gave FFCA in yields up to 80%.

Riisager et al. studied the potential feasibility of conducting HMF oxidation in ionic liquids using supported Ru based catalysts [93]. Various catalyst supports and ionic liquids were screened at 100°C–140°C and 1–30 bar O_2 to access the optimal

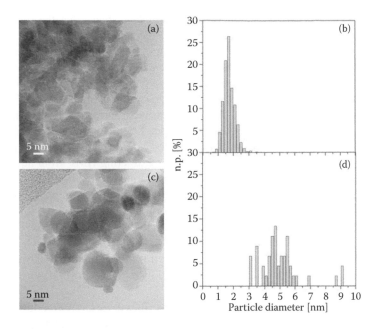

FIGURE 8.8   HRTEM images and particle size distributions of fresh Au/ZrO₂ (a, b) and Au/ ZrO₂ calcined at 600°C (c, d). (Reprinted from *Catal. Today,* The effects of gold nanosize for the exploitation of furfural by selective oxidation, 203, F. Pinna, A. Olivo, V. Trevisan, F. Menegazzo, M. Signoretto, M. Manzoli, B. Bocuzzi, 196–201, Copyright 2013, with permission from Elsevier.)

conditions required for successful oxidation. Full conversion of HMF was easily achieved; however, the selectivity to oxidation products was generally low, indicating the instability of HMF under these conditions. The best results were obtained for Ru/La₂O₃ in 1-ethyl-3-methylimidazolium acetate [EMIm]OAc, giving FDCA yields up to 48% and HMFCA yields up to 56%. A hot filtration test showed that up to 10% Ru leached into the ionic liquid. It appears therefore that further studies are required to improve both selectivity of the reaction and reusability of the catalyst. Nevertheless, these results demonstrate that the aerobic oxidation of HMF can be performed heterogeneously in ionic liquids.

8.3.1.4   Selective Oxidation of HMF

Next to the full oxidation of HMF to FDCA, the selective oxidation of HMF to 2,5-diformylfuran (DFF) has also been extensively studied. Like FDCA, DFF has a high potential for use as a building block in chemicals production. There have been numerous reports describing various useful applications [94] of 2,5-diformylfuran as a monomer for the preparation of aldehyde/urea-based resins [95] and for fine chemical synthesis [62a,63,96].

Several kinetic studies on the gold-catalyzed oxidation of HMF to FDCA revealed that HMFCA is more readily obtained as an intermediate oxidation product, suggesting that oxidation of the carbonyl function is more easy. However, in case of ruthenium and platinum-catalyzed reactions, both DFF and HMCA were present as intermediates, often in comparable amounts. This shows that the type of metal used as catalyst can have a large influence on the selectivity of the oxidation reaction. Nanostructured vanadium-based catalysts exhibit high selectivity in oxidation reactions and prevent overoxidation as is typically observed for other systems [97]. Therefore, these catalysts are commonly employed for the selective oxidation of HMF to DFF. For example, Tichit et al. reported on the oxidation of HMF to DFF catalyzed by supported V_2O_5/TiO_2 catalysts [98]. Reactions were performed in a batch reactor at 90°C, in toluene or methyl isobutyl ketone under 1.6 bar air and resulted in full conversion to DFF. Higher TOFs were obtained for MIBK, which is particularly useful due to the fact that MIBK is a good solvent for extracting HMF from aqueous solutions. This potentially allows coupling the dehydration and oxidation.

Grushin et al. screened the activity of a number of vanadium oxide–based catalysts for DFF synthesis via the aerobic oxidation of HMF in DMSO obtained from the dehydration of fructose [99]. Starting from HMF, DFF yields of 49%–61% were obtained for structured catalysts such as V_2O_5, $(VO)_2(P_2O_7)$, γ-$VOPO_4$, δ-$VOPO_4$, $VOHPO_4 \cdot H_2O$ and $VO(PO_3)_2$. Using a two-step protocol involving HMF synthesis from fructose with the same catalysts followed by oxidation, overall yields of 36%–45% DFF were obtained from fructose.

A similar study reported by Sbrana et al. describes the selective oxidation of HMF to DFF catalyzed by heterogeneous $VOPO_4 \cdot 2H_2O$-based catalysts [100]. However, in this case, attempts to synthesize DFF from fructose via dehydration and subsequent oxidation in water and water/MIBK solutions were unsuccessful. The best results for the oxidation of HMF to DFF were obtained in N,N-dimethylformamide (DMF) at 100°C and 1 atm air.

Corma et al. compared the activity of molecular Cu-, Pd-, and V-based catalysts immobilized on poly(vinylpyridine) (PVP) crosslinked with divinylbenzene and pyridine functionalized SBA-15 (Figure 8.9) [101]. Reactions were performed in DMSO and toluene at 130°C and 10 bar O_2. The highest conversions and selectivities to DFF were observed for the Cu/PVP and V/SBA catalysts in DMSO. In case of toluene, oxidation of the solvent occurred and led to lower conversions and selectivities. When trifluorotoluene was used as alternative, the reaction proceeded much more selectively giving 60% conversion with 99% selectivity to DFF.

Riisager et al. reported the use of acidic zeolites (e.g., H-beta, H-Y, H-mordenite, H-ZSM-5) impregnated with vanadia as catalysts for the selective oxidation of HMF to DFF [102]. Reactions were carried out in dimethylformamide at 100°C and 1 bar O_2. Under these conditions the catalysts gave 7%–19% DFF selectivity at 32%–60% conversion. In DMSO, selectivity was much higher and an optimal yield of 81% DFF was obtained at 92% conversion using 10 bar O_2. Catalyst reuse and hot filtration showed that considerable leaching of the supported vanadia occurs.

Next to these studies on vanadia-based catalysts, Ebitani et al. reported on the selective oxidation of HMF to DFF by Ru/HT catalysts [103]. Comparison of various supported Ru catalysts in dimethylformamide at 120°C, 1 atm, O_2 flow, showed that hydrotalcite-supported catalysts gave much higher conversion (92%) and selectivity (97%) to DFF than on Al_2O_3, $Mg(OH)_2$, and C. Furthermore, the authors demonstrated that DFF can be obtained from glucose via the stepwise addition of HT (for glucose isomerization to fructose), Amberlyst-15 (for the dehydration of fructose to HMF), and

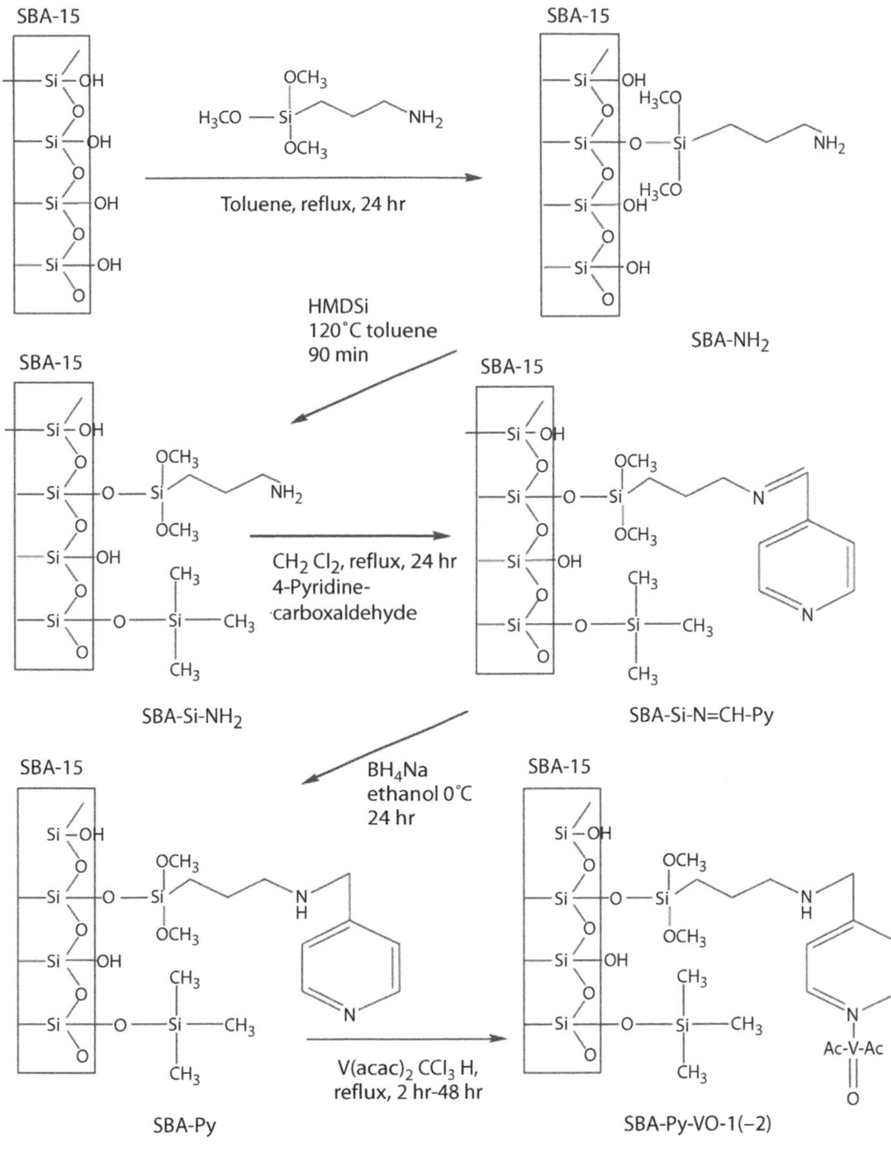

FIGURE 8.9 Preparation of pyridine modified SBA-15. (With kind permission from Springer Science+Business Media: *Top. Catal.*, Chemicals from Biomass: Aerobic Oxidation of 5-Hydroxymethyl-2-Furaldehyde into Diformylfurane Catalyzed by Immobilized Vanadyl-Pyridine Complexes on Polymeric and Organofunctionalized Mesoporous Supports 52, 2009, 304–314, O. Casanova, A. Corma, S. Iborra.)

Ru/HT (for the oxidation of HMF to DFF) (Figure 8.10). An overall yield of 49% DFF was obtained starting from fructose, whereas 25% was obtained starting from glucose. These studies provided a clear proof of principle for the viability of a multicompartment continuous flow process using basic, acidic, and oxidation catalysts in tandem.

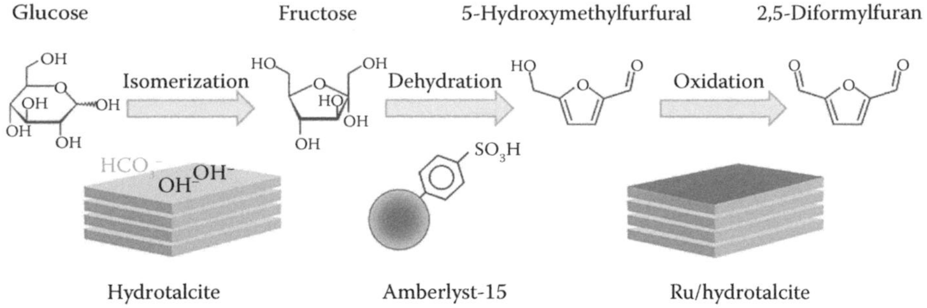

Glucose   Fructose   5-Hydroxymethylfurfural   2,5-Diformylfuran

Hydrotalcite   Amberlyst-15   Ru/hydrotalcite

FIGURE 8.10   **(See color insert.)** Reaction sequence to obtain DFF from glucose using different nano-structured catalysts. (Reprinted with permission from A. Takagaki, M. Takahashi, S. Nishimura, K. Ebitani, *ACS Catal.* 2011, 1, 1562–1565. Copyright 2011 American Chemical Society.)

8.3.1.5   Oxidation of Furfural

In analogy to the oxidation of HMF to FDCA, oxidation of furfural will yield furoic acid. Although examples of homogeneous and stoichiometric oxidation of FF have been described [104], heterogeneously catalyzed oxidation of FF remains relatively undocumented. Sha et al. proposed Ag_2O and CuO catalysts for the aqueous aerobic oxidation of FF to furoic acid [105]. As with FDCA, the presence of excess base was required to stabilize the catalyst and allow full conversion. It was shown that the yield strongly depended on temperature, pH, and catalyst amount. At 70°C and pH 13, an optimal yield of 92% was found for an Ag_2O/CuO catalyst.

Interestingly, when oxidation is carried out with hydrogen peroxide instead of molecular oxygen several new products are obtained. For example, Ebitani et al. reported the Amberlyst-15 catalyzed oxidation of FF to succinic acid using H_2O_2 as oxidant [106]. Reactions were performed for 24 h in 1.33 M H_2O_2 at 80°C. A screening using various homogeneous acids, metal oxides (e.g., Ni_2O_5, (sulphated) ZrO_2, ZSM-5), and acidic resins revealed that Amberlyst-15 selectively catalyzed the formation of succinic acid (74%) with 12% maleic acid as the major side-product. This shows that besides oxidation of the carbonyl function, decarboxylation and ring opening reactions can also be used to broaden the product spectrum obtainable from biogenic feeds such as FF (Scheme 8.7).

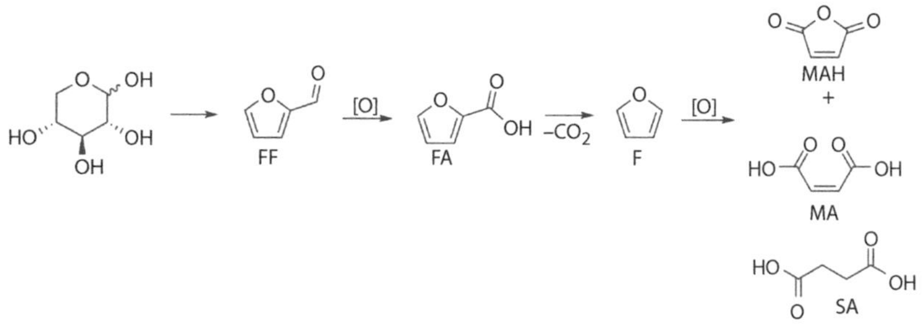

SCHEME 8.7   Oxidation of furfural.

This is further exemplified by the results of Ojeda et al. who reported on the gas phase oxidation of FF to maleic anhydride (MAH) catalyzed by VO_x/Al_2O_3 [107]. Catalysts with varying VO_x surface densities (at_V/nm^2) were prepared by incipient wetness impregnation of NH_3VO_3 on γ-alumina (134 m^2 g^{-1}) and calcination at 500°C. The resulting catalysts were screened at 280°C with a furfural/O_2 (1.6 kPa furfural, 2.5 kPa O_2) gas mixture and optimal conversion and selectivity were obtained for a catalyst with VO_x surface density of 10 at_V/nm^2. Further optimization of temperature (320°C) and O_2 pressure (5.7 kPa) resulted in 73% maleic anhydride yield at full conversion. The use of lower temperatures and/or O_2 pressures resulted in the additional formation of furan (F) with a maximum yield of 9%. Based on kinetic data of maleic anhydride and furan formation, it was proposed that Al_2O_3-supported polyvanadates are intrinsically more active than monovanadates (VO_4) and V_2O_5 crystals.

8.3.2 REDUCTIVE TRANSFORMATION OF PLATFORM CHEMICALS BY SOLID CATALYSTS

As indicated above, the reductive transformation of HMF and FF can give rise to a variety of interesting value-added products. In case of furfural (FF), selective hydrogenation of the aldehyde function results in the formation of furfuryl alcohol (FFA), which is an important intermediate in the manufacturing of lysine, vitamin C, plastics, and many furan-type resins such as various synthetic fibers and rubbers. It can also be used as an environmentally benign solvent for pigments or phenolic resins and as lubricant or dispersing agent. Common methods for the production of FFA from FF are the catalytic gas phase and liquid phase hydrogenation reaction. The resulting product spectrum largely depends on the employed catalysts and conditions. As shown in Scheme 8.8, further hydrogenation of the furan ring can result in 2-(hydroxymethyl)tetrahydrofuran (HMTHF). Hydrogenolysis of the alcohol functions of FFA and HMTHF can also lead to products such as 2-methylfuran (2-MF) and 2-methyltetrahydrofuran (2-MTHF). Besides this, decarbonylation of FF may result in the formation of furan (F), which can further react to tetrahydrofuran (THF). The development of catalysts, which exhibit high selectivities to different products is particularly useful when considering different applications starting from FF.

8.3.2.1 Gas Phase Hydrogenation of Furfural

Vannice et al. reported on the gas phase oxidation of FF to FFA over a commercial copper chromite catalyst from Engelhard (Cu-1800P) [108]. Reduction of the catalyst

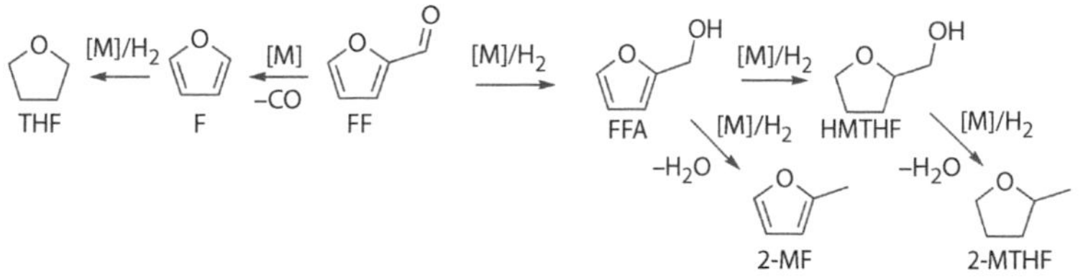

SCHEME 8.8 Reduction of furfural.

in H_2 atmosphere at 200°C–400°C was performed prior to the reaction to obtain active surface species such as Cu^+ (from $CuCrO_2$ phases) and Cu^0 (metallic phases). Hydrogenation of FF at 140°C (13 kPa H_2, 1.3 kPa FF) gave an optimal FFA selectivity of 70% at 55% conversion with 2-MF arising as the only side-product. However, continued operation showed that the catalyst slowly deactivates over time. The same group also reported the use of copper catalysts supported on active carbon, graphitic fibers, and synthetic diamond powder [109]. The active carbon support was pretreated with nitric acid to increase the concentration of oxygen-containing surface groups. 5 wt% Cu/C catalysts were prepared by wet impregnation with $Cu(NO_3)$ and reduction at 300°C–400°C. Hydrogenation of FF at 100°C–200°C showed that these catalysts exhibit high selectivity (>90%) at low conversions (<20%); however, at increased conversion, FFA selectivity drops considerably and 2-MF is also formed. Continued operation showed that the active carbon supported catalyst exhibits high stability (i.e., activity could be maintained for more than 10 h), whereas for the other catalyst severe deactivation led to complete loss of activity after 4 h.

Mikolajska et al. reported the preparation of Pt-based catalysts supported on SiO_2, γ-Al_2O_3, MgO, and TiO_2 materials, which are covered with a monolayer of TiO_2, V_2O_5, or ZrO_2 [110]. Figure 8.11 illustrates the formation of a TiO_2 monolayer on MgO by a surface reaction with $Ti(O^iPr)_4$. Subsequent heating of the material releases propene, thereby generating the active monolayer surface. As a result of a strong oxide–oxide interaction (SOOI), the reactivity and chemical properties of the deposited monolayers are considerably different than the corresponding bulk material. The Pt-loaded materials were obtained by dry impregnation with H_2PtCl_6, calcination at 550°C in dry air and dry N_2. Reduction at 350°C was carried out prior to the reaction, which was performed at 150°C–300°C under atmospheric pressure using an H_2/FF molar ratio of 2.

Compared with the catalysts supported on simple oxide materials, the monolayer modified catalysts showed much higher conversion and selectivity. Particularly, Pt/ZrO_2/TiO_2 and Pt/ZrO_2/SiO_2 performed very well, giving 88%–96.3% selectivity to FFA at 56.9%–61.5% conversion. These studies showed that both strong surface-metal interactions (SMSI) and SOOI effects lead to enhanced activity and selectivity.

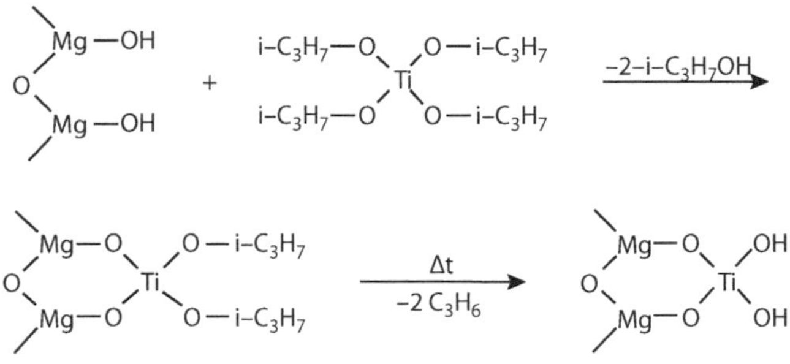

FIGURE 8.11 Deposition of a TiO_2 monolayer on MgO by surface grafting. (Reprinted from *Appl. Catal., A,* Platinum deposited on monolayer supports in selective hydrogenation of furfural to furfuryl alcohol, 233, J. Kijenski, P. Winiarek, T. Paryjczak, A. Lewicki, A. Mikolajska, 171–182, Copyright 2002, with permission from Elsevier.)

Using a coprecipitation method, Raju et al. prepared Cu/MgO catalysts for the vapor phase hydrogenation of furfural at 180°C and 1 atm [111]. High conversion, and high selectivity toward furfuryl alcohol (98%) were reported. Continuous operation showed that the catalyst remains stable for more than 5 hours on stream.

Cu–Ca/SiO$_2$ catalysts (20 wt% Cu, 161–166 m^2 g^{-1}) were prepared by sol–gel and impregnation techniques and calcination at 400°C [112]. After reduction at 250°C for 3 h, hydrogenation of FF was carried out in a fixed bed reactor at 130°C using a H$_2$/FF ratio of 5. The sol–gel catalyst demonstrated 98.7% selectivity at full conversion for more than 80 h, whereas the impregnation catalyst showed 100% selectivity with activity decreasing over time. The increased activity of the sol–gel catalyst was attributed to the stabilizing effect of Ca as a structural promotor.

Resasco et al. studied the gas phase hydrogenation of FF with monometallic Pd, Ni, Cu [113], and bimetallic Pd/Cu catalysts supported on SiO$_2$ [114]. Reactions were carried out at 220°C–290°C and 1 atm H$_2$ with a H$_2$/FF ratio of 25. Cu produced mainly furfuryl alcohol (99% selectivity at 230°C) and minor amounts of methylfuran. Over Pd, furan is the main product with 70% selectivity and THF and furfuryl alcohol are produced in nearly equal amounts (15%). Variation of temperature showed that decarbonylation increases with temperature. Over Ni, furan and furfuryl alcohol are the main products at lower temperatures whereas at higher-temperature selectivity shifts toward decarbonylation and ring opening reactions. For the bimetallic Pd/Cu catalysts, the selectivity of the reaction was found to depend strongly on the ratio of the metals, i.e., the decarbonylation, which is typically observed for Pd, is suppressed by increasing the Cu loading to 2%. As such, the selectivity to FFA can be increased to 70%.

Tomishige studied the use of Ni/SiO$_2$ catalysts for the gas phase total hydrogenation of FF to THFFA [115]. Catalysts were prepared by the deposition of Ni(NO$_3$)$_2$ on SiO$_2$ (500 m^2 g^{-1}) and reduction at 400°C–600°C or by calcination at 500°C followed by reduction. Characterization of the obtained catalysts showed that the direct reduction resulted in average particle sizes of 2.6–4.0 nm, whereas for the calcined and reduced catalysts, the particle size was much larger (12 nm). At 130°C, using an H$_2$/FF/N$_2$ ratio of 1:36:72, the main products of the reaction are FFA and HMTHF. At low conversions, FFA is selectively obtained, whereas at higher conversions, the selectivity to HMTHF increases. By optimization of conditions, up to 94% yield of HMTHF could be obtained.

Somorja et al. studied the effect of particle size on product selectivity in the Pt/MCF-17 (mesoporous silica) catalyzed gas phase hydrogenation of FF [116]. Catalysts with various particle sizes were prepared by the deposition of PVP stabilized nanoparticles. Reactions were performed at 200°C, 1 bar hydrogen pressure with a H$_2$/FF ratio of 9. As shown in Figure 8.12, the selectivity of the reaction strongly depends on the size of the supported particles. For smaller particle sizes, the reaction produced mainly furan indicating a high selectivity toward decarbonylation. In contrast for larger particle sizes, FFA is the main product and indicates an increased selectivity toward hydrogenation. Side reactions such as ring cracking and dehydration hardly occurred in both cases. This study clearly demonstrates that tailoring the physical properties of the supported metal particles can have a large effect on the product selectivity of the reaction.

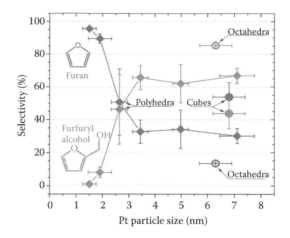

FIGURE 8.12 **(See color insert.)** Steady-state product selectivities to furan (blue) and fur-furyl alcohol (red) versus TEM-projected Pt particle sizes. (Reprinted with permission from V. V. Pushkarev, N. Musselwhite, K. An, S. Alayoglu, G. A. Somorjai, *Nano Lett.* 2012, 12, 5196–5201. Copyright 2012 American Chemical Society.)

8.3.2.2 Liquid Phase Hydrogenation of Furfural

Gaset et al. reported the hydrogenation of FF and FFA to THFFA over various commercial Pd-, Ru-, Rh-, Ni-, and Cu-based catalysts [117]. Comparison of the selected catalysts under different conditions led to the conclusion that mixtures of Ni- and Cu-based catalysts were required for the hydrogenation of furfural, whereas for the hydrogenation of furfuryl alcohol, a 59 wt% $Ni/SiO_2/Al_2O_3$ was most suited.

The activity of Raney nickel catalysts impregnated with various types of hetero-polyacid salts (M(HPA), M = Cu, Zn, Co, Fe, Ca, Na, HPA = $PMo_{12}O_{40}$, $SiMo_{12}O_{40}$, $PW_{12}O_{40}$, $SiW_{12}O_{40}$) was screened for the liquid phase hydrogenation of FF to FFA in methanol [118]. Comparison of the parent Raney Ni with the impregnated catalysts showed a clear beneficial effect of the presence of the HPA salt on the selectivity toward FFA. Optimization of the reaction conditions and variation of the supported HPA led to conversions up to 91%–98%, with FFA selectivities up to 93%–98%, with the supported $Cu_{1.5}PMo_{12}O_{40}$ being identified as the best in terms of conversion (98.1%) and selectivity (98.5%).

Bimetallic $PtSn/SiO_2$ catalysts with varying amounts of tin were obtained by the modification of 1 wt% Pt/SiO_2 using controlled surface reactions [119]. Liquid phase hydrogenation of furfural was carried out at 100°C, 10 bar H_2 in isopropanol. Screening of the catalysts showed that all exhibit high selectivity (96%–98%) and that the most efficient catalyst contained the lowest tin content ($PtSn_{0.3}$). Reuse of the catalyst showed that the bimetallic catalysts suffer deactivation over time.

Shimazu et al. also proposed the use of bimetallic tin-based catalysts for the liquid phase hydrogenation of FF to FFA [120]. $NiSn/Al(OH)_x$ catalysts were prepared by hydrothermal treatment of Raney Ni supported on $Al(OH)_3$ with $SnCl_2$ in EtOH/H_2O. Reactions were carried out in isopropanol at 150°C and 30 bar H_2. For the catalyst without Sn, full conversion to THFFA was found, whereas for the modified catalysts selectivity changed to FFA with a Ni/Sn ratio 1.4 giving full conversion to FFA after 1 h.

Lui et al. reported the preparation, and use of Cu/Ni/Mg/Al/O mixed hydrotalcite materials for the liquid and gas phase hydrogenation of FF to FFA [121]. Catalysts were prepared by the coprecipitation method and calcination at 500°C in O_2. The materials were activated prior to the reaction by reduction in H_2 atmosphere at 220°C–400°C. Characterization of the activated materials showed that as a result of the pretreatment Cu migrates to the surface and nickel toward the interior of the support. For the liquid phase reaction in ethanol, selectivities of 80%–86% FFA at 82%–90% conversion were found irrespective of the activation temperature. For the gas phase reaction, the selectivity to furfuryl alcohol was greatest when the catalysts were activated at 400°C (80% conversion with 85% selectivity).

8.3.2.3 Hydrogenation of HMF

The selective hydrogenation of HMF using supported metal catalysts can give rise to several interesting products. As shown in Scheme 8.9, hydrogenation of the aldehyde function of HMF yields 2,5-bis-(hydroxymethyl)furan (BHMF), whereas hydrogenation of the ring results in 2-formyl-5-(hydroxymethyl)tetrahydrofuran (FHMTHF). Further hydrogenation of either results in 2,5-bis(hydroxymethyl)tetrahydrofuran (BHMTHF). Depending on the employed catalyst, hydrogenolysis of the hydroxyl groups of BHMF and BHMTHF may occur leading to the production of 2,5-dimethylfuran (DMF) and 2,5-dimethyltetrahydrofuran (DMTHF), respectively. DMF can potentially serve as a liquid fuel, and its production from glucose and fructose is increasingly studied [122]. BHMF and BHMTHF may be used directly as novel monomers in the preparation of resins and polymers such as polyesters and polyamides [123]; however, of equal interest is the conversion of BHMTHF to obtain a biomass-based alternative for 1,6-hexanediol, which is used in the production of caprolactam and nylon [124].

The catalytic hydrogenation of HMF to BHMTHF using silica (kieselgur or diatomaceous earth) supported Ni (20%) catalysts with 3% Cu and 1% Fe, has been patented by Hales [125]. After 1.5 h at 100°C and 103–138 bar H_2, yields up to 97.5% were obtained after isolation by distillation. Mentech et al. studied the

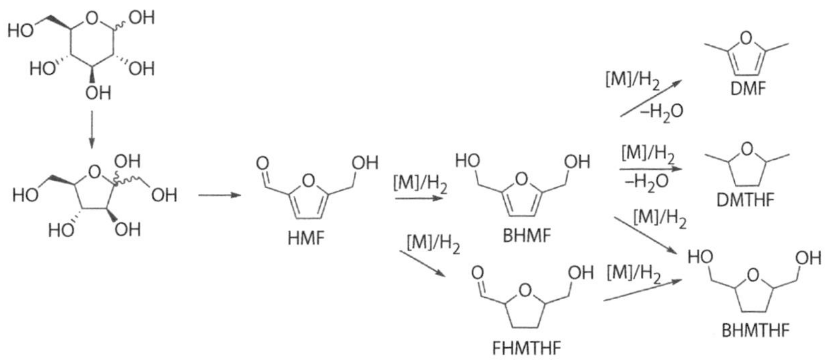

SCHEME 8.9 Reduction of HMF.

hydrogenation of HMF using various supported transition metals such as Ni, Cu, Pt, Pd, and Ru in aqueous medium [126]. In case of Cu- and Pt-catalyzed reactions, BHMF was obtained as the predominant product, whereas for Ni and Pd, BHMTHF was obtained.

Lilga et al. patented the total hydrogenation of HMF to BHMTHF via BHMF in aqueous medium using a fixed-bed continuous flow reactor [127]. A large number of commercial catalysts based on Ni, Co, Cu, Pd, Pt, Ru, Ir, Re, and Rh and various promoted Co, Ir, Ni, and Pd catalysts were tested in batch reductions that were typically carried out at 60°C and 34 bar H_2. Depending on the employed catalyst, various degrees of hydrogenation were observed after 2 h reaction with the combined selectivities to BHMF and BHMTHF generally being very high. For the continuous flow setup, two catalysts, comprised of Co and Ni, were used in series and allowed full conversion of HMF to BHMTHF. It was also noted that Pt(Ge)/C and Co/SiO$_2$ exhibited high selectivity for the conversion of HMF to BHMF in solutions that contain fructose.

Sandorn et al. patented the hydrogenation of HMF to BHMTHF using commercial Ni/SiO$_2$ catalysts promoted with ZrO$_2$ in ethanolic medium [128]. At 200°C and 103 bar H_2, full conversion of HMF was obtained after 1 h.

Sutherland et al. reported the use of commercial Pd/C and Raney Ni catalysts for the hydrogenation of HMF to BHMTHF [129]. Various solvents such as MeOH, EtOAc, toluene, and isopropanol were screened at 60°C and 4.1 bar H_2. After 16–18 h, full conversion of HMF to BHMTHF could be achieved.

Tomishige et al. studied the use of NiPd alloy catalysts supported on SiO$_2$ for the total hydrogenation of furan derivatives [129]. Catalysts with 2 wt% Pd were prepared by impregnation of SiO$_2$ (BET: 535 m^2 g^{-1}) with mixed solutions of PdCl$_2$ and Ni(NO$_3$)$_2$ and calcination at 500°C in air for 3 h. Activation of the catalysts by reduction was performed in H$_2$ atmosphere at 500°C. TEM data showed that alloy particles with an average size of 10.7 nm were formed. The formation of the Ni/Pd alloy was supported by XRD data. Reactions were carried out in water containing a small amount of acetic acid at 40°C and 80 bar H$_2$. Various alloy compositions were tested in the reaction and an Ni/Pd molar ratio of 7 was identified as giving the highest selectivities to BHMTHF. In contrast, Pd/SiO$_2$ and Ni/SiO$_2$ resulted mainly in the production of BHMF, showing that the hydrogenation of the furan ring is rather difficult at lower temperatures. Further optimization of the reaction conditions led to an optimal selectivity of 96% to BHMTHF at 99% HMF conversion. Similar selectivities and conversions were also obtained for the hydrogenation of FF and F.

Dumesic et al. reported the use of supported Ru, Pd, and Pt catalysts for the reduction of HMF to BHMTHF [130]. Catalysts were prepared by the impregnation of C, Al$_2$O$_3$, SiO$_2$, CeO$_x$, and MgZrO with the appropriate metal precursor and drying followed by reduction at 300°C in H$_2$ atmosphere for 8 h. Reactions were performed in water/1-butanol at 130°C and 27.5 bar H$_2$. For the oxide supported ruthenium catalysts, high selectivities to BHMTHF were obtained with BHMF being the major side-product. In contrast for the carbon supported and unsupported Ru, a large amount of ring-opened products such as 1,2,5-hexanetriol, 1,2,6-hexanetriol, 1,2,5,6-hexanetetrol were obtained next to BHMTHF.

8.4 CONCLUSION

The adaption of known solid catalysts and the development of new ones are crucial for future biorefinery schemes. The efficient conversion of biomass as a renewable resource strongly depends on the employed catalyst species. Since conventional refinery processes are based on high-temperature gas-phase transformations involving compounds with a low polarity, a significant change in the process technologies is required. In contrast, polymeric biomass feeds such as cellulose can be transformed into polar platform molecules that often exhibit only limited temperature stability. Thus, an efficient conversion of such resources requires process parameters under comparable mild conditions entailing preferably liquid phase processes in polar solvents as shown in this chapter by processes mostly carried out in water or ionic liquids to enable optimal interactions with the polar substrates. Nevertheless, also the types of required reactions have changed with the compounds involved from classical hydrocarbons to biogenic carbohydrates, especially involving all kinds of oxygen-containing organic functional groups.

In recent years, an exponentially increasing interest in catalytic processes for biomass conversion is obvious. In terms of the overall efficiency of the processes involving the downstream processing, the separation of the catalyst has to be taken into account. In this regard, the utilization of solid catalysts is preferred. Although lots of different types of solid catalysts are known, the investigation and adaption of conventional catalysts is not always sufficient. A targeted development of novel solid catalysts based on a detailed understanding of the substrate-catalyst interaction is a challenging target.

Especially the conversion of cellulose using solid acid catalysts for hydrolytic depolymerization or bifunctional catalysts for hydrolytic hydrogenation/hydrogenolysis is a bottleneck for the development of efficient processes. The main challenge is to increase the contact interface between the solid substrate and the solid catalyst particles. Thus, in most reported examples, catalyst amounts as high as the substrate amount or even excess amounts of the catalyst have been used which is far from an efficient process design. In case solvents are used that actually dissolve the cellulose, the problem is only reduced since the polymeric chains have to reach the catalytic active surface in the often microporous and mesoporous materials. More promising in this matter are nanoparticulate materials with a high external surface area for enhanced interaction with a solid polymeric substrate. In fact, the most promising approach seems to involve mechanical treatment such as ball milling of the cellulose in combination with the solid acid as shown for the abundant acidic mineral kaolinite resulting in water-soluble monomeric and oligomeric species. In the case of metallic catalysts involving hydrogen, an interesting approach was demonstrated applying reshaped Ni nanoparticles on the tip of carbon nanofibers resulting in a high selectivity for hydrogenation and nearly avoiding hydrogenolysis of C–C bonds. However, the effect is not yet understood, nicely demonstrating that a lot of research is necessary to deal with the great challenge of an efficient biomass conversion.

Catalytic oxidative and reductive transformations of lignocellulose-derived platform chemicals such as HMF and FF can result in a wide variety of different value-added products. The overview presented above demonstrates how the various properties of nanostructured catalysts can play an important role in influencing catalyst activity, selectivity, and stability. Particularly for the oxidation of HMF to FDCA, it was shown that the nature of the support greatly influences catalyst stability as well as selectivity. Also, a clear particle size effect was established, stressing the need for the preparation of tailored nanostructured catalysts. This is especially true when considering the potential integration of different catalytic technologies to convert raw materials such as glucose toward value-added products, in situ or using continuous flow setups. For the hydrogenation reactions, it was demonstrated that modifications of catalyst surfaces as well as the use of bimetallic catalysts can greatly alter the selectivity of reactions. These are necessary steps to develop a diverse library of catalysts, which can be implemented in future biorefinery schemes to enable the production of a wide range of products starting from a limited amount of platform chemicals.

ACKNOWLEDGMENTS

This work has been funded by the Robert Bosch Junior professorship for sustainable utilization of renewable natural resources. It was performed as part of the Cluster of Excellence "Tailor-Made Fuels from Biomass" funded by the Excellence Initiative by the German federal and state governments to promote science and research at German universities.

REFERENCES

1. (a) A. Corma, S. Iborra, A. Velty, *Chem. Rev.* 2007, *107*, 2411–2502; (b) M. J. Climent, A. Corma, S. Iborra, *Green Chem.* 2011, *13*, 520–540.
2. B. Kamm, P. R. Gruber, M. Kamm, *Biorefineries—Industrial Process and Products, Vol. 1*, Wiley-VCH, Weinheim, 2006.
3. M. Balat, M. Balat, E. Kirtay, H. Balat, *Energy Convers. Manage.* 2009, *50*, 3158–3168.
4. R. Rinaldi, F. Schüth, *ChemSusChem* 2009, *2*, 1096–1107.
5. C. Judson King, in *Ullmann's Encyclopedia of Industrial Chemistry, Vol. 6*, electronic ed., Wiley-VCH, Weinheim, 2001.
6. N. Mosier, C. Wyman, B. Dale, R. Elander, Y. Y. Lee, M. Holtzapple, M. Ladisch, *Bioresour. Technol.* 2005, *96*, 673–686.
7. (a) X. Zhao, K. Cheng, D. Liu, *Appl. Microbiol. Biotechnol.* 2009, *82*, 815–827; (b) Y. Sun, J. Cheng, *Bioresour. Technol.* 2002, *83*, 1–11.
8. (a) A. T. W. M. Hendriks, G. Zeeman, *Bioresour. Technol.* 2009, *100*, 10–18; (b) G. P. van Walsum, S. G. Allen, M. J. Spencer, M. S. Laser, M. J. Antal, L. R. Lynd, *Appl. Biochem. Biotechnol.* 1996, *57–58*, 157–170; (c) M. Laser, D. Schulman, S. G. Allen, J. Lichwa, M. J. J. Antal, L. R. Lynd, *Bioresour. Technol.* 2002, *81*, 33–44.
9. (a) H. Wang, G. Gurau, R. D. Rogers, *Chem. Soc. Rev.* 2012, *41*, 1519–1537; (b) T. Vancov, A.-S. Alston, T. Brown, S. McIntosh, *Renewable Energy* 2012, *45*, 1–6; (c) S. Zhu, Y. Wu, Q. Chen, Z. Yu, C. Wang, S. Jin, Y. Ding, G. Wu, *Green Chem.* 2006, *8*, 325–327.
10. R. Rinaldi, *Chem. Commun.* 2011, *47*, 511–513.
11. N. Meine, R. Rinaldi, F. Schüth, *ChemSusChem* 2012, *5*, 1449–1454.
12. S. V. de Vyver, J. Geboers, P. A. Jacobs, B. F. Sels, *ChemCatChem* 2011, *3*, 82–94.

13. C. E. A. Kirschhock, E. J. P. Feijen, P. A. Jacobs, J. A. Martens, in *Handbook of Heterogeneous Catalysis, Vol. 2* (Eds.: G. Ertl, H. Knözinger, F. Schüth, J. Weitkamp), Wiley-VCH, Weinheim, 2008.

14. (a) C. Moreau, R. Durand, J. Duhamet, P. Rivalier, *J. Carbohydrate Chemistry* 1997, *16*, 709–714; (b) R. Shukla, X. E. Verykios, R. Mutharasan, *Carbohydr. Res.* 1985, *143*, 97–106; (c) A. Abaddi, K. F. Gotlieb, H. van Bekkum, *Starch/Stärke* 1998, *50*, 23–28; (d) P. L. Dhepe, M. Ohashi, S. Inagaki, M. Ichikawa, A. Fukuoka, *Catal. Lett.* 2005, *102*, 163–169.

15. R. Rinaldi, R. Palkovits, F. Schüth, *Angew. Chem. Int. Ed.* 2008, *47*, 8047–8050.

16. Z. Zhang, Z. K. Zhao, *Carbohydr. Res.* 2009, *344*, 2069–2072.

17. H. Cai, C. Li, A. Wang, G. Xu, T. Zhang, *Appl. Catal., B* 2012, *123–124*, 333–338.

18. P. Lanzafame, D. M. Temi, S. Perathoner, A. N. Spadaro, G. Centi, *Catal. Today* 2012, *179*, 178–184.

19. B. C. Gates, in *Handbook of Heterogeneous Catalysis, Vol. 2* (Eds.: G. Ertl, H. Knözinger, F. Schüth, J. Weitkamp), Wiley-VCH, Weinheim, 2008.

20. Available at www.amberlyst.com

21. R. Rinaldi, N. Meine, J. vom Stein, R. Palkovits, F. Schüth, *ChemSusChem* 2010, *3*, 266–276.

22. J. Ahlkvist, S. Ajaikumar, W. Larsson, J.-P. Mikkola, *Appl. Catal., A* 2013, *454*, 21–29.

23. M. Benoit, A. Rodrigues, Q. Zhang, E. Fourré, K. De Oliveira Vigier, J.-M. Tatibouet, F. Jérôme, *Angew. Chem. Int. Ed.* 2011, *50*, 8964–8967.

24. R. Schlögl, in *Handbook of Heterogeneous Catalysis, Vol. 2* (Eds.: G. Ertl, H. Knözinger, F. Schüth, J. Weitkamp), Wiley-VCH, Weinheim, 2008.

25. (a) A. Onda, T. Ochi, K. Yanagisawa, *Green Chem.* 2008, *10*, 1033–1037; (b) A. Onda, T. Ochi, K. Yanagisawa, *Top. Catal.* 2009, *52*, 801–807.

26. M. Hara, T. Yoshida, A. Takagaki, T. Takata, J. N. Kondo, S. Hayashi, K. Domen, *Angew. Chem. Int. Ed.* 2004, *43*, 2955–2958.

27. (a) S. Suganuma, K. Nakajima, M. Kitano, D. Yamaguchi, H. Kato, S. Hayashi, M. Hara, *J. Am. Chem. Soc.* 2008, *130*, 12787–12793; (b) M. Kitano, D. Yamaguchi, S. Satoshi, K. Nakajima, H. Kato, S. Hayashi, M. Hara, *Langmuir* 2009, *25*, 5068–5075; (c) D. Yamaguchi, M. Kitano, S. Suganuma, K. Nakajima, H. Kato, M. Hara, *J. Phys. Chem. C* 2009, *113*, 3181–3188; (d) M. Hara, D. Yamaguchi, *WO2009099218*, 2009; (e) K. Fukuhara, K. Nakajima, M. Kitano, H. Kato, S. Hayashi, M. Hara, *ChemSusChem* 2011, *4*, 778–784.

28. R. Schlögl, in *Handbook of porous solids, Vol. 3* (Eds.: F. Schüth, K. Sing, J. Weitkamp), Wiley-VCH, Weinheim, 2002, p. 1863.

29. L. Vanoye, M. Fanselow, J. D. Holbrey, M. P. Atkins, K. R. Seddon, *Green Chem.* 2009, *11*, 390–396.

30. Y. Wu, Z. Fu, D. Yin, Q. Xu, F. Liu, C. Lu, L. Mao, *Green Chem.* 2010, *12*, 696–700.

31. J. Lee, J. Kim, T. Hyeon, *Adv. Mater.* 2006, *18*, 2073–2094.

32. S. Jun, S. H. Joo, R. Ryoo, M. Kruk, M. Jaroniec, Z. Liu, T. Ohsuna, O. Terasaki, *J. Am. Chem. Soc.* 2000, *122*, 10712–10713.

33. J. Pang, A. Wang, M. Zheng, T. Zhang, *Chem. Commun.* 2010, *46*, 6935–6937.

34. H. Kobayashi, T. Komanoya, K. Hara, A. Fukuoka, *ChemSusChem* 2010, *3*, 440–443.

35. T. Komanoya, H. Kobayashi, K. Hara, W.-J. Chun, A. Fukuoka, *Appl. Catal., A* 2011, *407*, 188–194.

36. S. V. de Vyver, L. Peng, J. Geboers, H. Schepers, F. de Clippel, C. J. Gommes, B. Goderis, P. A. Jacobs, B. F. Sels, *Green Chem.* 2010, *12*, 1560–1563.

37. D. Lai, L. Deng, J. Li, B. Liao, Q. Guo, Y. Fu, *ChemSusChem* 2011, *4*, 55–58.

38. A. Takagaki, M. Nishimura, S. Nishimura, K. Ebitani, *Chem. Lett.* 2011, *40*, 1195–1197.

39. A. Takagaki, C. Tagusagawa, S. Hayashi, M. Hara, K. Domen, *Energy Environ. Sci.* 2010, *3*, 82–93.

40. A. Takagaki, C. Tagusagawa, K. Domen, *Chem. Commun.* 2008, 5363–5365.
41. C. Tagusagawa, A. Takagaki, A. Iguchi, K. Takanabe, J. N. Kondo, K. Ebitani, T. Tatsumi, K. Domen, *Chem. Mater.* 2010, *22*, 3072–3078.
42. S. M. Hick, C. Griebel, D. T. Restrepo, J. H. Truitt, E. J. Buker, C. Bylda, R. G. Blair, *Green Chem.* 2010, *12*, 468–476.
43. (a) W. Deng, Y. Wang, Q. Zhang, Y. Wang, *Catal. Surv. Asia* 2012, *16*, 91–105; (b) H. Kobayashi, H. Ohta, A. Fukuoka, *Catal. Sci. Technol.* 2012, *2*, 869–883; (c) P. L. Dhepe, A. Fukuoka, *Catal. Surv. Asia* 2007, *11*, 186–191.
44. A. M. Ruppert, K. Weinberg, R. Palkovits, *Angew. Chem. Int. Ed.* 2012, *51*, 2564–2602.
45. L.-N. Ding, A.-Q. Wang, M.-Y. Zheng, T. Zhang, *ChemSusChem* 2010, *3*, 818–821.
46. N. Ji, M. Zheng, A. Wang, T. Zhang, J. G. Chen, *ChemSusChem* 2012, *5*, 939–944.
47. (a) H. Kobayashi, Y. Ito, T. Komanoya, Y. Hosaka, P. L. Dhepe, K. Kasai, K. Hara, A. Fukuoka, *Green Chem.* 2011, *13*, 326–333; (b) R. Palkovits, K. Tajvidi, J. Procelewska, R. Rinaldi, A. M. Ruppert, *Green Chem.* 2010, *12*, 972–978.
48. J. W. Han, H. Lee, *Catal. Commun.* 2012, *19*, 115–118.
49. P. Jacobs, H. Hinnekens, *US4950812*, 1990.
50. J. Geboers, S. V. de Vyver, K. Carpentier, P. Jacobs, B. Sels, *Chem. Commun.* 2011, *47*, 5590–5592.
51. (a) R. Palkovits, K. Tajvidi, A. M. Ruppert, J. Procelewska, *Chem. Commun.* 2011, *47*, 576–578; (b) J. Geboers, S. V. de Vyver, K. Carpentier, K. de Blochouse, P. Jacobs, B. Sels, *Chem. Commun.* 2010, *46*, 3577–3579; (c) Y. Ogasawara, S. Itagaki, K. Yamaguchi, N. Mizuno, *ChemSusChem* 2011, *4*, 519–525.
52. (a) T. Okuhara, N. Mizuno, M. Misono, *Appl. Catal., A* 2001, *22*, 63–77; (b) T. Okuhara, *Catal. Today* 2002, *73*, 167–176.
53. J. Geboers, S. V. de Vyver, K. Carpentier, P. Jacobs, B. Sels, *Green Chem.* 2011, *13*, 2167–2174.
54. M. Liu, W. Deng, Q. Zhang, Y. Wang, Y. Wang, *Chem. Commun.* 2011, *47*, 9717–9719.
55. H. Hattori, T. Yamada, T. Shishido, *Res. Chem. Intermed.* 1998, *24*, 439–448.
56. (a) A. Fukuoka, P. L. Dhepe, *Angew. Chem. Int. Ed.* 2006, *45*, 5161–5163; (b) V. Jollet, F. Chambon, F. Rataboul, A. Cabiac, C. Pinel, E. Guillon, N. Essayem, *Green Chem.* 2009, *11*, 2052–2060; (c) C. Luo, S. Wang, H. Liu, *Angew. Chem. Int. Ed.* 2007, *46*, 7636–7639.
57. N. Yan, C. Zhao, C. Luo, P. J. Dyson, H. Liu, Y. Kou, *J. Am. Chem. Soc.* 2006, *128*, 8714–8715.
58. Y. Zhu, Z. N. Kong, L. P. Stubbs, H. Lin, S. Shen, E. V. Anslyn, J. A. Maguire, *ChemSusChem* 2010, *3*, 67–70.
59. W. Deng, X. Tan, W. Fang, Q. Zhang, Y. Wang, *Catal. Lett.* 2009, *133*, 167–174.
60. H. Wang, L. Zhu, S. Peng, F. Peng, H. Yu, J. Yang, *Renewable Energy* 2012, *37*, 192–196.
61. S. V. de Vyver, J. Geboers, M. Dusselier, H. Schepers, T. Vosch, L. Zhang, G. van Tendeloo, P. A. Jacobs, B. F. Sels, *ChemSusChem* 2010, *3*, 698–701.
62. (a) J. Lewkowski, *ARKIVOC* 2001, *i*, 17–54; (b) P. Gallezot, *Chem. Soc. Rev.* 2012, *41*, 1538–1558.
63. C. Moreau, M. N. Belgacem, A. Gandini, *Top. Catal.* 2004, *27*, 11–30.
64. L. Hu, G. Zhao, W. Hao, X. Tang, Y. Sun, L. Lin, S. Liu, *RSC Adv.* 2012, *2*, 11184–11206.
65. G. Petersen, T. Werpy, *Vol. 1*, US Department of Energy, 2004, pp. 26–28.
66. P. Benecke, A. W. Kawczak, D. B. Garbak, *US207847*, 2008.
67. J. L. King II, A. W. Kawczak, H. P. Benecke, K. P. Mitchell, M. C. Clingerman, *US81883*, 2008.
68. M. Toshinari, K. Hirokazu, K. Takenobu, M. Hirohide, *US232815*, 2007.
69. W. Partenheimer, V. V. Grushin, *Adv. Synth. Catal.* 2001, *343*, 102–111.
70. S. E. Davis, M. S. Ide, R. J. Davis, *Green Chem.* 2013, *15*, 17–45.
71. E. I. Leupold, M. Wiesner, M. Schlingmann, K. Rapp, *US4977283*, 1990.

72. P. Vinke, H. E. van Dam, H. van Bekkum, *Stud. Surf. Sci. Catal.* 1990, *55*, 147–158.
73. M. A. Lilga, R. T. Hallen, J. Hu, J. F. White, M. J. Gray, *US20080103318*, 2008.
74. (a) N. Merat, P. Verdeguer, L. Rigal, L. Gaset, M. Delmas, *FR2669634*, 1992; (b) P. Verdeguer, N. Merat, A. Gaset, *J. Mol. Catal. A: Chem.* 1993, *85*, 327–344.
75. Y. Y. Gorbanev, S. K. Klitgaard, J. M. Woodley, C. H. Christensen, A. Riisager, *ChemSusChem* 2009, *2*, 672–675.
76. O. Casanova, S. Iborra, A. Corma, *ChemSusChem* 2009, *2*, 1138–1144.
77. S. E. Davis, L. R. Houk, E. C. Tamargo, A. K. Datye, R. J. Davis, *Catal. Today* 2011, *160*, 55–60.
78. N. K. Gupta, S. Nishimura, A. Takagaki, K. Ebitani, *Green Chem.* 2011, *13*, 824–827.
79. T. Pasini, M. Piccinini, M. Blosi, R. Bonelli, S. Albonetti, N. Dimitratos, J. A. Lopez-Sanchez, M. Sankar, Q. He, C. J. Kiely, G. J. Hutchings, F. Cavania, *Green Chem.* 2011, *13*, 2091–2099.
80. B. N. Zope, S. E. Davis, R. J. Davis, *Top. Catal.* 2012, *55*, 24–32.
81. S. E. Davis, B. N. Zope, R. J. Davis, *Green Chem.* 2012, *14*, 143–147.
82. Y. Y. Gorbanev, S. Kegnaes, A. Riisager, *Top. Catal.* 2011, *54*, 1318–1324.
83. Y. Y. Gorbanev, S. Kegnaes, A. Riisager, *Catal. Lett.* 2011, *141*, 1752–1760.
84. B. Saha, D. Gupta, M. M. Abu-Omar, A. Modak, A. Bhaumik, *J. Catal.* 2013, *299*, 316–320.
85. A. Modak, M. Nandi, J. Mondal, A. Bhaumik, *Chem. Commun.* 2012, *48*, 248–250.
86. M. Kröger, U. Prüße, K.-D. Vorlop, *Top. Catal.* 2000, *13*, 237–242.
87. M. L. Ribeiro, U. Schuchardt, *Catal. Commun.* 2003, *4*, 83–86.
88. J. M. R. Gallo, D. M. Alonso, M. A. Mellmer, J. A. Dumesic, *Green Chem.* 2013, *15*, 85–90.
89. E. Taarning, I. S. Nielsen, K. Egeblad, R. Madsen, C. H. Christensen, *ChemSusChem* 2008, *1*, 75–78.
90. O. Casanova, S. Iborra, A. Corma, *J. Catal.* 2009, *265*, 109–116.
91. (a) F. Pinna, A. Olivo, V. Trevisan, F. Menegazzo, M. Signoretto, M. Manzoli, B. Bocuzzi, *Catal. Today* 2013, *203*, 196–201; (b) M. Signoretto, F. Menegazzo, L. Contessottoa, F. Pinna, M. Manzoli, B. Bocuzzi, *Appl. Catal., B* 2013, *129*, 287–293.
92. B. Saha, S. Dutta, M. M. Abu-Omar, *Catal. Sci. Technol.* 2012, *2*, 79–81.
93. T. Stahlberg, E. Eyjóflsdóttir, Y. Y. Gorbanev, I. Sádaba, A. Riisager, *Catal. Lett.* 2012, *142*, 1089–1097.
94. F. W. Lichtenthaler, *Acc. Chem. Res.* 2002, *35*, 728–737.
95. A. Gandini, M. N. Belgacem, *Prog. Polym. Sci.* 1997, *22*, 1203–1379.
96. (a) K. T. Hopkins, W. D. Wilson, B. C. Bender, D. R. McCurdy, J. E. Hall, R. R. Tidwell, A. Kumar, M. Bajic, D. W. Boykin, *J. Med. Chem.* 1998, *41*, 3872–3878; (b) M. Baumgarten, N. Tyutyulkov, *Chem. Eur. J.* 1998, *4*, 987–989; (c) A. Gandini, *Green Chem.* 2011, *13*, 1061–1083; (d) J. Ma, Z. Du, J. Xu, Q. Chu, Y. Pang, *ChemSusChem* 2011, *4*, 51–54.
97. A. Datta, M. Agarwal, S. Dasgupta, *Proc. Indian. Acad. Sci. (Chem. Sci.)* 2002, *114*, 379–390.
98. C. Moreau, R. Durand, C. Pourcheron, D. Tichit, *Stud. Sur. Sci. Catal.* 1997, *108*, 399–406.
99. G. A. Halliday, R. J. Young, V. V. Grushin, *Org. Lett.* 2003, *5*, 2003–2005.
100. C. Carlini, P. Patrono, A. M. Raspolli, G. Sbrana, V. Zima, *Appl. Catal., A* 2005, *289*, 197–204.
101. O. Casanova, A. Corma, S. Iborra, *Top. Catal.* 2009, *52*, 304–314.
102. I. Sádaba, Y. Y. Gorbanev, S. Kegnaes, S. Sankar Reddy Putluru, R. W. Berg, A. Riisager, *ChemCatChem* 2013, *5*, 284–293.
103. A. Takagaki, M. Takahashi, S. Nishimura, K. Ebitani, *ACS Catal.* 2011, *1*, 1562–1565.
104. (a) R. J. Harrisson, M. Moyle, *Org. Synth.* 1956, *36*; (b) J. A. Moore, E. M. Partain, *Org. Prep. Proced. Int.* 1985, *17*, 203–205; (c) K. G. Sekar, *Int. J. Chem. Sci.* 2003, *1*, 227–232.
105. Q. Tian, D. Shi, Y. Sha, *Molecules* 2008, *13*, 948–957.
106. H. Choudhary, S. Nishimura, K. Ebitani, *Chem. Lett.* 2012, *41*, 409–411.

107. N. Alonso-Fagúndez, M. L. Granados, R. Mariscal, M. Ojeda, *ChemSusChem* 2012, *5*, 1984–1990.

108. R. Rao, A. Dandekar, R. T. K. Baker, M. A. Vannice, *J. Catal.* 1997, *171*, 406–419.

109. R. S. Rao, R. T. K. Baker, M. A. Vannice, *Catal. Lett.* 1999, *60*, 51–57.

110. J. Kijenski, P. Winiarek, T. Paryjczak, A. Lewicki, A. Mikolajska, *Appl. Catal., A* 2002, *233*, 171–182.

111. B. M. Nagaraja, V. S. Kumar, V. Shasikala, A. H. Padmasri, B. Sreedhar, B. D. Raju, K. S. R. Rao, *Catal. Commun.* 2003, *4*, 287–293.

112. J. Wu, Y. Shen, C. Liu, H. Wang, C. Geng, Z. Zhang, *Catal. Commun.* 2005, *6*, 633–637.

113. S. Sitthisa, D. E. Resasco, *Catal. Lett.* 2011, *141*, 784–791.

114. S. Sitthisa, T. Pham, T. Prasomsri, T. Sooknoi, R. G. Mallinson, D. E. Resasco, *J. Catal.* 2011, *280*, 17–27.

115. Y. Nakagawa, H. Nakazawa, H. Watanabe, K. Tomishige, *ChemCatChem* 2012, *4*, 1791–1797.

116. V. V. Pushkarev, N. Musselwhite, K. An, S. Alayoglu, G. A. Somorjai, *Nano Lett.* 2012, *12*, 5196–5201.

117. N. Merat, C. Godawa, A. Gaset, *J. Chem. Tech. Biotechnol.* 1990, *48*, 145–159.

118. L. Baijun, L. Lainhai, W. Bingchun, C. Tianxi, K. Iwatani, *Appl. Catal., A* 1998, *171*, 117–122.

119. A. B. Merlo, V. Vetere, J. F. Ruggera, M. L. Casella, *Catal. Commun.* 2009, *10*, 1665–1669.

120. Rodiansono, T. Hara, N. Ichikuni, S. Shimazu, *Chem. Lett.* 2012, *41*, 769–771.

121. C. Xu, L. Zheng, D. Deng, J. Liu, S. Liu, *Catal. Commun.* 2011, *12*, 996–999.

122. (a) Y. Román-Leshkov, C. J. Barrett, Z. Y. Liu, J. A. Dumesic, *Nature* 2007, *447*, 982–986; (b) J. B. Binder, R. T. Raines, *J. Am. Chem. Soc.* 2009, *131*, 1979–1985; (c) M. Chidambaram, A. T. Bell, *Green Chem.* 2010, *12*, 1253–1262.

123. (a) J. A. Moore, J. E. Kelly, *Macromolecules* 1978, *11*, 568–573; (b) W. J. Pentz, *GB2131014*, 1984; (c) M. Durant-Pinchard, *FR2556344*, 1985.

124. (a) T. Utne, R. E. Jones, J. D. Garber, *US3070633*, 1962; (b) T. Buntara, S. Noel, P. Huat Phua, I. Melián-Cabrera, J. G. de Vries, H. J. Heeres, *Angew. Chem. Int. Ed.* 2011, *50*, 7083–7087.

125. R. A. Hales, *US3040062*, 1962.

126. V. Schiavo, G. Descotes, J. Mentech, *Bull. Soc. Chim. Fr.* 1991, 704.

127. M. A. Lilga, R. T. Hallen, T. A. Werpy, J. F. White, J. E. Holladay, J. G. Frye Jr., A. H. Zacher, *US2007287845*, 2007.

128. A. J. Sanborn, P. D. Bloom, *US7393963*, 2008.

129. Y. Nakagawa, K. Tomishige, *Catal. Commun.* 2010, *12*, 154–156.

130. R. Alamillo, M. Tucker, M. Chia, Y. Pagán-Torres, J. Dumesic, *Green Chem.* 2012, *14*, 1413–1419.

9 Chemocatalytic Processes for the Production of Bio-Based Chemicals from Carbohydrates

Jan C. van der Waal and Ed de Jong

CONTENTS

9.1   INTRODUCTION

The production of bio-based chemicals-is it the new future or are we looking at the return of an old strategy? Certainly, the world is changing its chemistry backbone as it has become apparent that the current fossil-based source of almost all carbon will run out on the long term. Although there may still be in the order of a century for oil and natural gas and several centuries for coal, other factors such as the developing insights in the role of CO_2 on the changing climate of our planet may speed up the transgression to a fully bio-based society. Many countries have set programs in place for the sustainable growth, harvesting, and conversion of locally produced biomass into renewable fuels and chemicals. Not all biomass is the same, and it is typically classified into three major groups: lignocellulosics or woody biomass (the nonedible portion of biomass, e.g., bagasse, corn stover, grasses, wood), amorphous sugars (e.g., starch, glucose), and triglycerides (e.g., vegetable oil). Other important groups are rosins, crude tall oils, terpenes, and proteins, which also attract interest.

The most abundant, fastest-growing, and cheapest form of land-based biomass is lignocellulosic biomass, which is composed of three primary components: cellulose, hemicellulose, and lignin.[1] Although lignocellulosic biomass is a desirable feedstock, it is not yet economically viable to convert it into liquid transportation fuels and chemicals. It has been stated that "the central and surmountable impediment to more widespread application of biocommodity engineering (or biorefining) is the general absence of low-cost processing technology."[2] Several challenges must be overcome for biomaterials and biofuels from lignocellulosic biomass to be economically competitive with fossil-based fuels.[3] Since lignocellulosic biomass has generally a high carbohydrate content (>66%), it is to be expected that many of the processes will use enzymatic and/or fermentative steps in the conversion routes. However, several new technologies are emerging that clearly show this is not the only way. Classical conversion techniques such as hydrothermal upgrading, pyrolysis, gasification, and chemocatalytic processes are among these new approaches taken.

Here, the focus will be an industrial perspective on several chemocatalytic process technology routes using carbohydrate feedstocks. The most important research efforts and strategies will be discussed. However, the change to a biomass-based chemical industry still faces several challenges as inevitably new chemistry will be needed as new platform feedstocks, such as lignocellulosic glucose and chemicals become available, which can differ radically from those currently obtained from petroleum feedstocks. It is clear that a considerable research effort still lays ahead of us. Chemocatalysis will go side-by-side with biological conversions and will most certainly play an equally important role in this biobased economy as it does now for the petrochemical industry.

It is exemplary to consider the petrochemical industry. It has taken close to a century for this industry to fully adopt and maximize the use of petroleum for the production of fuels and chemicals, and even today, this research effort is continuing. The chemistry typically involves cracking and breaking of carbon–carbon bonds using heterogeneous acid catalysis and hydrotreating of nearly pure hydrogen-rich carbon based feedstocks. In contrast, biomass is generally wet, chemically diverse, and oxygen-rich, which is not very compatible with existing refinery operations. It is clear that new catalytic processes will need to be developed, but as Kamm et al.[4] have pointed out, we should not forget the major lessons from the petroleum-based industry, e.g., employing continuous operation procedures, using scale-of-economics, employing heat integration, minimizing concentration and purifications steps, and applying the conversion of the entire feed into multiple products.

Several chemocatalytic industrial and pilot-plant processes that all have sugars as starting points will be discussed. Glucose is one of the main biomass-based platform chemicals for further chemical, catalytic, and biotechnological conversion.[5] The chemistries discussed here are the Quaker Oats process for furfural, the Avantium furan dicarboxylic acid process, GEVO's route to *p*-xylene, Virent's aqueous phase reforming (APR) of lignocellulosic biomass to fuels and *p*-xylene and Annellotech's catalytic pyrolysis of lignocellulosic biomass, the Braskem ethylene process, and finally Roquette's process to isosorbide. This chapter focuses mostly on the current industrial practice starting from sugar as the feedstock, although if this involves homogeneous catalysts, the perspective for the introduction of heterogeneous catalysts will be discussed.

9.2 EMERGING RENEWABLES LANDSCAPE

Recently, many companies have become now active in the renewables arena.[6] These companies span a wide range of chemicals as shown in Table 9.1, some with strong growth potential in the current chemical industry and some clearly still further down the pipeline. In this respect, a commonly cited report by Werpy and Petersen[9] identified in 2004 and updated in 2010[7] a list of 12 high potential platform chemicals from carbohydrate sources (Table 9.2). Comparing both tables clearly shows that although some overlap exists, clearly industry has developed a much wider focus, resulting from considerations on using all forms of biomass available.

When considering which of the chemicals in Tables 9.1 and 9.2 are produced in a chemocatalytic manner, it is often good to realize that enzymatic and/or fermentation processes are used earlier on in the process, and that catalysis only started to play a role once the biomass has been transformed in more suitable platform chemicals. For example, ethylene is produced by solid-acid–catalyzed dehydration of ethanol, which in turn was obtained by fermentation of sugars. The latter could be obtained from sugar- and starch-containing crops, but currently, significant research efforts are undertaken in pretreatment and (enzymatic) hydrolysis methods of lignocellulosic crops and waste streams. This teaming up of biocatalysis and chemocatalysis to produce products in an integrated process is one of the examples of a so-called biorefinery in analogy with today's petrorefineries.[8] IEA Bioenergy Task 42 has defined a biorefinery as "Biorefinery is the sustainable processing of biomass into a spectrum of marketable products."[10] And if the biorefinery of the future will become true mimics of today's petrorefineries, chemocatalysis will certainly play an important role, as it is more amendable to large volume processes, which will be key to obtain the required scale-of-economic and heat integration. An indication of this can be found in the announced new processes up to 2016. Our own research has shown currently almost all biochemicals are made via fermentation, with bioethanol being the biggest

TABLE 9.1
Production of Bio-Based Chemicals

	Products with Strong Growth Potential		Bio-Based Chemicals in the Pipeline	
C#	**Chemical**	**Company**	**Chemical**	**Company**
1	Methanol	BioMCN, Chemrec	Formic acid	Maine BioProducts
2	Ethylene	Braskem, DOW/Mitsui, Songyuan Ji'an Bio-chemical	Ethyl acetate	Zeachem
	Ethanol	Many	Glycolic acid	Metabolix Explorer
	Ethylene glycol	India Glycols Ltd, Greencol Taiwan, JBF Industries	Acetic acid	Wacker
3	Lactic acid	Purac, Henan Jindan, BBCA NatureWorks, Galactic	Acrylic acid	Cargill, Perstorp, OPXBio, DOW, Arkema
	Glycerol	Many	Propylene	Braskem/Toyota Tsusho, Mitsubishi Chemical, Mitsui Chemicals
	Epichlorohydrin	Solvay, DOW	3-Hydroxypropionic acid	Cargill
	1,3-Propanediol	DuPont/Tate&Lyle	n-Propanol	Braskem
	Ethyl lactate	Vertec BioSolvents		
	Propylene glycol	ADM, Dupont	Isopropanol	Genomatica, Mitsui Chemicals
4	n-Butanol	Cathay Industrial Biotech, Butamax, Butalco, Cobalt/Rhodia	1,4-Butanediol	Genomatica/M&G, Genomatica/Mitsubishi Chemical, Genomatica/ Tate & Lyle
	iso-Butanol	Butamax, Gevo	Methyl methacrylate	Lucite/Mitsubishi Rayon, Evonik/ Arkema
	Succinic acid	BioAmber, Myriant, BASF/Purac, Reverdia (DSM/Roquette), PTT Chem/Mitsubishi CC	Iso-butene	Gevo/Lanxess
5	Furfural	Many	Itaconic acid	Itaconix, Qingdao Kehai Biochemistry Co,
	Xylitol	Lenzing	Isoprene	Goodyear/Genencor, GlycosBio, Amyris
	Glutamic acid	Global Biotech, Meihua, Fufeng, Juhua	Levulinic acid	Maine BioProducts, Avantium, Segetis, Circa Group
6	Sorbitol	Roquette, ADM,	Adipic acid	Verdezyne, Rennovia, BioAmber, Genomatica

(continued)

TABLE 9.1 (Continued)
Production of Bio-Based Chemicals

	Products with Strong Growth Potential		Bio-Based Chemicals in the Pipeline	
C#	Chemical	Company	Chemical	Company
	Isosorbide	Roquette	2,5-Furandicarboxylic acid	Avantium
	Lysine	Global Biotech, Evonik/ RusBiotech, BBCA, Draths[a], Ajinomoto	Glucaric acid	Rivertop renewables, Genencor
	Citric acid	Cargill, DSM, BBCA, Ensign, TTCA, RZBC	Caprolactam	DSM
n	PHA	Metabolix, Meridian plastics, Tianjin Green Biosience Co.	para-Xylene	Gevo, Draths,[a] Annellotech, Virent, UOP, Sabic
	Fatty acid derivatives	Croda, Elevance,	Farnesene	Amyris

Source: Data from M. Dohy, E. De Jong, H. Jørgensen et al. 2009. Adding value to the sustainable utilization of biomass, RRB5 conference, Ghent.

[a] Draths was recently acquired by Amyris.

by far, and catalysis is applied almost exclusively for production of biodiesel from vegetable oils.[11] However, current research efforts aimed at new production processes that have been announced by industry show an almost equal split between fermentation and chemocatalytic processes (see Figure 9.1). With catalytic processes mostly using platform chemicals derived from fermentation processes, in particular ethanol and butanol, but the conversion of sugars is also in development. Similar conclusions can be drawn from data supplied by *Biofuels Digest* (June 2012).[12] These numbers can change dramatically; recently, JBF Industries and the Coca Cola Company announced plans for a 500 kT/a ethylene glycol plant in Brazil based on sugarcane ethanol.

TABLE 9.2
The Top Carbohydrate-Derived Building Blocks Outlined by DOE

Succinic, Fumaric, and Maleic	Itaconic acid
2,5-Furandicarboxylic acid	Levulinic acid
3-Hydroxyl propionic acid	3-Hydroxybutyrolactone
Aspartic acid	Glycerol
Glucaric acid	Sorbitol
Glutamic acid	Xylitol/Arabinitol

Source: Data from T. Werpy and G. Petersen, 2004. Top value added chemicals from biomass, Pacific Northwest National Laboratory, and the National Renewable Energy Laboratory.

(a)

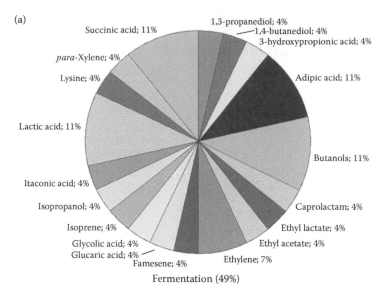

Fermentation (49%)

(b)

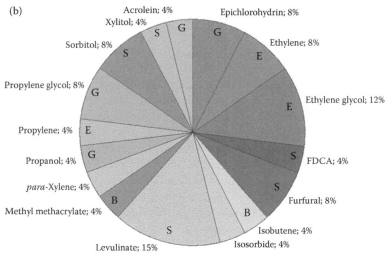

Catalytic (46%)

FIGURE 9.1   **(See color insert.)** Announced commercial processes and research projects up to 2016, as announced June 2012. (a) Fermentation and (b) chemocatalytic. Feedstocks used: S, sugar; E, ethanol; B, butanol; G, glycerol.

9.3 QUAKER OATS PROCESS FOR THE PRODUCTION OF FURFURAL

The production of furfural from oat husks by Quaker Oats Co. started in 1922 and can be considered one of the earliest examples of a catalytic process employing major amounts of biomass for the production of a commodity chemical.[13] The basic chemistry is that of dehydration of the abundant C5 sugars from the hemicellulose part of the straw and husks of the plant. The predominant feedstock is the hemicellulose part, also called pentosan or xylan part, of the plant, which mostly consists of polymers of xylose. Arabinose and galacturonic acid, also present in the biomass, are reported to be also converted to the desired furfural (see Figure 9.2).

The formation of furfural consists of a two-step catalytic reaction. In the first step, the hemicellulose is hydrolyzed in the presence of an acid to form the monomeric sugars. In the second step, these monomeric sugars are dehydrated to give furfural. The mechanism of this dehydration is generally considered to be a Brønsted acid catalyzed reaction, and although the exact mechanism is still unknown, it is generally assumed to consist of the consecutive removal of 3 molecules of water. Two general types of dehydration mechanism are considered. The first mechanism was proposed by Zeitsch[14] and assumes the dehydration to proceed on the acyclic structure, with the furan ring being formed in the last step (see Figure 9.3). The other mechanism is assuming that the pentose sugars dehydrate from its cyclic furanoside form (see Figure 9.4) was proposed by Antal.[15] Recently, work by Marcotullio[16] showed the beneficial role of halide ions, chloride in particular, on the dehydration of xylose and improvements in yield up to 81% were reported. Here it also worthy to note that the formation of diacetyl and 2,3-pentadione as valuable by-products was extremely important to the overall economics of a commercial plant. Both have a buttery flavor and were used in the coloring of butter and margarine. About 14 and 1.2 kg, respectively, are produced per ton of furfural produced in the ROSENLEW process. As the market price of the diacetyl was 14USD/kg and the 2,3-pentadione was 300USD/kg, these products, if recovered, would add 0.54$ value produced per kilogram of furfural.

The catalysts typically used in the process are cheap inorganic acids such as sulfuric acid in the original Quaker Oats[13] process and phosphoric acid and superphosphates in the Petrol-Chime process.[14,17] Although the use of these acids leads to difficulties in separation and recycling and are generally corrosive and toxic, their main drawback is that extensive side-reactions occur due to the long residence times in the reactor. In general, the yield of furfural in commercial units does not

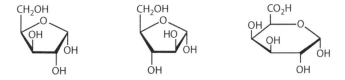

FIGURE 9.2 Monomeric sugars for furfural production: (a) xylose, (b) arabinose, and (c) galacturonic acid.

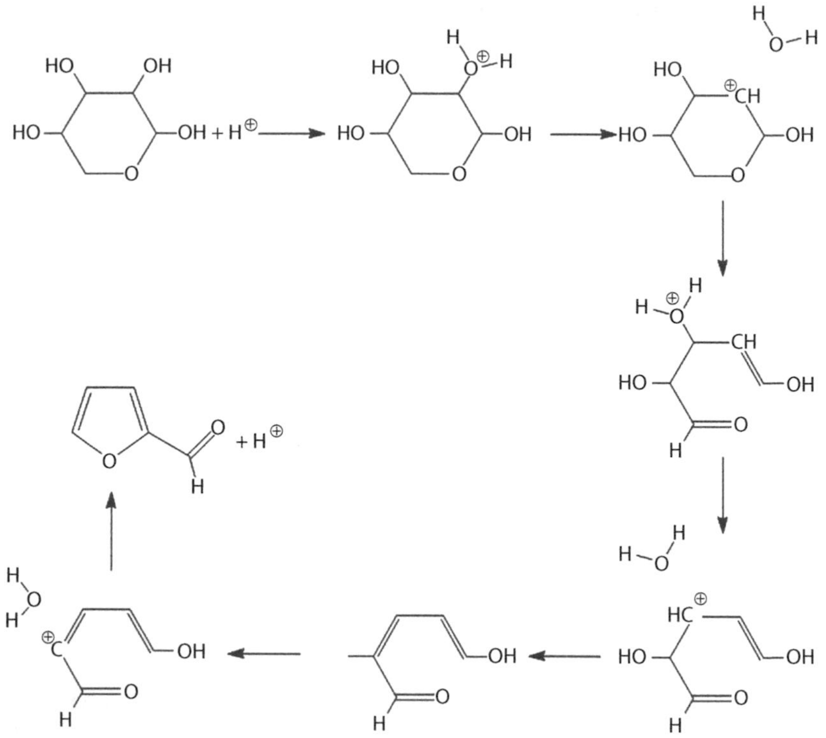

FIGURE 9.3 Acyclic dehydration pathway of C5 sugars to furfural. (Adapted from K.J. Zeitsch, 2000. *The Chemistry and Technology of Furfural and Its Many By-Products*, 1st ed.; Elsevier: Amsterdam.)

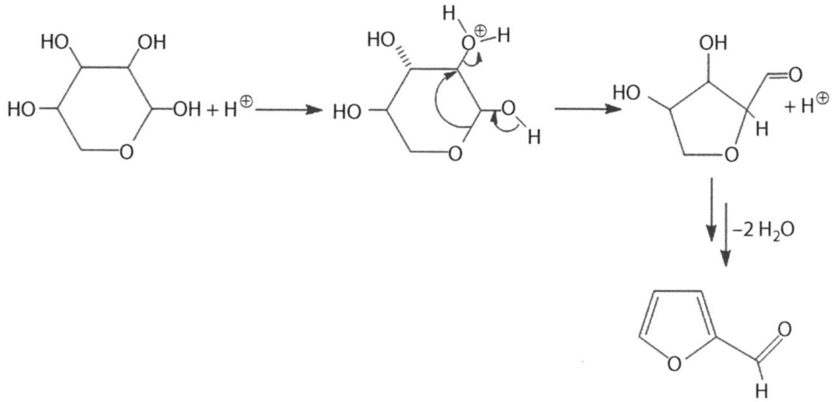

FIGURE 9.4 Cyclic dehydration pathway of C5 sugars to furfural. (Adapted from M.J. Antal, G.N. Richards. *Carbohydr. Res.* 1991, 217, 71.)

exceed 50%–55% base on carbon content. This is in contrast to well-established analytic methods, which determine the presence of C5 sugars by analysis of furfural from a selective, quantitative dehydration.[14]

In the past decades, several studies toward the use of heterogeneous catalysts have been published. In 1998, Moreau et al.[18] was the first to report a heterogeneous catalyst system using acid zeolites faujasite and mordenites in the dehydration of xylose in a biphasic system with high selectivity to furfural of 90%–96% at low conversions of 27%–37%. Interesting results were obtained by Dias[19] using MCM-41–type acid niobates; at nearly full conversion, furfural yields around 50% were obtained. The same group also reported the use of sulfated zirconia and micromesoporous sulfonates.[20,21] Good results were also obtained using the micromesoporous sulfonates with 73.8% furfural yield at conversions over 90%. Another interesting new development is the use of nonaqueous solvents as described by Binder.[22] The best catalyst system found was $CrCl_3$ in dimethylacetamide with a bromide salt as additive, which suggests that rather than the classical Brønsted acid, a Lewis acid–sugar complex may be involved. Although the best reported yield in furfural of 56% does not significantly change from the Brønsted acid catalyzed commercial process.

Several production processes are described by Zeitsch in great detail,[14] and both batch and continuous processes have been developed for the production of furfural. Currently, batch operation is still the predominant production method as it is inexpensive and does not suffer from the high maintenance costs and frequent shutdowns typically associated with continuous processes due to the high corrosivity of the reaction mixture and sensitivity of the moving parts to the abrasion by sand in the feed.

Irrespective of the type of process used, it should be considered that major parts of the operational costs are due to the steam stripping of the furfural in the reactor and due to the purification of the crude furfural stream from the reactor. The reactor effluent mainly consists of water but also has small amounts of the diacetyls mentioned above and usually organic acids such as acetic acid and formic acid are present as well. As water and furfural form an azeotropic mixture, considerable effort is needed to separate the two if pure furfural is to be obtained. Zeitsch describes a typical purification process, in which the furfural-water azeotrope is distilled over the top and the crude furfural is obtained by cooling the mixture until liquid–liquid phase separation occurs (see Figure 9.5).[14] The furfural is still not very pure and requires further distillation. One of the lessons of petrorefineries is that extensive purification and concentration has to be avoided, and because pure furfural will not always be the final product, it is suggested that the furfural is converted from less pure streams in the process.

It is obvious that heterogeneous catalyst systems offer advantages over the corrosive, toxic homogeneous systems currently in use, although considerable research is still required to obtain materials that are active and selective. In addition, most research uses xylose as the starting material, and it should be realized that the feedstock of the future is most likely polymeric hemicelluloses such as the xylans in the plant.

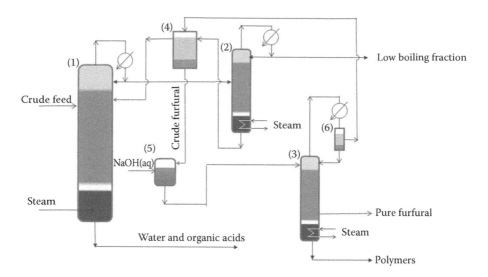

FIGURE 9.5 Typical purification section of a furfural plant. (1) Main reactor, (2) distillation column, (3) azeotropic distillation column, (4) decanter, (5) acid neutralizer, (6) decanter. (Adapted from K.J. Zeitsch, 2000. *The Chemistry and Technology of Furfural and Its Many By-Products*, 1st ed.; Elsevier: Amsterdam.)

9.3.1 OUTLOOK FOR NEW FURFURAL CATALYSTS

Historically, the conversion of hemicellulose to furfural is done with homogeneous acids, as heterogeneous catalysts are typically not the best solution for solid materials. With the rise of lignocellulosic materials as the starting biomass, the availability of huge amounts of dissolved, monomeric C5 sugars from the hemicellulose part of the plant will become available. These new feeds will be much more compatible with the use of solid acids.

New catalytic materials will have a good chance of beating the classical homogeneous acids, if they can give improved yields in furfural. The current best reported results (see Table 9.3) has already a yield in furfural of 47.2 wt%/wt% xylose, which is considerably higher than the typical 32–35 w%/w% xylose, which is typically observed in current industrial practice. However, there is still considerable room in further improving these results.

TABLE 9.3

Theoretical and Overall Yield for the Conversion of Xylose to Furfural

Step	Theoretical Yield[a] Weight/Weight Substrate	Best Reported Selectivity	Combined Best Yields on Xylose	Overall Carbon Efficiency
Dehydration of xylose	64.0 w/w%	82%[b]	47.2 w/w xyl%	73.8%

[a] Assuming a 100% molar yield and selectivity.
[b] At 90% conversion.

9.4 PRODUCTION OF FURAN DICARBOXYLIC ACID VIA FURANIC INTERMEDIATES

Where the acid-catalyzed dehydration of pentoses gives rise to furfural as described above, the similar dehydration of hexoses gives rise to a group of chemicals called furanics, of which hydroxymethyl furfural (HMF) and levulinic acid (LA) are the most commonly observed compounds. Both compounds have been identified by Werpy and Petersen[9] as important building blocks of the future. HMF has recently received a lot of attention as it can be oxidized to obtain furan dicarboxylic acid (FDCA), as starting monomer in the production of polyethylene furandicarboxylate, or PEF, a fully bio-based material suitable for water and carbonated soft drink bottles. Recently collaboration between the Coca Cola Company and Danone announced their collaborations with Avantium Chemicals on this new material.[23,24]

The general approach for the production of FDCA from glucose is a three-step process consisting of an isomerization of glucose to fructose, followed by dehydration to hydroxymethyl furfural (HMF) while suppressing formation of levulinic acid (LA) and finally, an oxidation to the desired FDCA (see Figure 9.6). Direct conversion of glucose is also possible, and may or may not proceed via fructose as an intermediate.[25] The enzymatic isomerization of glucose to fructose is a well-established process that is predominantly used in the conversion of corn starch hydrolyzates into high-fructose corn syrups (HFCS) and is already cost-effectively being done. It will not be discussed here. The next step, dehydration to HMF, is considered to be the most critical process. The HMF is a very unstable molecule that easily converts further to levulinates and to (partly insoluble) polymers called humins. The formation of HMF from fructose was discovered by Mulder in 1840.[26] Several solutions have been proposed, and two pilot plants for HMF have been operated using a biphasic reaction media. Recently, two alternative strategies have emerged that do not aim at the direct production of HMF but rather a more stable product thereof. Avantium Chemicals aims to in situ produce the alkoxy methyl furfural ethers (RMF),[27] and Mascal produces the chloromethyl furfural (CMF).[28]

9.4.1 CONVERSION OF FRUCTOSE TO HMF

The dehydration of fructose to HMF is a typical acid-catalyzed reaction and has been studied extensively since its discovery in 1840.[26] By far, the most commonly studied catalyst systems are the homogeneous acids, in particular, sulfuric acid. For the interested reader, a recent review by van Putten[29] has comprehensively collected all available catalytic, kinetic, and process data available. Typical yields at full fructose conversions are around 55% in water, approximately 90% in organic solvents such as DMSO and DMA, and over 99% yields in ionic liquids.[28] In many cases, an insoluble product, referred to as humans, is formed as a major by-product by condensation of the HMF formed with unconverted sugar and sugar dehydration products. In light of the purpose of this book, we will only discuss the advances made by employing heterogeneous catalysts.

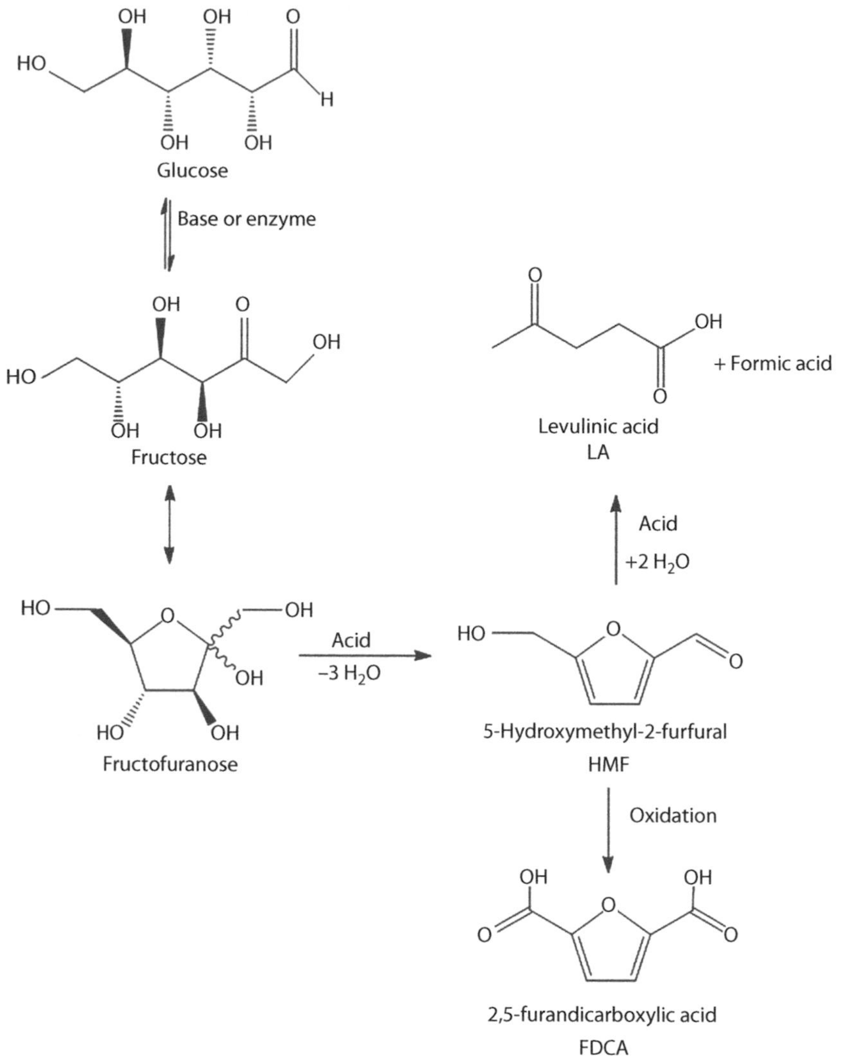

FIGURE 9.6   Conversion of glucose and fructose via HMF to FDCA.

The use of heterogeneous catalysts for the dehydration of fructose has been studied extensively, see Tables 9.4 and 9.5 for fructose and Table 9.6 for glucose, the most relevant results are tabulated. Considering the heterogeneous acids result in aqueous solutions, Table 9.4 clearly shows that the best reported yields are actually very similar to those obtained for the homogeneous Brønsted acids. The only two exceptions seem to be related to how the process is operated, i.e., by continuous extraction of the HMF formed and when considerable amounts of an organic solvent are present. Both these strategies are also known to enhance HMF selectivity. This strongly seems to suggest that a heterogeneous catalyst that contains Brønsted acid

TABLE 9.4

Selected Results for Heterogeneous Catalysts in the Dehydration of Fructose in Aqueous Solutions

Catalyst	Temperature (°C)	Reaction Time	Yield[a] (%)	Selectivity (%)	Reference
AlVOP	80	2 h	58	76	31
FeVOP	80	1 h	60	84	30
Dowex 50	150	15 min	73[b]	81	32
TiO_2	200[a]	5 min	38	45	31,35
TiP	100	1 h	67[c]	95	37
Niobic acid–H_3PO_4	100	0.5 h	22	61	33,34
NiP_2O_7	100	3 h	30	59	32,35
ZrO_2	200[a]	5 min	31	48	31,36
ZrP	240	2 min	49	61	37
ZrP_2O_7	100	2 h	43	81	38
SO_4^{2-}/ZrO_2	200[d]	5 min	26	44	39
Mordenite SAR 5.5	165	30 min	70	92	40

Source: See table for multiple data references.

[a] Yield defined on carbon basis.

[b] 70 wt% acetone as co-solvent.

[c] HMF extraction with MIBK after 0.5 and 1 h.

[d] Best yield in HMF reported only in case of multiple references.

activity will be enough for the dehydration and that research should be aimed at new materials with enhanced solid–sugar or solid–HMF interactions to prevent the undesired humins by-product formation. This seems to be in-line with the arguments put forward why some organic solvents and especially ionic liquids are such selective solvents.[30] They form a clathrate around the sugar and/or HMF that effectively shields them from reacting with each other or from further hydration to levulinic acid.

Thus far, heterogeneous catalysts have been tested in batch experiments only. However, this method of testing does not or gives only to a very limited extent information on the deactivation of the catalysts under industrially more relevant continuous operation.

9.4.2 Conversion of Glucose to HMF

One aspect in which solid acids may have an edge over their homogeneous counter parts is in having different catalytic sites present physically separated on their surface. Considering Figure 9.6, the route to obtain HMF from glucose first requires an isomerization that is base catalyzed, followed by an acid catalyzed dehydration of the fructose formed. Several authors[31,35,48–50] have reported multifunctional catalysts as shown in Table 9.6. Yields in HMF are in general lower, but they show the

TABLE 9.5

Selected Results[a] for Heterogeneous Catalysts in the Dehydration of Fructose in Organic Solutions

Solvent	Catalyst	Temperature (°C)	Reaction Time	Yield[b] (%)	Selectivity (%)	Reference
Acetone/DMSO (70:30 w/w)	Dowex 50WX8-100	150[c]	20 min	88	90	31
Acetone/DMSO (70:30 w/w)	SO_4^{2-}/ZrO_2	180[c]	5 min	63	74	38
Acetone/DMSO (70:30 w/w)	ZrO_2	180[c]	5 min	41	57	38
DMA[e]	Amberlyst	105	5 h	62	83	41
DMF	Lewatit SPC 108	96	5 h	80	—	42
DMSO	Amberlyst15 powder	120	2 h	100[d]	100	43,43
DMSO	$Cs_{2.5}H_{0.5}PW_{12}O_{40}$	120	2 h	91[d]	91	44
DMSO	Diaion PK-216	80	500 min	90	—	45
DMSO	$FePW_{12}O_{40}$	120	2 h	97[d]	97	43
DMSO	PTA/MIL-101	130	30 min	63	77	46
DMSO	Si-3-IL-HSO_4	130	30 min	63	63	47
DMSO	SiO_2-SO_3H	100[b]	4 min	63	66	48
DMSO	SO_4^{2-}/ZrO_2-Al_2O_3	130	4 h	57	57	49,43
DMSO	WO_3/ZrO_2	120	2 h	94[c]	94	43
DMSO	Zeolite H-beta	130	30 min	63	63	46,43
NMP[e]	Amberlyst 35	115	5 h	81	86	40

Source: See table for multiple data references.

[a] Best yield in HMF reported only in case of multiple references.

[b] Yield defined on carbon basis.

[c] Heating by microwave irradiation

[d] Continuous water removal.

[e] Cornsweet 90 high fructose syrup as the substrate.

TABLE 9.6

Selected Results[a] for Heterogeneous Catalysts in the Dehydration of Glucose in Aqueous Solutions

Catalyst	Catalyst Loading (wt%)	Temperature (°C)	Reaction Time (min)	Yield (%)	Conversion (%)	Selectivity (%)	Reference
HY-zeolite	50	160	3	8	83	10	50
TiO_2	100	200	5	20[b]	81	25	31,35,51
TiO_2-ZrO_2	100	250	5	29[b]	44	67	50
ZrO_2	100	250	5	17[b]	38	46	31,35,50

Source: See table for multiple data references.

[a] Best yield in HMF reported only in case of multiple references.

[b] 2%–2.5% fructose yield.

concept. Interesting is the work from Yan[48] who also used fructose as a substrate and observed lower yields in HMF than for glucose. This was ascribed to severe conditions required for the glucose isomerization that are detrimental to fructose if present in high concentrations. It also clearly shows that further optimization of these catalyst systems with respect to matching the isomerization and dehydration activity is needed.

9.4.3 OXIDATION OF FURANICS TO FDCA

The selective oxidation of HMF results in 2,5-furandicarboxylic acid (FDCA, Figure 9.6), which was identified by the U.S. Department of Energy to be a key bio-derived platform chemical.[9] Recently, Avantium Chemicals have announced collaborations with the Coca Cola Company, with Alpla and with Danone on the FDCA-based ethylene glycol polyester as a full bio-based material for use in water and carbonated soft drink bottles. Although the required selective oxidation of the alcohol and the aldehyde to corresponding carboxylic acid is a well-known reaction, the oxidation of HMF in particular has several features that need to be taken into consideration. First, as already mentioned, HMF is a rather unstable molecule. Prolonged heating at elevated temperature or in the presence of acids or bases should be avoided. Second, the solubility of the obtained FDCA in common solvents is very low, which is similar to terephtalic acid. Lastly, the oxidation needs to be performed at very high selectivity, as even small amounts of intermediates or decarboxylated products would be detrimental for the polymerization chemistry to follow. Several oxidation methods have been described in the literature so far and are shown in Table 9.7. In analogy with the oxidation of *para*-xylene to terephthalic acid, the Co/Mn/Br system has also been claimed by ADM[52] and Avantium Chemicals[53] for the oxidation of HMF and MMF, respectively.

TABLE 9.7
Oxidation of HMF to FDCA

Reaction Conditions	HMF Conversion	FDCA Yield	Reference
Co/Mn/Br, air (HMF)	91	61	51
Co/Mn/Br, air (MMF)			52
Pt/Al$_2$O$_3$ (basic conditions)		100	55
Co(acetylacetonate)-SiO$_2$ (from fructose), air	72	99	58
Pb-Pt/C (NaOH), air	100	81	59
Pt/C, air	100	95	60
Pt-ZrO$_2$; Pt/Al$_2$O$_3$, air	100	98	61
PtBi/C (from fructose, solid acid, H$_2$O/MIBK), air	50	25	62
Au/TiO$_2$ (basic conditions/O$_2$)	100	71	53
Au (hydrotalcite), air	100	100	56
Pt/C, Pd/C, Au/C, Au/TiO$_2$, air	100	79	63
NiO$_2$/OH anode (electrochemical oxidation)		71	64

Source: See table for multiple data references.

Heterogeneous catalysts can also be used in the oxidation of HMF to FDCA with molecular oxygen. Supported platinum catalysts were first used in the presence of base, resulting in near quantitative FDCA yields. Recently, two examples were reported using supported gold catalysts for aqueous HMF oxidation. Gorbanev et al. demonstrated that Au/TiO_2 could oxidize HMF into FDCA in 71% yield at near-room temperature[54] and Casanova et al. showed that Au/CeO_2 was even more active and selective.[55] However, it should be noted that these noble catalyst systems all require at least stoichiometric amounts of base to keep the FDCA formed in aqueous solution as the di-alkaline salt[56] and often high oxygen pressures (10–20 bar). This underlines the low solubility of the free acid in water even under the elevated reaction conditions. Recently, Gupta et al. reported the base-free oxidation over gold catalysts supported on hydrotalcites.[57] Since the FDCA formed is an acid, it can be expected that a reaction between the basic hydroxyl groups of the hydrotalcite and the produced FDCA will take place.

An interesting approach is to combine both the formation of HMF from fructose with the oxidation to FDCA in one pot. Similar to the approach via CMF and MMF, the aim is to convert the unstable HMF before it can react further to levulinates or degrade into humins. A FDCA yield of 25% was reported using Pt-Bi/C in combination with a solid acid in water/MIBK, constituting an overall 50% selectivity to FDCA on fructose.[61] Ribeiro reported the direct conversion of fructose to FDCA with high conversion and excellent selectivity (99%) using $Co(acac)–SiO_2$ as a bifunctional catalyst at 160°C and 20 bar air pressure.[58]

9.4.4 INDUSTRIAL PRODUCTION OF HMF

Full-scale industrial production of HMF has so far proven to be very difficult. Due to the unstable character of HMF, it is not only difficult to obtain HMF in high yields and selectivity in the reactor, it has often also been found that the subsequent work-up to the pure HMF is challenging in itself. In the past, two pilot-scale plants have been operated, the first by Roquette Frères in 1986[65] and the second by Suddeutsche Zucker-aktiengesellschaft in 1988.[66] Both plants have since closed. In 2011, Avantium Chemicals started up a pilot plant for the production of MMF and FDCA.

The process operated by Roquette Frères in France was based on the biphasic approach using methylisobutyl ketone (MIBK) or dimethoxyethane (DME) to continuously extract the HMF from the reaction mixture.[64] Cationic Lewatit SPC 108 was used at relatively low temperatures of 70°C–95°C. The organic phase was rich in HMF and also contained some levulinic acid, adding up to an HMF yield of 38% at 51% fructose conversion. Further details on the isolation of HMF from the organic phase and subsequent HMF purification were not described. However, considering that 36 l of MIBK was needed to be evaporated per 266 g of HMF produced and further purification of the crude HMF is still needed, one can assume that this process was not very economical. No economic data are available for this process, but for a similar process layout, Dumesic calculated the minimum cost price of HMF at 1.08$/kg.[67] However, the production data figures given add up to over 100% selectivity while the actual data from the best biphasic process reported indicates expected yields closer to 69%.[73] A low cost of 300$/tonne for pure fructose was also taken into account, and it would be fairer to at least double the minimum cost price for HMF.

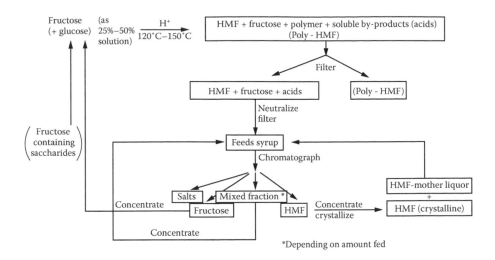

FIGURE 9.7 HMF production process as developed by Südzucker. (Adapted from G. Fleche, A. Gaset, J.-P. Gorrichon, E. Truchot, P. Sicard, 1982. Process for manufacturing 5-hydroxymethylfurfural, US4339387, to Roquette Frères.)

The second approach by Südzucker was based on water as a diluent and conditions that maximized HMF selectivity, rather than maximizing HMF yield or hexose conversion.[66] The process involved the use of a homogeneous catalyst such as sulfuric acid and an ion-exchange resin to obtain 99% pure HMF in crystalline form (Figure 9.7). For HMF production from fructose in an aqueous reaction system patented by Südzucker AG, an HMF manufacturing price of 6 DM/kg, equivalent to approximately €3/kg, was reported when fructose is available at a price of 0.5 DM/kg, and similar to the process described by Roquette. The major factor contributing to the cost price is the extensive purification rather than the catalytic conversion step.

9.4.5 PRODUCTION OF CHLOROMETHYLFURFURAL

Recently, Mascal reported a promising approach to obtaining high isolated CMF yields ranging from 75 to 89 wt% using glucose, sucrose, and cellulose as the feed in a biphasic reactor system.[28] The reactions were performed in a batch reactor using highly concentrated aqueous HCl with 1,2-dichloroethane (DCE) as the extracting solvent and 5 to 9 wt% levulinic acid as major by-product. The direct use of glucose is very surprising, although a study by Braholz[68] showed yields of 81% and 58% for fructose and glucose, respectively. Since concentrated HCl is used in the formation of the CMF, a change to a heterogeneous acid may be difficult to envisage. The hydration of CMF to HMF is also reported by Brasholz[67] with yields in HMF of 71% and 20% to the valuable levulinic acid as a major side-product. Similarly, Mascal reported 86% and 10% yields for HMF and levulinic acid, respectively.[69]

However, CMF is shown to have a considerable higher mutagenic and cytotoxic activity in bacteria compared with HMF,[71] which must be considered if CMF will be used as a platform chemical. However, Mascal already has shown that it is reasonably easy to convert CMF to HMF, 5-alkoxymethylfurfurals or levulinates. Although no process details have been given, the indicative price of 0.53USD/l CMF when starting from lignocellulosic feedstocks was reported by Mascal,[28] which would make this a very interesting platform chemical.

9.4.6 PRODUCTION OF 5-ALKOXYMETHYLFURFURALS

5-Alkoxymethylfurfurals were identified in the reaction mixture when reacting fructose in the presence of an alcohol (Figure 9.8). Bicker et al.[70] reported the formation of 5-methoxymethylfurfural (MMF) when the reaction was performed in supercritical methanol using sulfuric acid as the catalyst. The maximum selectivity of MMF was 84 mol% for a 1 wt% fructose solution in methanol, although at higher fructose loading the maximum selectivity of MMF dropped to 54 mol%. The formation of 5-ethoxymethylfurfural (EMF) was reported when ethanol was used as the solvent.[27,41] 5-Alkoxymethylfurfurals have been identified as promising HMF derivatives for biofuel applications. Avantium examined various types of alcohols for the reaction, ranging from low-chain alcohols such as methanol and ethanol, branched alcohols, higher-chain alcohols, and even mixtures of various alcohols.[27]

Since 2011, Avantium practices a continuous production of 5-alkoxymethyl-furufal at a pilot scale, and they claim that this is possible using heterogeneous catalysts, giving examples with zeolites. Some details of the process are discussed by Eerhart et al.[71] who performed a cradle-to-grave life-cycle analysis for PEF via MMF as the intermediate. Eerhart assumes a process that runs at 200°C–220°C, 50 bar, and has an 8-column distillation section for the recovery of MMF and methyl levulinate as the major products. Another aspect of the process was shown by Dias,[72] showing that starting with methyl glycosides as feedstock, much higher sugar concentrations in alcohol could be achieved compared with the normal solubility of fructose or glucose in those solvents. The combined yields of MMF and levulinates of over 75% using up to 25 wt% sugar solutions were presented.[72] Van Aken reported expected prices for FDCA of €0.6–0.9/kg, later updated as a

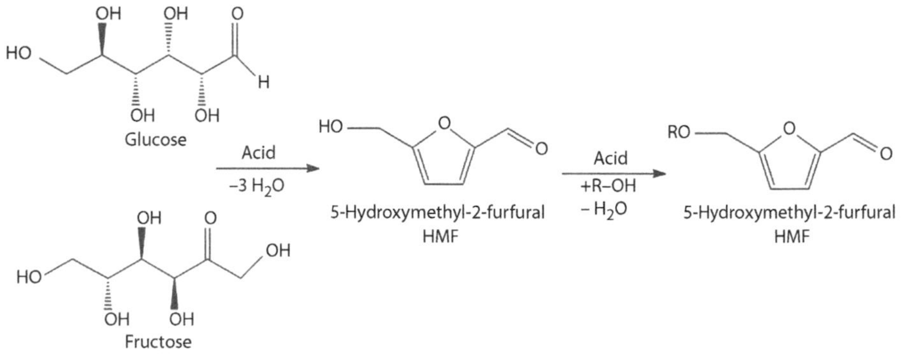

FIGURE 9.8 Formation of 5-alkoxymethylfurfural from fructose.

result of increased feedstock prices to €0.6–1.2/kg by de Jong for a 500-ktonne/year plant.[73]

9.4.7 Outlook for New Furanics Catalysts

The field of furanic chemistry has long been recognized as a potentially important way of generating platform chemicals; however, the low yields in HMF using fructose as a feedstock combined with its unstable nature made economic production unfeasible. In Table 9.8, the HMF, MMF and CMF routes are presented. All three processes are similar in their feedstock use, but as the limited economic data on these processes already suggested, the decisive factor is in finding a cost-effective purification. Although further improvements of catalysts for the production of HMF and/or MMF are still needed, the real challenge is in the development of catalysts that can convert much cheaper carbohydrate mixtures derived from lignocellulosic feedstocks. These aldose-rich feedstocks require both acid and base active sites, which cannot be obtained in a homogeneous catalyst system.

TABLE 9.8
Theoretical and Overall Yield for the Conversion of Sugars to FDCA via HMF, MMF and CMF Respectively

HMF Steps	Theoretical Yield[a] Weight/ Weight Fructose	Combined Theoretical Yield on Fructose	Best Reported Selectivity[b]	Combined Best Yields on Fructose	Overall Carbon Efficiency
Dehydration to HMF[c]	70.0 w/w%	70 w/w%	69%	48.3 w/w%	69.0%
Oxidation to FDCA	123.8 w/w%	86.7 w/w%	98%[b]	58.6 w/w%	67.6%
MMF Steps					
Dehydration to MMF	77.8 w/w%	77.8 w/w%	62.5%	48.6 w/w%[d]	62.5%
Oxidation to FDCA	111.4 w/w%	86.7 w/w%	98%[e]	53.1 w/w%	61.2%[e]
CMF Steps					
Dehydration to CMF	89.1w/w%	89.1 w/w%	90%	80.1 w/w%	90%[f]
Conversion in HMF	87.1 w/w%	70.0 w/w%	86%	60.0 w/w%	77.4%
Oxidation to FDCA	123.8 w/w%	86.7 w/w%	98%	72.8 w/w%	75.8%

Source: See table for multiple data references.

a Assuming a 100% molar yield and selectivity.
b On carbon basis.
c Assuming a biphasic process as used by Roquette, catalytic data from Kuster.[74]
d Combined MMF and HMF, considerable amounts, approximately 14.8 w/w fru% of levulinates (LA) are co-produced.[71]
e No catalytic data on continuous operation known. Eerhart assumes similar selectivities for HMF/MMF oxidation as for *para*-xylene oxidation.[70]
f Levulinic acid is coproduced in the first step (4.0 w/w%) and second step (5.9 w/w%).

9.5 PRODUCTION OF *para*-XYLENE FROM ISO-BUTANOL

Where FDCA and its subsequent PEF plastic are emerging as fully bio-based alternatives to terephthalic acid (TA) and its subsequent PET plastic, others in the field have focused on obtaining *para*-xylene, the starting material for the production of TA, from biomass. Recently, the Coca Cola Company announced collaborations with Virent and Gevo Inc. on the development of fully bio-based TA.[23] Virent's technology is based on APR and will be discussed later. The technology route developed by Gevo Inc. will be discussed first.

Gevo Inc. was founded in 2005 and went for an initial public offering (IPO) in February 2011. In the beginning of the company, Gevo licensed intellectual property developed by Jim Liao at UCLA and Frances Arnold at Caltech. Gevo claims that its fermentation processes can be retrofitted into existing ethanol plants with a limited amount of capital. The first of these retrofits was at St. Joseph, Missouri, in 2009, with an organism capable of making butanol. Of the three butanols that their technology can produce, iso-butanol is particularly attractive because it can be readily dehydrated into isobutylene and offers an opportunity as a renewable drop-in substitute into existing processes as shown in Figure 9.9. The product spectrums that can be obtained cover a wide range of renewable chemicals and renewable fuels such as jet fuel, high-octane gasoline, solvents, renewable terephthalic acid for PET bottles, and butyl rubbers. In 2010, Gevo announced a joint venture with Lanxess for the production of butenes, and in 2011, a collaboration with the Coca Cola Company on PET for beverage bottles.

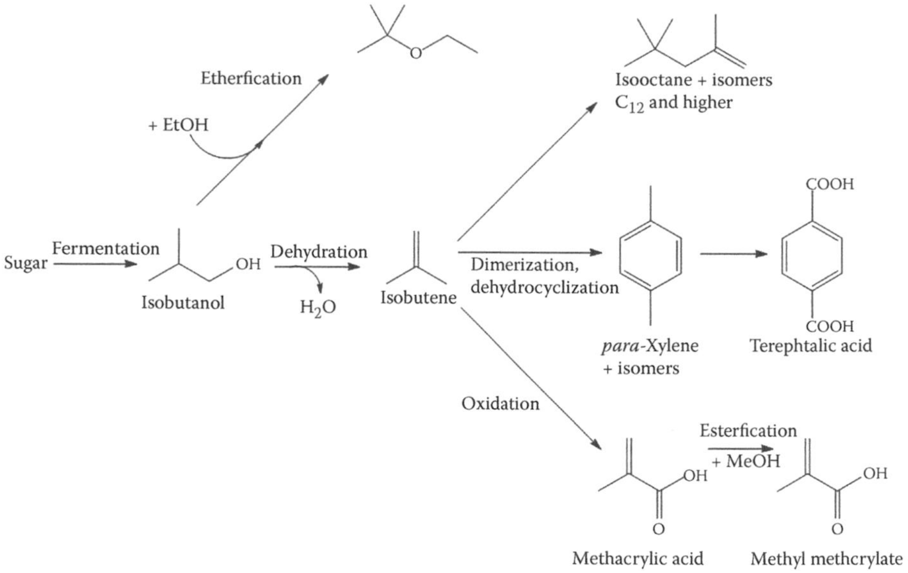

FIGURE 9.9 Some derivatives available from bio-isobutanol. (From J.D. Taylor, M.M. Jenni, M.W. Peters, 2010. Dehydration of fermented isobutanol for the production of renewable chemicals and fuel, *Top. Catal.* 53, 1224; B. Bernacki, 2011, *Biobutanol potential for biobasedmonomers and polymers at Bioplastics*, Las Vegas, NV, USA.)

So far, very little is disclosed in literature about the actual process of iso-butanol toward *para*-xylene. Clearly, the chemistry is based on the fermentation technology to iso-butanol developed by Gevo Inc.[76] The theoretical stoichiometric route for obtaining iso-butanol via the fermentation pathway developed by GEVO is

$$1 \text{ glucose} \rightarrow 1 \text{ } i\text{BuOH} + 2 \text{ } CO_2 + H_2O$$

Two moles of CO_2 are lost in this pathway, and even with a 100% theoretical molar yield of the reaction above, only a 41% theoretical weight yield of isobutanol per weight of glucose can be obtained. A practical isobutanol yield of approximately 82%–90% of theoretical was reported in the examples using corn starch as the starting sugar, which would be equivalent to 0.37 g isobutanol/g of glucose.[76,77]

An example of the fermentative production of isobutanol and its conversion to PX is given in the patent literature[79,80] and illustrated below. After fermentation, the next step is the dehydration of isobutanol. This can be achieved using a variety of commercial zeolite catalysts, and this can be tuned to give all different isomers of butene, both isobutene and double-bond isomers of *n*-butene.[79] The best catalyst reported was the BASF AL3996, which, at 99.8% conversion, gave a 96% yield to the desired iso-butene and 2.5% to the 2-butene (*cis* and *trans*) and 1% to 1-butene. The values for the 2- and 1-butenes are well below the thermodynamic equilibrium values,[75] suggesting that the H-ZSM-5 catalysts kinetically control the reaction. Fermentative routes to iso-butene are also explored.[78]

The dehydration is followed by dimerization of the isobutene,[78] although this step can be combined with the dehydration.[79,80] The oligomerization subsequently takes place in the presence of suitable zeolite catalysts. Avoiding the formation of trimer and higher oligomers is critical in this reaction. As such, the process is not run at full conversion and thus will require a recycle of unconverted butenes. Results for a HZSM-5 zeolite (Zeolyst CBV 2314) are given, which show a selectivity of 98.5 to octenes and 1.5% to trimers at 70% conversion. Since the linear butenes are less reactive, they will accumulate in the recycle stream. In another example, it was shown that after isomerization of the linear butenes to the thermodynamic equilibrium and combining with fresh iso-butene, this recycled feed gave 89% of the desired octenes and 10% of the trimers at 99% conversion. The major octenes formed are the 2,5-dimethyl hexanes. The trimers and higher oligomers are not considered waste products, as they are highly branched and would be excellent fuel compounds.

The next step is the dehydrocyclization reaction, which is catalyzed by a chromium oxide catalyst, BASF-1145E at 550°C. Each octane is converted in xylene, and additional H_2 is produced, which is captured for use with other processes. The highest yield in xylenes reported is 42% with a *p*-xylene fraction of 90%; however, isomerization of xylenes is a well-established process, and all xylenes produced are considered valuable products. It is also mentioned that unconverted octenes and butenes formed can be recycled in the process, and thus higher overall yields can be expected. As significant amounts of H_2 are formed in the dehydrocyclization, fully hydrogenated alkanes are formed as well, which cannot be recycled and only have an outlet as a fuel component. The entire process yields are reported by Dodds to yield 18.7 kg *p*-xylene per 100 kg glucose.[82]

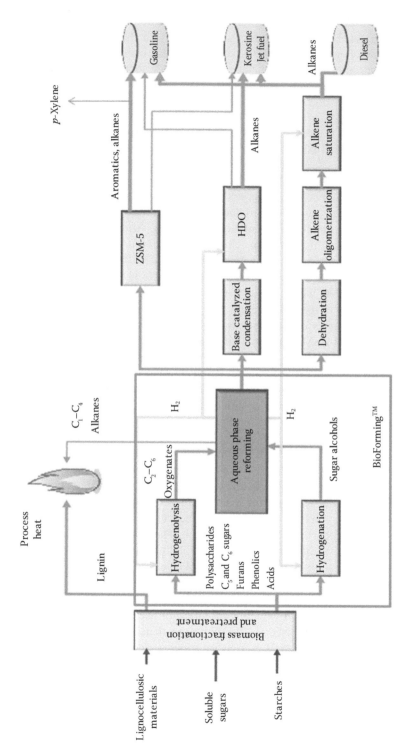

FIGURE 9.10 Production scheme for drop-in biofuels using Virent BioForming™ approach. HDO, hydrodeoxygenation; MTO/MTG, methanol to olefins/gasoline. (From Randy D. Cortright, 2012, *Aqueous-Phase Reforming Process*, 2nd Annual Wisconsin Bioenergy Summit.)

TABLE 9.9
Theoretical and Overall Yield for the Conversion of Glucose to *p*-Xylene

Step	Theoretical Yield[a] Weight/ Weight Substrate	Combined Theoretical Yield on Glucose	Best Reported Selectivity[b]	Combined Best Yields on Glucose	Overall Carbon Efficiency
Fermentation of glucose to isobutanol	41.1 w/w%	41.1 w/w%	90%	27.0 w/w%	35.2%
Isobutanol to isobutene	75.7 w/w%	31.1 w/w%	96%	26.8 w/w%	33.8%
Oligomerization to octenes	100 w/w%	31.1 w/w%	89.9%[c]	24.1 w/w%	30.4%
Cyclization of octenes to *p*-xylene	98.2 w/w%	30.5 w/w%	77.7%[d]	18.7 w/w%	23.6%
Oxidation of *p*-xylene to TA	156.6 w/w%	47.8 w/w%	98%	28.1 w/w%	23.1%

[a] Assuming a 100% molar yield and selectivity.
[b] For the reaction described assuming near 100% conversion by recycling unconverted starting material.
[c] Assuming isomerization and recycle of the linear butenes.
[d] Best reported selectivity calculated from the 18.7% yields reported by Dodds.[81]

The final step in the process would be the oxidation of the *p*-xylene to terephthalic acid. Worldwide, the BP/Amoco process is the most commonly used, which has a reported selectivity of 98% at 99.8% conversion to the purified terephthalic acid.[82]

9.5.1 Outlook for the GEVO Process

In Table 9.9 the best reported process yields for each process and for the overall process are tabulated. The best reported overall yield for TA on glucose of 28.1 w/w% means that 3.55 kg of glucose is needed per kg of TA. With historic prices for TA in the range of €0.6–1.2/kg in the 2005–2009 period,[72] and not taking processing costs into account, very low sugar prices will be needed to make this an economically viable process.

9.6 AROMATIC CHEMICALS VIA BTX BIOREFINING

Another route to *para*-xylene, and from there into terephthalic acid, is by BTX biorefining. This is the direct production of benzene, toluene, and xylenes (BTX) in an analogous route compared with a petroleum refinery without the involvement of a biological step. As petroleum refining is fully developed for the conversion and production of hydrocarbons, these technologies are based on the initial formation of low-oxygen–containing hydrocarbons from the oxygen-rich biomass. The attraction of this approach is obvious: once the biomass is converted into biohydrocarbons, it can be used in the existing, fully integrated, continuous refining processes, by either using existing infrastructure or making use of fully developed catalytic processes. Starting from lignocellulosic material offers even the possibility to burn the lignin to cover a majority of the energy needs. The major drawback of this approach is due to the high oxygen content of biomass, which requires externally sourced hydrogen, or the sacrifice of part of the biomass to generate hydrogen (and CO_2) via the water-gas shift reaction.

At present, several companies are working to commercialize biorefinery technology; here we discuss in more detail the technologies of Virent Inc. and Anellotech Inc.

9.6.1 VIRENT BIOFORMING™ PROCESS

Virent is developing technology for the conversion of biomass-derived sugars into hydrocarbons that are useful as fuels and industrial chemicals, including aromatics, using the BioForming™ process technology initially established by Cortright and Dumesic.[83,84] Recently, Virent was announced to be the third partner in a collaboration with the Coca Cola Company on the development of fully biobased terephthalic acid.[85] The BioForming™ process uses an aqueous sugar feed-stream and is able to convert it in a wide range of products, including aromatics.[86] Typically, three key steps can be identified.

In the first step, an aqueous carbohydrate stream is subjected to catalytic hydrogenation and/or hydrogenolysis to produce sugar alcohols and low-molecular-weight oxygenates, respectively. The typical reaction temperature is in the range of 100°C–150°C. The hydrogenolysis process relies on heterogeneous catalysts, with patent examples including a supported ruthenium-based hydrogenation catalyst and a supported platinum/rhenium catalyst. The hydrogenation of sugars to sorbitol is long known and can already be achieved with very high selectivity using Ni-based catalysts.[87] The required hydrogen preferentially is generated on site, for which Virent is using the aqueous phase reforming (APR) process, as outlined below.

In the second step, the oxygenates from the hydrotreating step then enter a process known as APR. The process reduces the oxygen content of the feed. Supported platinum catalysts modified with rhenium are typically used for this step. Virent claims that this step is the key to their process and relies on proprietary catalyst technology, which in a presentation by Cortright appears to be a NiSn-based material.[88] The process temperature is typically in the range of 177°C–302°C, depending on feedstock. Products from the process include "mono-oxygenates," smaller hydrocarbons, and significant amounts of molecular hydrogen (see Figure 9.11). The formation of hydrogen is critical as this is needed in the first step to hydrogenate/hydrogenolyse the sugar stream. Product distribution (paraffin vs. oxygenate; type and molecular weight of oxygenate) can be controlled by the reaction conditions and the catalyst composition. Excess hydrogen production is accompanied by biomass consumption and co-production of CO_2. The excess H_2 is fed back to the first hydrogenation/hydrogenolysis reactor to generate the isosorbide or used elsewhere on the plant site.

To come to BTX, a third and last step is needed to convert the combined monooxygenates and hydrocarbon stream into aromatics. The patent examples show that the product distributions depend on feed composition, catalyst, and reaction conditions, but yields as high as 75% of the desired aromatics C8–C9 range can be obtained using an acid ZSM-5 catalyst at a typical reaction temperature of 375°C.[89]

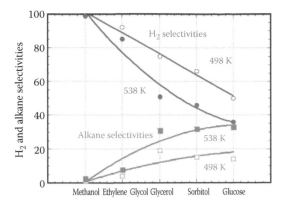

FIGURE 9.11 (See color insert.) Aqueous phase reforming of 1 wt% methanol, ethylene glycol, glycerol, sorbitol, and glucose. (From R.D. Cortright, 2012. *Aqueous-Phase Reforming Process*, 2nd Annual Wisconsin Bioenergy Summit.)

A number of the Virent patent applications and issued patents describe variations in the overall process to produce mixtures of aromatics in the C6 to C9 range.[89] Of particular interest is that the process can produce BTX along with ethylbenzene (precursor to styrene), all of which are commercially important, high-volume aromatic chemicals.[90] The overall reaction for the production of *para*-xylene produces 2.5 CO_2 per mole of *para*-xylene formed, which would indicate a maximum carbon efficiency of 76.2% and a loss of 23.8% of the carbon to CO_2.[82] As the production of CO_2 is directly linked to the amount of hydrogen present in the final products and significant amounts of lower (C1–C5) paraffins are produced, the true carbon losses to CO_2 will be significantly higher. Based on sucrose the equation for *para*-xylene is

$$0.875\ C_{12}H_{22}O_{11} \rightarrow C_8H_{10} + 2.5\ CO_2 + 4.6\ H_2O$$

Simplifying the reactions using a general $C(H_2O)$ carbohydrate as a source of carbon and as a source for hydrogen with CO_2 co-production, the chemistry can be described by three equations. Each product can be seen as a combination of C, H, and O.

$$C(H_2O) \rightarrow C + H_2O$$

$$C(H_2O) + H_2O \rightarrow 4\ H + CO_2$$

$$H_2O \rightarrow O + 2\ H \text{ (equivalent to } -0.5\ CO_2)$$

The theoretical CO_2 for an average product composition $C_nH_xO_y$ obtained can be deducted to be

$$\text{Loss } CO_2 = 100 \times (0.25x - 0.5y)/(n + 0.25x - 0.5y)$$

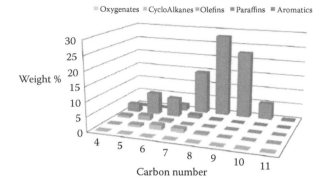

FIGURE 9.12 Example of paraffin, olefin, and aromatics distribution. (Patents with specific examples of aromatics generation via the BioformingTM Process: US8017818; US2011/0257448; US8053615; US2008/0300434, WO2011/082222; US7977517 all to Virent Energy Systems, Inc.)

A typical characteristic of the Virent process is the fact that it is not selective in the production of *para*-xylene. A wide range of products is claimed (see Figure 9.12), and in Table 9.10, compositional data for a sucrose/xylose mixture and sorbitol are given. Despite claims that lignocellulosic biomass can be used, no example has been found in the extensive patent literature covering the Virent's BioForming™ technology.[83,90] Although the example given does not specifically mention CO_2, it can be calculated using the above outlined approach. The results are shown in Table 9.10 indicate that close to 36% of the carbon present in the starting carbohydrates is converted into CO_2.

TABLE 9.10

Conversion of Sucrose/Xylose (93/7) and Sorbitol to Paraffin and Aromatics

	Product Yield (wt% of Feed Carbon)[a]				
Feed	Total C1–C3	C5+ Paraffin	C5+ Olefins	Aromatics	Estimated CO_2 Loss[b]
Sucrose/xylose (93/7)	22.3	20	0.8	25.0	35.1
Sorbitol	18.1	11.3	7.8	22.3	29.8 (35.9)[c]

Source: Data from G.W. Huber, J.N. Chheda, C.J. Barrett, J.A. Dumesic, 2005. Processing of biomass-derived carbohydrates production of liquid alkanes by aqueous-phase. *Science*, 308, 1446; see also US5736478; US6441241; US6964758; US7977517; US8075642; WO2009129019.

[a] Data from Example 55 of US Patent 7,977,517.[83]

[b] Calculated assuming ethane, pentane, butene, and *para*-xylene as representative examples for each of the product classes.

[c] In brackets the value compensated for the additional H_2 needed in the production of sorbitol.

Plans to build a 37,800-tonne/year aromatics plant have been announced, and it can be expected that the *para*-xylene will be used for PET production.[91] To secure the availability of sufficient sustainable C6 and C5 sugar streams in the future, Virent has recently announced a partnership with HCl CleanTech, now Virdia, a company that is developing technology for large-scale conversion of lignocellulosic material from nonfood plant sources to mixed C5–C6 sugars.[92]

9.6.2 ANELLOTECH "BIOMASS TO AROMATICS™" PROCESS

Using technology originally developed by Huber,[93] Anellotech is further developing technology for the direct conversion of biomass to chemicals, including the BTX aromatics, using a proprietary process called "Biomass to Aromatics™." The basic process is the catalytic pyrolysis of dried, ground biomass using zeolites as the catalyst. The process is described in detail and can use simple sugars as well as raw biomass.[94]

Three basic steps in the process can be identified. In the first step, the biomass source is prepared by drying and grinding it to the desired particle size range. All forms of biomass have been claimed, and patent examples given include lignocellulosic, wood, xylitol, cellobiose, sugarcane, and hemicellulose, corn stover.[96]

In the second step, the pretreated biomass is catalytically pyrolyzed into a fluidized bed reactor. By very rapid heating of the biomass, the process is geared to produce mostly gaseous materials in favor over coke formation, and this gives a much better output for downstream processing.

In the third step following the pyrolysis, the gas rapidly enters a reactor bed with a catalyst that is specifically chosen to favor formation of aromatic products. Similar to the processes by Gevo and Virent, acidic zeolites, H-ZSM-5 in particular, are used. A high catalyst to biomass feed ratio, and lower weight hourly space velocity are preferred conditions for high conversion to aromatics. Alternatively, the pyrolysis temperature can be set to ca. 600°C to favor aromatics formation in the pyrolysis step, although more coke is formed under these conditions.

An example of the distribution of products obtained from the process is shown in Figure 9.13.[95] Yields of aromatics can be as high as 30% based on biomass carbon recovered as products. Purer sugars give higher amounts of the desired aromatic fraction compared with raw lignocellulosic biomass, which is most likely due to the high amounts of lignin present in the latter feedstock. The coke can easily be recovered, and it is suggested that it is burned to generate the required heat (and power) for the pyrolysis process.

The yield and distribution of aromatic chemicals from pyrolysis of wood using the Annelotech process is shown in Figure 9.14. It is clear from this figure that a considerable amount of heavy aromatic, i.e., the naphthalene fraction, is present, and that from the 30% carbon yield, about half is toward the desire desired C6–C8, i.e., BTX, and ethylbenzene, fraction. The typical catalyst for such a conversion would be ZSM-5, as other zeolites have been reported but show very high amounts of coke formed.[96]

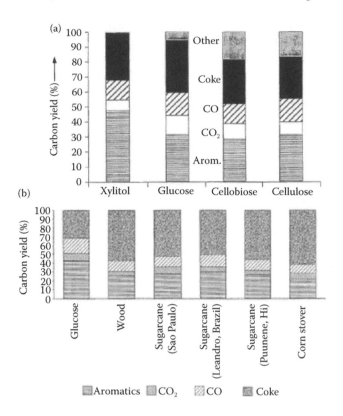

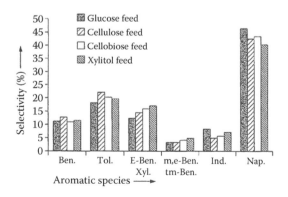

FIGURE 9.13   Product distribution in the Anellotech process for several biomass sources. (From G.W. Huber, J. Jae, T.P. Vispute, T.R. Carlson, G. Tompsett, Y.-T. Cheng, 2009. *Catalytic Pyrolysis of Solid Biomass and Related Biofuels, Aromatic, and Olefin Compounds*, US20090227823 and WO2009111026, to University Massachusetts.)

FIGURE 9.14   Yields of aromatic chemicals from pyrolysis of various biomass sources. (From G.W. Huber, J. Jae, T.P. Vispute, T.R. Carlson, G. Tompsett, Y.-T. Cheng, 2009. *Catalytic Pyrolysis of Solid Biomass and Related Biofuels, Aromatic, and Olefin Compounds*, US20090227823 and WO2009111026, to University Massachusetts.)

9.6.3 Outlook for Bio-Aromatic from Direct Catalytic Routes

Both the Virent BioForming™ process and the Annellotech "Biomass to Aromatics™" process are capable of producing the desired BTX from biomass. Both technologies are at an early stage of development and still must address the challenge of the high oxygen content of biomass compared with hydrocarbons. At the current state of technology reported, neither process seems to be capable of producing aromatics exclusively, and further catalysis research is needed to improve the process and the products yields. Based on biomass carbon consumed, both technologies as currently described in the patent literature are capable of generating 25%–30% yields on carbon of aromatics, although not all of this is in the commercially very important C6–C9 range, including BTX and ethylbenzene.

Comparing the aromatics yields for both processes is shown in Table 9.11. For both processes, the best reported examples using C6 sugars are used. For Virent's BioForming™, this is a glucose/xylose 97/3 mixture, and for Annellotech's Biomass to Aromatics™ process, this is pure glucose. In case of Annellotech, examples are also available using raw lignocellulosic biomass. The yield in aromatics decreases significantly, which is probably due to the lignin present in these materials, giving increased amounts of coke. Since the coke is burned in the process for energy production, one should also include a credit for electricity produced in these plants. Unfortunately, no data are available to estimate these credits. Both processes appear to produce a relatively low amount of *para*-xylene (starting molecule for terephthalic acid) compared with those discussed earlier. But this is offset by a much more diverse product palette, and one could consider both these processes to be true representatives of biorefineries, where as the other described processes would be bioequivalents of the chemical industry.

TABLE 9.11

Theoretical and Overall Yield for the Conversion of Pure Sugars by the Virent BioForming™ and Annellotech Biomass to Aromatics™ Process

Step	Theoretical Yield[a] Weight/ Weight Glucose	Best Reported Selectivity[b]	Combined Best Yields on Fructose	Overall Carbon Efficiency
BioForming™ to *p*-xylene	33.7 w/w%	25%	11.0 w/w%	24.9%
Oxidation of *p*-xylene to TA	52.7 w/w%	98%	16.9 w/w%[c]	24.4%
Biomass to Aromatics™ Steps				
Pyrolysis to *p*-xylene		22.8%	10.1 w/w%	22.8%
Oxidation of *p*-xylene to TA		98%	15.5 w/w%[d]	22.4%

[a] Assuming a 100% molar yield and selectivity.
[b] On carbon basis assuming all aromatics to be benzene or xylenes.
[c] The BioForming™ also produces about 20.7 w/w% hydrocarbons.
[d] The Biomass to Aromatics™ co-produces ~13.7 w/w% of naphthalene.

9.7   PRODUCTION OF ETHYLENE AND GLYCOLS FROM ETHANOL

With a total annual production exceeding 50 Mtonne annually, ethanol is at this moment the most abundant nonfood chemical produced from biomass. Ethanol is a liquid at room temperature, easy to volatilize, and, compared with the sugars it is made of, has a low oxygen content. Not surprisingly, many industries have considered that this would be an ideal biofeedstock to use (Figure 9.15). Braskem, JGC, and Solvay have announced plans for ethylene production; India Glycol produces ethylene glycol, and Teijin and Greencol Taiwan have announced plans; Braskem reported propylene and propanol as products; Lucite reports a completely new route to methylmethacrylaat (MMA) based on ethylene, methanol, and carbon monoxide.[6] It does not take much imagination to see many other products such as acetic acid, acetaldehyde, diethyl ether, ethyl acetate, and vinyl acetate as possibilities.[96]

Here the synthesis of ethylene from ethanol is discussed. Braskem has 100- and 200-ktonne/year plants operational in Brazil and recently announced a 600-ktonne/year plant in Mexico to be operational by 2015. Solvay has a small 60 ktonne/year plant operational. The technology is far from a new process. The first papers go back to 1783–97[97], and the first plant was operated by Elektrochemische Werke Gmbh in Germany in 1913. Ethylene was not the primary product, rather ethane for refrigeration use. During the Second World War, ethylene was almost exclusively made by dehydration in Germany, Great Britain, and the United States.[98]

Ethylene is produced by acid-catalyzed dehydration of ethylene, with diethyl ether formation as the main side reaction (see Figure 9.16). This intramolecular dehydration reaction to ethylene is highly endothermic: $\Delta H = +10.82$ kcal/g-mol. Other products are formed as well. Table 9.12 gives an overview of the typical product distribution

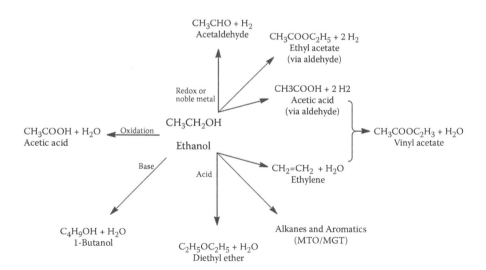

FIGURE 9.15   Ethanol as a renewable source for chemicals production. (Adapted from B. Voss, 2011, Value-added chemicals from biomass by heterogeneous catalysis, PhD thesis, Department of Chemical and Biochemical Engineering, Technical University of Denmark.)

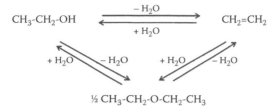

FIGURE 9.16 Main reactions in ethanol dehydration.

TABLE 9.12
Typical Product Distribution in Ethanol Dehydration at 99% Conversion for a Commercial Catalyst

Product	Selectivity
Ethylene	96.8
Ethane	0.5
Propylene	0.06
Butylenes	2.4
Acetaldehyde	0.2

Source: Data from Ethylene from Ethanol. Brochure Business Area Biomass Chemicals of the Chematur Engineering Group, http://www.chema tur.se/sok/download/Ethylene_rev_0904.pdf, accessed July 2012.

---

observed. Morschbacker cites a number of by-products besides diethyl ether that have been observed in ethanol dehydration: acetic acid, ethyl acetate, acetone, methanol, methane, ethane, propane, propylene, *n*-butane, butylene isomers, hydrocarbons with 5 carbons or more, carbon monoxide, and carbon dioxide.[98] By-products formed from ethanol can in some cases still be converted to the desired ethylene, for example diethyl ether (reaction 3, Figure 9.16). Other side products such as the oligomerization to butylenes and the dehydrogenation to acetaldehyde and hydrogen are losses that will need to be avoided. Another important side product from ethanol or ethylene is coke. Coke is particularly unwanted as it causes catalyst deactivation.

The last group of by-products come from impurities in the feedstock. These include the dehydration of the higher alcohols present in the ethanol. These are co-produced during the fermentation and called fusel alcohols. For example, propene can result from the dehydration of the 1-propanol.

9.7.1 CATALYST FOR THE DEHYDRATION OF ETHANOL

In general, heterogeneous catalysts are used for the ethanol dehydration. It can be said that there are two drivers behind catalysts selection. The first is related to the relatively high cost of ethanol, which requires highly selective catalysts and in particularly coke formation is to be avoided. The second finds its origin in the strongly endothermic nature of the reaction. This requires a reactor design to deal with the decrease in temperature (adiabatic operation) or the striving for isothermal operation.

Adiabatic operation results in a temperature profile along the catalysts bed. The temperature drop in the catalyst bed can easily reach 100°C and the reactor feed must be heated so that the exit temperature of the catalytic bed is still high enough to convert the last remaining traces of ethanol. The much higher temperature at the top of the reactor, however, also means that the catalyst in this area is very active, and care should be taken that unselective reactions, especially coke formation, are not taking place.

Both fixed bed and fluidized bed have been used as isothermal reactor systems. Fixed bed reactors are usually described as isothermal since the heat transfer is supposed to maintain a nearly constant reaction temperature, although the low heat transfer rates in the catalyst bed typically result in a drop in the actual bed temperature sometimes up to 100°C–110°C as mentioned by Kagyrmanova et al.[100] The temperature then rises slowly and reaches essentially the feed temperature at the reactor exit. Lower temperature dips of about 25°C have been reported by Morais et al. for a multitubular reactor system.[101] They observed the same fast temperature drop at the beginning, followed by a slow recovery. Thermal shock resistance is thus an important factor in designing the best catalyst. Although fluidized beds have the advantage of a constant temperature and thus better control of the reaction, operating fluidized bed reactors at the industrial scale is a costly operation, and the constantly moving catalyst particles suffer from considerable attrition. No commercial exploitation of fluidized bed reactor technology has been reported.

Many catalysts have been proposed for this reaction, although, in general, the main dehydration reaction is not that demanding and any type of acidity can do the trick given time and temperature. Evaluation of recent literature for highly active materials, i.e., over 90% activity, shows that catalyst systems based on modified H-ZSM-5, SAPO-34, and alumina are the three materials used. Modifications of the materials with metal salts and/or inorganic acid are generally not to enhance selectivity, but rather to increase the stability of the catalyst. As mentioned, coke formation, coupled with deactivation of the catalysts should be the primary focus of research. Pilot-plant studies have shown that over 2000-h time on stream can be obtained[102] and over 8000 h have been claimed[99] for ZSM-5–based catalysts, but research on new materials rarely mentions testing beyond 50-h time on stream.

Once ethylene is obtained, it seems that the conversion to ethylene oxide and subsequent ethylene glycol is straightforward. Both technologies have long been known, and catalyst systems have been researched and optimized over many decades. However, at the time of this writing, only Indian Glycol had developed a process for ethylene glycol, based on bioethylene from Braskem, and GreenCol has announced plans. The Coca Cola Company recently announced a partnership with JBF Industries Ltd. to further expand production of monoethyleneglycol. Construction on the new facility is expected to begin at the end of 2012 and last for 24 months. At full capacity, it is estimated the facility will produce 500,000 metric tons of MEG per year.[103]

TABLE 9.13

Theoretical and Overall Yield for the Conversion of Pure Sugars to Ethylene via Ethanol Dehydration

Step	Theoretical Yield[a] Weight/ Weight Glucose	Combined Theoretical Yield on Glucose	Best Reported Selectivity[a]	Combined Best Yields on Glucose	Overall Carbon Efficiency
Ethanol fermentation	51.1 w/w%	51.1 w/w%	94.5%	48.3 w/w%	63.0%
Dehydration of ethanol to ethylene	60.9 w/w%	31.1 w/w%	98.5%	30.0 w/w%	62.1%
Oxidation to ethylene oxide	157 w/w%	48.9 w/w%	82%	37.3 w/w%	50.9%
Hydration of EO to ethylene glycol	140.9 w/w%	68.9 w/w%	99.7%	52.4 w/w%	50.8%

[a] Assuming a 100% molar yield and selectivity.

9.8 OUTLOOK OF BIOETHYLENE IN ETHYLENE GLYCOL PRODUCTION

The production of bioethylene cannot be fully considered without taking the fermentation to ethanol in account. Theoretically, one CO_2 is produced for each molecule of ethanol formed. Thus, the theoretically maximum recoverable amount of ethanol is 51.1 w%/w% on glucose. Current-day efficiencies in the fermentation of pure glucose report numbers, which are in the lower 80s, of the theoretical recoverable ethanol, i.e., around 483 kg ethanol per tonne of sugar used.[104] In the future, a shift to lignocellulosic ethanol is foreseen. Badger reported a total yield/tonne of about 227 l (179 kg) of ethanol per tonne of corn dry stover used fermenting both the glucose and the xylose present.[105] If one could attain 95% for both efficiencies, then the yield would be approximately 350 l (277 kg) of ethanol/tonne of biomass. Table 9.13 shows that with the current status of technology, approximately 4 kg of pure glucose or any other fermentable sugar are needed to obtain 1 kg of ethylene. For the conversion to ethylene glycol, via ethylene oxide, this value improves to slightly more than 2 kg glucose per kg of glycol due to the incorporation of oxygen and water in the ethylene.

9.9 ROQUETTE PROCESS TO ISOSORBIDE

Another good candidate from glucose is isosorbide, which is the 1,4:3,6-dianhydro-hexitol, which can be derived from glucose by hydrogenation to sorbitol and twofold dehydration via sorbitan. Currently, it is mainly used in pharmaceutical formulation, but its use in polymers (PEIT and PolySorb™ plastizicer) is starting and application as fuel additive has been reported.[106] The production of isosorbide from sorbitol relies on mineral acids as catalysts and more environmentally benign methods are desirable. Fermentative routes to produce sorbitol are also suggested, but it is unlikely that these routes can replace the technically mature catalytic hydrogenation process.[108]

Based on research performed by Fleche and Huchette in the 1980s,[108] Roquette Frères have commercialized the production of isosorbide from sorbitol (Figure 9.17). There are currently two industrial processes, by Roquette and by Dupont, and both use sulfuric acid as the catalyst. In a similar manner to the dehydration of sugars to HMF, the reaction just requires Brønsted acid sites. The chemistry of the reaction is well understood and consists of various equilibrium reactions toward the desired product. This is opposite to what is the case with HMF formation. As a consequence of these equilibria, it is required to remove the water as soon as it is formed to drive the reaction to completion. The commercial processes are claimed to run at 80%–85% yields of sorbitol at full conversion.[110]

If water is not removed sufficiently fast, both further dehydration reactions occur into colored products, especially above 135°C,[108] and rehydration reactions take place with potential inversion on either of the chiral centers C2 to C5. In the latter case, two additional sorbides are formed, e.g., the *exo–exo* configuration (isoiodide) and the *endo–endo* configuration (isomannide). The removal of water is typically done by vacuum, which implicates that batch or continuous stirred tank reactors are preferentially used in the process.

An interesting new approach is investigated by Moulijn in collaboration with Petrobras.[110] In their process, cellulose is directly converted to sorbitol by dissolving the cellulose in $ZnCl_2$ with some HCl as acid and a supported Ru hydrogenation

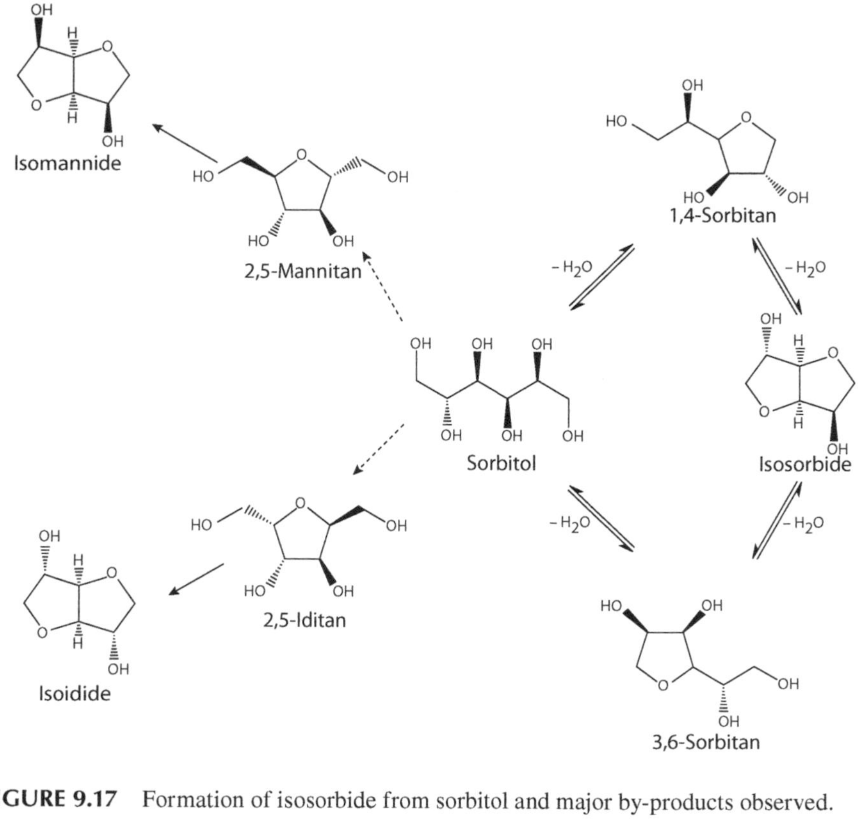

FIGURE 9.17 Formation of isosorbide from sorbitol and major by-products observed.

catalyst. The ruthenium catalyst tolerates the acidic media while maintaining high sorbitol selectivity, but further catalyst research is needed to overcome the reported loss in activity. The acid catalyst is both responsible for the hydrolysis of cellulose in glucose and for the dehydration of the sorbitol to isosorbide. Since the hydrolysis consumes half of the water liberated in the isosorbide dehydration, combining both reactions in one pot partially solves the need for fast water removal as described above.

9.9.1 HETEROGENEOUS CATALYSTS

Heterogeneous catalysts have been reported for isosorbide production. Zeolites have also been used in the past,[111] but new materials have recently been described as well (see Table 9.14). In general, they report lower yields in the desired product. Recently, an effort in the USA has started that is aimed at finding more selective heterogeneous catalysts for the process with an improved yield over 90%.[109] However, considering the catalysis that is involved in this reaction, one can easily identify several constraints. Considering the homogeneous reaction as the most relevant background, clearly the active site should be a strong Brønsted acidic site. In the commercial process, water is removed immediately. The presence of water not only slows down the reaction rate, the rehydration can also give rise to inversion of the chiral centers and thus lowers the yield in the desired product. Any heterogeneous catalyst need to be very apolar in order not to retain water too strongly close to its active sites. Either a dispersed slurry reaction in liquid phase or a high-temperature gas-phase reaction would be most suited. Another potentially interesting route that heterogeneous catalysts offers would be the one of catalytic membranes, designed to remove the water very fast from the active sites.

TABLE 9.14
Heterogeneous Catalysts for Isosorbide Formation from Sorbitol

Catalyst	Conditions	Yield at >90% Conversion	Reference
Cs-HPA			113
Amberlyst-35	Melt, 135°C, 9–12 Torr	73.4% Sorbitol 18.6% Sorbitan	114
Sn-PO$_4$, Zr-PO$_4$, Ti-PO$_4$	Gas phase 300°C	47%–52% Sorbitol	115
HPA-ZrO$_2$	Gas phase 250°C	71.4% Sorbitol	116
NiO-C	250°C	28.9% Sorbitol 38.0% Sorbitan	117
Sulfated-CuO	Fixed bed 200°C	67.3% Sorbitol	118
Nb$_2$O$_5$-H$_3$PO$_4$	Fixed bed	62.5% Sorbitol 0.5% Sorbitan	119

Source: See table for multiple data references.

TABLE 9.15

Theoretical and Overall Yield for the Conversion of Pure Sugars to Isosorbide Through Sorbitol

Step	Theoretical Yield Weight/Weight Substrate	Combined Theoretical Yield on Glucose	Best Reported Selectivity	Combined Best Yields on Glucose
Hydrogenation of glucose to sorbitol and mannitol	101.1 w/w%	101.1 w/w%	99%	100.1 w/w%
Dehydration to isosorbide	80.2 w/w%	81.1 w/w%	85%	68.3 w/w%

Source: Data from Bhatia, K.K. 2005, Process innovations for simultaneous reaction, separation, and purification of isosorbide, World Congress of Chemical Engineering, 7th, Glasgow, United Kingdom, p. 10.

9.9.2 OUTLOOK OF ISOSORBIDE

Despite the overall yield of isosorbide in the current homogeneous process, which is already close to the theoretical maximum, further improvements of the catalysts are highly sought after. The reason for this finding originates from the complexity of the purification of this process. The common industrial practice is to perform a melt crystallization, or less common, a high vacuum distillation. Both processes are very expensive and reduction of impurities would result in considerable gains in process costs (Table 9.15).

Isosorbide is already produced industrially for pharmaceutical purposes. Due to the high purity specifications required, melt crystallization is the preferred workup procedure, and the current price of isosorbide is above 5$/kg. Studies by Dupont have indicated that production prices of polymer-grade isosorbide of 1.10 to 2.20$/kg can be achieved by changing to a high-vacuum distillation process.[119]

9.10 GENERAL CONCLUSIONS

The sugar-based chemistries presented have shown that catalytic routes toward valuable biochemicals are not only possible but hold great promise to be cost competitive with chemicals derived from petrochemical sources. All chemistries evaluated require significant amounts of sugar per kilogram of chemical. This is directly related to the high oxygen content of biomass in general and carbohydrates in particular, which needs to be reduced either by the elimination of CO_2 or loss of H_2O via dehydration. In all cases, this oxygen elimination is accompanied by significant loss of mass of starting material.

The yield of product per kilogram of carbohydrate used is relatively low, and it is thus clear that the feedstock price will become an important driver for the economic production of biochemicals. Further reduction of the cost price of these biochemicals should come from improved yields by using more selective catalysts. The use of heterogeneous catalysts has been suggested for several of the processes as most likely candidates to achieve this. Heterogeneous catalysts can not only improve selectivity and yields for the current processes but will also allow a shift to cheaper but less refined, i.e., lignocellulosic, carbohydrate sources.

REFERENCES

1. G.W. Huber, S. Iborra, A. Corma, 2006. Synthesis of transportation fuels from biomass: chemistry, catalysts, and engineering. *Chem. Rev.*, 106, 4044, Cited in reference 3.
2. L.R. Lynd, C.E. Wyman, T.U. Gerngross, 1999. Biocommodity engineering. *Biotechnol. Prog.*, 15, 777. Cited in reference 3.
3. Y.-C. Lin, G.W. Huber, 2009. The critical role of heterogeneous catalysis in lignocellulosic biomass conversion. *Energy Environ. Sci.*, Published on 20 November 2008 on http://pubs.rsc.org doi: 10.1039/B814955K.
4. B. Kamm, P.R. Gruber, M. Kamm, 2007. Biorefineries, industrial processes and products. In: *Ullmann's Encyclopedia of Industrial Chemistry*, 7th ed.
5. M.J. Climent, A. Corma, S. Iborra, 2011. Converting carbohydrates to bulk chemicals and fine chemicals over heterogeneous catalysts. *Green Chem.*, 13, 520; A. Corma, S. Iborra, A. Velty, 2007. Heterogeneous catalysts for the one-pot synthesis of chemicals and fine chemicals. *Chem. Rev.*, 107, 2411.
6. E. de Jong, A. Higson, P. Walsh, M. Wellisch, 2012. Product developments in the bio-based chemicals arena. *Biofuels, Bioprod. Bioref.*, doi: 10.1002/bbb.
7. J.J. Bozell, G.R. Petersen, 2010. Technology development for the production of bio-based products from biorefinery carbohydrates—the US Department of Energy's "Top 10" revisited. *Green Chem.*, 12, 539.
8. M. Dohy, E. De Jong, H. Jørgensen et al. 2009. *Adding Value to the Sustainable Utilisation of Biomass*, RRB5 conference, Ghent.
9. T. Werpy and G. Petersen, 2004. *Top Value Added Chemicals from Biomass*, Pacific Northwest National Laboratory, and the National Renewable Energy Laboratory.
10. F. Cherubini, G. Jungmeier, M. Wellisch, T. Willke, I. Skiadas, R. van Ree, E. de Jong, 2009. Toward a common classification approach for biorefinery systems. *Biofuels, Bioprod. Bioref.*, 3(5), 534.
11. L.E. Manzer, J.C. van der Waal, P. Imhof, 2013. The industrial playfield for the conversion of biomass to renewable fuels and chemicals. In: *Catalytic Process Development for Renewable Materials*. John Wiley and Sons.
12. Advanced Biofuels & Chemicals Project Database from www.biofuelsdigest.com, Release 2.0 (11/16/11), accessed July 2012; http://www.biofuelsdigest.com/bdigest/2011/11/16/advanced-biofuels-chemicals-capacity-to-reach-5-11b-gallons-by-2015-207-projects-new-database/; Biofuels Venture Value Calculator from www.biofuelsdigest.com, Release 1.2, Mar 2012, accessed July 2012, http://www.biofuelsdigest.com/bdigest/2011/02/08/excel-xls-compatible-version-of-biofuels-venture-valutation-tool-released/.
13. H.E. Hoydonckx, W.M. Van Rhijn, W. Van Rhijn, D.E. De Vos, P.A. Jacobs, 2007. *Furfural and Derivatives in Kirk-Othmer Encyclopedia of Chemical Technology*, 4th ed., John Wiley and Sons: New York, doi: 10.1002/14356007.a12119.pub2.
14. K.J. Zeitsch, 2000. *The Chemistry and Technology of Furfural and its Many By-Products*, 1st ed., Elsevier: Amsterdam.
15. M.J. Antal, G.N. Richards, 1991. Mechanism of formation of 2-furaldehyde from d-xylose. *Carbohydr. Res.*, 217, 71.
16. G. Marcotullio, W. de Jong, 2010. Chloride ions enhance furfural formation from d-xylose in dilute aqueous acidic solutions. *Green Chem.*, 12, 1739.
17. B. Sain, A. Chaudhuri, J.N. Borgohain, B.P. Baruah, J.L. Ghose, 1998. Furfural and furfural-based industrial chemicals. *J. Sci. Ind. Res.*, 41, 431.

18. C. Moreau, R. Durand, D. Peyron, J. Duhamet, P. Rivalier, 1998. Selective preparation of furfural from xylose over microporous solid acid catalyst. *Ind. Crops Prod.*, 7, 95.
19. A.S. Dias, M. Pillinger, A.A. Valente, 2005. Liquid phase dehydration of D-xylose in the presence of Keggin-type heteropolyacids. *Appl. Catal. A: Gen.*, 285, 126; A.S. Dias, S. Lima, P. Brandao, M. Pillinger, J. Rocha, A.A. Valente, 2006. Liquid-phase dehydration of D-xylose over microporous and mesoporous niobium silicates. *Catal. Lett.*, 108, 179.
20. A.S. Dias, S. Lima, M. Pillinger, A.A. Valente, 2007. Modified versions of sulfated zirconia as catalysts for the conversion of xylose to furfural. *Catal. Lett.*, 114, 3.
21. A.S. Dias, M. Pillinger, A.A. Valente, 2005. Dehydration of xylose into furfural over micro-mesoporous sulfonic acid catalysts. *J. Catal.*, 229, 414.
22. J.B. Binder, J.J. Blank, A.V. Cefali, R.T. Raines, 2010. Synthesis of furfural from xylose and xylan. *ChemSusChem*, 3, 1268.
23. W. Neuman, 2011. Race to greener bottles could be long, printed on December 16, 2011, on page b1 of the New York Edition; press release 15 December 2011. http://www.the coca-colacompany.com/dynamic/press_center/2011/12/plantbottle-partnerships.html.
24. Danone se lance dans les bouteilles en plastique bio, accessed July 2012; http://www. lemonde.fr/economie/article/2012/03/22/danone-se-lance-dans-les-bouteilles-en-plas tique-bio_1674241_3234.html. http://avantium.com/news/Avantium-and-Danone-sign-development-partnership-for-Next-Generation-bio-based-plastic-PEF.html, accessed July 2012.
25. Y. Zhang, E.A. Pidko, E.J.M. Hensen, 2012. On the mechanism of Lewis acid catalyzed glucose transformations in ionic liquids. *ChemCatChem*, 4, 1263.
26. J.J. Mulder, 1840. *J. Prakt. Chem.*, 21, 229.
27. G.J.M. Gruter, F. Dautzenberg, 2010. Method for the synthesis of 5-alkoxymethyl furfural ethers and their use, WO2007104514 and US8133289; G.J.M. Gruter, 2010. Hydroxymethylfurfural ethers from sugars, or HMF and branched alcohol, WO2009030506; G.J.M. Gruter, L.E. Manzer, 2010. Hydroxymethylfurfural ethers from sugars, or HMF and higher alcohols, WO2009030507; G.J.M. Gruter, L.E. Manzer, 2010. Hydroxymethylfurfural ethers from sugars or HMF and mixed alcohols, WO2009030508; all to Furanix Technologies B.V.
28. M. Mascal, Chemical conversion of biomass into new generations of renewable fuels, polymers, and value-added products, White Paper, published on linkedin, accessed March 2012; S. Caratzoulas, D.G. Vlachos, 2011. Converting fructose to 5-hydroxymethylfurfural: a quantum mechanics/molecular mechanics study of the mechanism and energetic. *Carbohydr. Res.*, 346, 664; M. Mascal, E.B. Nikitin, 2009. Dramatic advancements in the saccharide to 5-(Chloromethyl)furfural conversion reaction. *ChemSusChem*, 2, 859; M. Mascal, 2010. High-yield conversion of cellulosic biomass into furanic biofuels and value-added products, US7829732, to the regents of the University of California.
29. R.-J. van Putten, J.C. van der Waal, E. de Jong, C.B. Rasrendra, H.J. Heeres, J.G. de Vries, 2013. Hydroxymethylfurfural, a versatile platform chemical made from renewable resources. *Chem. Rev.*, 113,1499.
30. E.A. Pidko, V. Degirmenci, R.A. van Santen, E.J.M. Hensen, 2010. Glucose activation by transient $Cr2+$ dimers. *Angew. Chem. Int. Ed.* 49, 2530; E.A. Pidko, V. Degirmenci, R.A. van Santen, E.J.M. Hensen, 2010. Coordination properties of ionic liquid-mediated chromium (II) and copper (II) chlorides and their complexes with glucose. *Inorg. Chem.*, 49, 10081.
31. C. Carlini, P. Patrono, A.M. Raspolli Galletti, G. Sbrana, 2004. Heterogeneous catalysts based on vanadyl phosphate for fructose dehydration to 5-hydroxymethyl-2-furaldehyde. *Appl. Catal. A: Gen.*, 275, 111.
32. X. Qi, M. Watanabe, T.M. Aida, R.L. Smith Jr., 2008. Catalytical conversion of fructose and glucose into 5-hydroxymethylfurfural in hot compressed water by microwave heating. *Catal. Commun.*, 9, 2244.

33. T. Armaroli, G. Busca, C. Carlini, M. Giuttari, A.M. Raspolli Galletti, G. Sbrana, 2000. Acid sites characterization of niobium phosphate catalysts and their activity in fructose dehydration to 5-hydroxymethyl-2-furaldehyde. *J. Mol. Catal. A: Chem.*, 151, 233.

34. P. Carniti, A. Gervasini, S. Biella, A. Auroux, 2006. Niobic acid and niobium phosphate as highly acidic viable catalysts in aqueous medium: fructose dehydration reaction. *Catal. Today*, 118, 373.

35. C. Carlini, M. Giuttari, A.M. Raspolli Galletti, G. Sbrana, T. Armaroli, G. Busca, 1999. Selective saccharides dehydration to 5-hydroxymethyl-2-furaldehyde by heterogeneous niobium catalyst. *Appl. Catal. A: Gen.*, 183, 295.

36. M. Watanabe, Y. Aizawa, T. Iida, T.M. Aida, C. Levy, K. Sue, H. Inomata, 2005. Glucose reactions with acid and base catalysts in hot compressed water at 473 K. *Carbohydr. Res.*, 340, 1925; M. Watanabe, Y. Aizawa, T. Iida, R. Nishimura, H. Inomata, 2005. Catalytic glucose and fructose conversions with TiO_2 and ZrO_2 in water at 473 K: relationship between reactivity and acid–base property determined by TPD measurement. *Appl. Catal. A: Gen.*, 295, 150.

37. F.S. Asghari, H. Yoshida, 2006. Dehydration of fructose to 5-hydroxymethylfurfural in sub-critical water over heterogeneous zirconium phosphate catalysts. *Carbohydr. Res.*, 341, 2379.

38. F. Benvenuti, C. Carlini, P. Patrono, A.M. Raspolli Galletti, G. Sbrana, M.A. Massucci, P.A. Galli, 2000. Heterogeneous zirconium and titanium catalysts for the selective synthesis of 5-hydroxymethyl-2-furaldehyde from carbohydrates. *Appl. Catal. A: Gen.*, 193, 147.

39. X. Qi, M. Watanabe, T.M. Aida, R.L. Smith Jr., 2009. Sulfated zirconia as a solid acid catalyst for the dehydration of fructose to 5-hydroxymethylfurfural. *Catal. Commun.*, 10, 1771.

40. P. Rivalier, J. Duhamet, C. Moreau, R. Durand, 1995. Development of a continuous catalytic heterogeneous column reactor with simultaneous extraction of an intermediate product by an organic solvent circulating in countercurrent manner with the aqueous phase. *Catal. Today*, 24, 165; C. Moreau, R. Durand, S. Razigade, J. Duhamet, P. Faugeras, P. Rivalier, P. Ros, G. Avignon, 1996. Dehydration of fructose to 5-hydroxymethylfurfural over H-mordenites. *Appl. Catal. A: Gen.*, 145, 211.

41. A.J. Sanborn, 2008. Processes for the preparation and purification of hydroxymethylfuraldehyde and derivatives, US7,317,116, to Archer Daniels Midland.

42. D. Mercadier, L. Rigal, A. Gaset, J.P. Gorrichon, 1981. Synthesis of 5-hydroxymethyl-2-furancarboxaldehyde. *J. Chem. Tech. Biotechnol.*, 31, 489.

43. M. Ohara, A. Takagaki, S. Nishimura, K. Ebitani, 2010. Syntheses of 5-hydroxymethylfurfural and levoglucosan by selective dehydration of glucose using solid acid and base catalysts. *Appl. Catal. A: Gen.*, 383, 149.

44. K.-I. Shimizu, R. Uozumi, A. Satsuma, 2009. Enhanced production of hydroxymethylfurfural from fructose with solid acidcatalysts by simple water removal methods. *Catal. Commun.*, 10, 1849.

45. Y. Nakamura, S. Morikawa, 1980. The dehydration of D-fructose to 5-hydroxymethyl-2-furaldehyde. *Bull. Chem. Soc. Jpn.*, 53, 3705.

46. Y. Zhang, V. Degirmenci, C. Li, E.J.M. Hensen, 2011. Phosphotungstic acid encapsulated in metal–organic framework as catalysts for carbohydrate dehydration to 5-hydroxymethylfurfural. *ChemSusChem*, 4, 59.

47. K.B. Sidhpuria, A.L. Daniel-da-Silva, T. Trindade, J.A.P. Coutinho, 2011. Supported ionic liquid silica nanoparticles (SILnPs) as an efficient and recyclable heterogeneous catalyst for the dehydration of fructose to 5-hydroxymethylfurfural. *Green Chem.*, 13, 340.

48. Q. Bao, K. Qiao, D. Tomida, C. Yokoyama, 2008. Preparation of 5-hydroxymethylfurfural by dehydration of fructose in the presence of acidic ionic liquid. *Catal. Commun.*, 9, 1383.

49. H. Yan, Y. Yang, D. Tong, X. Xiang, C. Hu, 2009. Catalytic conversion of glucose to 5-hydroxymethylfurfural over SO_4^{2-}/ZrO_2 and SO_4^{2-}/ZrO_2-Al_2O_3 solid acid catalysts. *Catal. Commun.*, 10, 1558.

50. K. Lourvanij, G.L. Rorrer, 1993. Reactions of aqueous glucose solutions over solid-acid Y-zeolite catalyst at 110–160°C. *Ind. Eng. Chem. Res.*, 32, 11.

51. A. Chareonlimkun, V. Champreda, A. Shotipruk, N. Laosiripojana, 2010. Reactions of C5 and C-sugars, cellulose, and lignocellulose under hot compressed water (HCW) in the presence of heterogeneous acid catalysts. *Fuel*, 89, 2873; A. Chareonlimkun, V. Champreda, A. Shotipruk, N. Laosiripojana, 2010. Catalytic conversion of sugarcane bagasse, rice husk and corn cob in the presenceof TiO_2, ZrO_2 and mixed-oxide TiO–ZrO_2 under hot compressed water (HCW) condition. *Bioresour. Technol.*, 101, 4179.

52. A.J. Sanborn, 2009. Oxidation of furfural compounds, WO2010132740, to Archer Daniels Midland company; see also W. Partenheimer, V.V. Grushin, 2001. Synthesis of 2,5-diformylfuran and furan-2,5-dicarboxylic acid by catalytic air-oxidation of 5-hydroxymethylfurfural. Unexpectedly selective aerobic oxidation of benzyl alcohol to benzaldehyde with metal/bromide catalysts. *Adv. Synth. Catal.*, 343, 102.

53. C. Munoz de Diego, W.P. Schammel, M.A. Dam, G.J.M Gruter, 2011. Method for the preparation of 2,5-furandicarboxylic acid and esters thereof, WO2010n150653, to Furanix Technologies B.V.

54. Y.Y. Gorbanev, S.K. Klitgaard, J.M. Woodley, C.H. Christensen, A. Riizager, 2009. Gold-catalyzed aerobic oxidation of 5-hydroxymethyl-furfural in water at ambient temperature. *ChemSusChem*, 2, 672.

55. O. Casanova, S. Iborra, A. Corma, 2009. Biomass into chemicals: one pot-base free oxidative esterificationof 5-hydroxymethyl-2-furfural into 2,5-dimethylfuroate with gold on nanoparticulated ceria. *J. Catal.*, 265, 109.

56. P. Vinke, W.V. Poel, H. van Bekkum, 1991. On the oxygen tolerance of noble metal catalysts in liquid phase alcohol oxidations the influence of the support on catalyst deactivation. *Stud. Surf. Sci. Catal.*, 59, 385.

57. N.K. Gupta, S. Nishimura, A. Takagaki, K. Ebitani, 2011. Hydrotalcite-supported gold-nanoparticle-catalyzed highly efficient base-free aqueous oxidation of 5-hydroxymethylfurfural into 2,5-furandicarboxylic acid under atmospheric oxygen pressure. *Green Chem.*, 13, 824.

58. M.L. Ribeiro, U. Schuchardt, 2003. Cooperative effect of cobalt acetylacetonate and silica in the catalytic cyclization and oxidation of fructose to 2,5-furandicarboxylic acid. *Catal. Commun.*, 4, 83.

59. N. Merat, P. Verdeguer, L. Rigal, A. Gaset, M. Delmas, 1992. Process for the manufacture of furan-2,5-dicarboxylic acid, FR2669634, to Furchim SARL; P. Verdeguer, N. Merat, A. Gaset, 1993. Oxydation catalytique du HMF en acide 2,5-furane dicarboxylique. *J. Mol. Catal. A: Chem.*, 85, 327.

60. B.W. Lew, 1967. Method of producing dehydromucic acid, US3,326,944, to Atlas Chemical Industries, Inc.; E.I. Leupold, M. Wiesner, M. Schlingmann, K. Kapp, 1990. Process for the oxidation of 5-hydroxymethylfurfural, EP0356703, to Hoechst AG.

61. M.A. Lilga, R.T. Hallen, J.F. White, G.J. Frye, 2008. Hydroxymethyl furfural oxidation methods, US20080103318, to Batelle Memorial Institute; M.A. Lilga, R.T. Hallen, M. Gray, 2010. Production of oxidized derivatives of 5-hydroxymethylfurfural. *Top. Catal.*, 53, 1264.

62. M. Kröger, U. Prüsse, K.-E. Vorlog, 2000. A new approach for the production of 2,5-furandicarboxylic acid by in situ oxidation of 5-hydroxymethylfurfural starting from fructose. *Top. Catal.*, 13, 237.

63. S.E. Davis, L.R. Houk, E.C. Tamargo, A.K. Datye, R.J. Davis, 2011. Oxidation of 5-hydroxymethylfurfural over supported Pt, Pd and Au catalysts. *Catal. Today*, 160, 55.

64. G. Grabowski, J. Lewkowski, R. Skowroński, 1991. The electrochemical oxidation of hydroxymethylfurfural with the nickel oxide/hydroxide electrode. *Electrochim. Acta*, 36, 1995.
65. G. Fleche, A. Gaset, J.-P. Gorrichon, E. Truchot, P. Sicard, 1982. Process for manufacturing 5-hydroxymethylfurfural, US4339387, to Roquette Frères.
66. K.M. Rapp, 1987. Process for preparing pure 5-hydroxymethylfurfuraldehyde, US4740605, to Süddeutsche Zucker-Aktiengesellschaft.
67. F.K. Kazi, A.D. Patel, J.C. Serrano-Ruiz, J.A. Dumesic, R.P. Anex, 2011. Techno-economic analysis of dimethylfuran (DMF) and hydroxymethylfurfural (HMF) production from pure fructose in catalytic processes. *Chem. Eng. J.*, 169, 329.
68. M. Brasholz, K. von Känel, C.H. Hornung, S. Saubern, J. Tsanaktsidis, 2010. Highly efficient dehydration of carbohydrate to 5-(chloromethyl)furfural (CMF), 5-(hydroxymethyl)furfural (HMF) and levulinic acid by biphasic continous flow processing. *Green Chem.*, doi: 10.1039/c1gc15107j.
69. M. Mascal, E.B. Nikitin, 2010. High-yield conversion of plant biomass into key value-added feedstocks 5-(hydroxymethyl)furfural, levulinic acid, and levulinic esters via s-(chloromethyl)furfural. *Green Chem.*, 12, 370.
70. M. Bicker, D. Kaizer, L. Ott, H.J. Vogel, 2005. Dehydration of D-fructose to hydroxymethylfurfural in sub- and supercritical fluids. *Supercrit. Fluids*, 36, 118.
71. A.J.J.E. Eerhart, A.P.C. Faaij, M.K. Patel, 2012. Replacing fossil based pet with biobased PEF; process analysis, energy and GHG balance. *Energy Environ. Sci.*, 5, 6407.
72. A.S. Dias, G.J.M. Gruter, R.-J. van Putten, 2012. Process for the conversion of a carbohydrate-containing feedstock, WO2012091570, to Furanix Technologies B.V.
73. T. van Aken, 2010. Avantium, at GPEC2010 Conference, Orlando, USA; E. de Jong, 2010. Furanics: versatile molecules applicable for biopolymers and biofuels applications, at Frontiers in Biorefining Conference, Georgia, USA.
74. B.F.M. Kuster, H.J.C. van der Steen, 1977. Preparation of 5-hydroxymethylfurfural. *Starch/Stärke*, 29, 99.
75. J.D. Taylor, M.M. Jenni, M.W. Peters, 2010. *Dehydration of Fermented Isobutanol for the Production of Renewable Chemicals and Fuel*, Top Catal 53, 1224; B. Bernacki, 2011, *Biobutanol Potential for Biobasedmonomers and Polymers at Bioplastics*, Las Vegas, USA.
76. A.C. Hawkins, D.A. Glassner, T. Buelter, J. Wade, P. Meinhold, M.W. Peters, P.R. Gruber, W.A. Evanko, A.A. Aristidou, 2009. *Methods for the Economical Production of Biofuel Precursor that is also a Biofuel from Biomass*, US20090215137, to Gevo Inc.
77. D.A. Glassner, 2009. *Hydrocarbon Fuels from Plant Biomass*, US Department of Energy, Advanced Biofuels II, Downloaded Aug 2012 from http://www1.eere.energy.gov/biomass/pdfs/Biomass_2009_Adv_Biofuels_II_Glassner.pdf.
78. B.N.M. van Leeuwen, A.M. van der Wulp, I. Duijnstee, A.J.A. van Maris, A.J.J. Straathof, 2012. Fermentative production of isobutene. *Appl. Microbiol. Biotechnol.*, 93, 1377–1387.
79. M.W. Peters, J.D. Taylor, M. Jenni, L.E. Manzer, D.E. Henton, 2011. *Integrated Process to Selectively Convert Renewable Isobutanol to p-Xylene*, WO2011044243 (A1), to Gevo Inc.
80. M.W. Peters, J.D. Taylor, M. Jenni, L.E. Manzer, 2011. *Integrated Methods of Preparing Renewable Chemicals*, WO2011085223 (A1), to Gevo Inc.
81. D.R. Dodds, B. Humphreys, 2012. Production of aromatic chemicals from biobased feedstock. In: *Catalytic Process Development for Renewable Materials*. John Wiley and Sons, in press.
82. F.G. Belmonte, D.L. Sikkenga, O.S. Ogundiran, K.J. Abrams, L.K. Leung Linus, C.G. Meller, D.A. Figgins, A.B. Mossman, 2005. Staged countercurrent oxidation, WO2005051881 (a1) to BP Corporation North America; an overview of alternative TA oxidation strategies is given. In L. Loydd (Ed.), 2011. *Handbook of Industrial Catalysis*, Springer, 294 v.v.

83. G.W. Huber, J.N. Chheda, C.J. Barrett, J.A. Dumesic, 2005. Processing of biomass-derived carbohydrates production of liquid alkanes by aqueous-phase. *Science*, 308, 1446; see also US5736478; US6441241; US6964758; US7977517; US8075642; WO2009129019.

84. Available at http://www.biofuelstp.Eu/Downloads/Virent_Technology_Whitepaper.Pdf, accessed June 2012.

85. The Coca Cola Company, 2011. The Coca-Cola Company announces partnerships to develop commercial solutions for plastic bottles made entirely from plants. http://www.thecoca-colacompany.com/dynamic/press_center/2011/12/plantbottle-partnerships.html.

86. R.D. Cortright, 2012. *Aqueous-Phase Reforming Process*, 2nd Annual Wisconsin Bioenergy Summit.

87. B. Kusserow, S. Schimpf, P. Claus, 2003. Hydrogenation of glucose to sorbitol over nickel and ruthenium catalysts. *Adv. Synth. Catal.*, 345(1–2), 289.

88. R.D. Cortright, Virent Energy Systems, 2005. *Hydrogen Generation from Biomass-Derived Carbohydrates via the Aqueous-Phase Reforming (APR) Hydrogen Program Process.* DOE Review: Arlington, Virginia. http://hydrogendoedev.nrel.gov/pdfs/review05/pd7_cortright.pdf, accessed July 2012.

89. Patents with specific examples of aromatics generation via the BioformingTM Process: US8017818; US2011/0257448; US8053615; US2008/0300434, WO2011/082222; US7977517, all to Virent Energy Systems, Inc.

90. R.D. Cortright, P.G. Blommel, 2011. US20110245543, to Virent Energy Systems Inc. R.D. Cortright, catalytic conversion of lignocellulosic biomass to conventional liquid fuels and chemicals, downloaded Aug 2012 from http://www.ccrhq.org/publications_docs/CAT_Cortright.pdf.

91. D. Komula, 2011. Completing the puzzle: 100% plant-derived PET. *Bioplastics Magazine*, 04/11, 6, 14.

92. Available at http://www.virent.com/news/virent-and-hcl-clean-tech-receive-grant-to-demonstrate-cellulosic-sugars-as-feedstocks-for-drop-in-biofuels-and-bioproducts/, accessed July 2012.

93. Y.-T. Cheng, G.W. Huber, 2011. *Catalysis*, 1(6), 611; A. Corma, G.W. Huber, L. Sauvanaud, P. O'Connor, 2008. *J. Catalysis*, 257(1), 163; T.P. Vispute, H. Zhang, A. Sanna, R. Xiao, G.W. Huber, 2010. *Science*, 330(6008), 1222.

94. T.R. Carlson, T.P. Vispute, G.W. Huber, 2008. *Green Gasoline by Catalytic Fast Pyrolysis of Solid Biomass-Derived Compounds. ChemSusChem* 1, 397.

95. G.W. Huber, J. Jae, T.P. Vispute, T.R. Carlson, G. Tompsett, Y.-T. Cheng, 2009. *Catalytic Pyrolysis of Solid Biomass and Related Biofuels, Aromatic, and Olefin Compounds*, US20090227823 and WO2009111026, to University of Massachusetts.

96. B. Voss, 2011. Value-added Chemicals from Biomass by Heterogeneous Catalysis, PhD Thesis, Dep. Chemical and Biochemical Engineering, Technical University of Denmark.

97. Fourcroy, 1797. *Annales De Chimie*, 48; J. Priestley, K. Banks, 1783. Experiments relating to phlogiston, and the seeming conversion of water into air. *Philos. Trans. Royoy. Soc. Lond.*, 73, 398.

98. A. Morschbacker, 2009. Bio-ethanol based ethylene. *J. Macromol. Sci. C*, 49, 79.

99. *Ethylene from Ethanol*. Brochure Business Area Biomass Chemicals, of the Chematur Engineering Group, http://www.chematur.se/sok/download/Ethylene_rev_0904.pdf, accessed July 2012.

100. A.P. Kagyrmanova, V.A. Chumachenko, V.N. Korotkikh, V.N. Kashkin, A.S. Noskov, 2011. Catalytic dehydration of bioethanol to ethylene: pilot-scale studies and process simulation. *Chem. Eng. J.*, 176–177, 188.

101. E.R. Morais, B.H. Lunelli, R.R. Jaimes, I.R.S. Victorino, M.R.W. Maciel, R. Maciel Filho, 2011. Development of an Industrial Multitubular Fixed Bed Catalytic Reactor as CAPE-OPEN Unit Operation Model Applied to Ethene Production by Ethanol Dehydration Process Chem. Eng. Trans.

102. Ethanol dehydration to ethylene over submicron zsm-5 zeolite: study on deactivation, regeneration and kinetics. http://www.engineering-tech.com/organic-chemical-engineer ing/ethanol-dehydration-to-ethylene-over-submicron-zsm-5-zeolite-study-on-deacti vation-regeneration-and-kinetics.html, accessed June 2012.
103. The Coca Cola Company, 2012. Construction on the new facility is expected to begin at the end of this year and last for 24 months. At full capacity, it is estimated the facility will produce 500,000 metric tons of material per year. http://www.thecoca-colacom pany.com/dynamic/press_center/2012/09/partnership.html.
104. J. Arrizon, A. Gschaedler, 2002. Increasing fermentation efficiency at high sugar concentrations by supplementing an additional source of nitrogen during the exponential phase of the tequila fermentation process. *Can. J. Microbiol.*, 48, 965.
105. P.C. Badger, 2002. Ethanol from cellulose: a general review. In: J. Janick, A. Whipkey (Eds.), *Trends in New Crops and New Uses*. Ashs Press: Alexandria, p 14.
106. Available at http://www.specialchem4polymers.com/sf/roquette/index.aspx?id=polysorb-additive, accessed July 2012.
107. O. Akinterinwa, R. Khankal, P.C. Cirino, 2008. Metabolic engineering for bioproduction of sugar alcohols. *Curr. Opin. Biotechnol.*, 19(5), 461.
108. G. Fleche, M. Huchette, 1986. Isosorbide: preparation, properties and chemistry. *Starch/Stärke*, 1, 26.
109. R. Williamson, 2005. Continuous isosorbide production from sorbitol using solid acid catalysis, Iowa Corn Promotion Board, Doe–Products Platform Stage Gate Review Meeting, http://www.slideserve.com/london/continuous-isosorbide-production-from-sorbitol-using-solid-acid-catalysis-doe-products-platform-stage-gate-review-mee, accessed July 2012.
110. R.M. De Almeida, J. Li, C. Nederlof, P. O'Connor, M. Makkee, J.A. Moulijn, 2010. Cellulose conversion to isosorbide in molten salt hydrate media. *ChemSusChem*, 3, 325.
111. G.J. Dozeman, 1989. *Catalytic Conversion of D-Glucitol and Isosorbide Using Zeolite Catalysts*. Michigan State University, Department of Chemical Engineering.
112. X. Zhang, Y. Wang, J. Hu, X. Li, J. Holladay, J. White, T.A. Werpy, 2003. *Sorbitol Conversion Over Solid Acid Catalysts*. Poster presentation at 18th NAM conference, Cancun, Mexico.
113. M. Kevin, A.J. Sanborn, 2002. *Process for the Production of Anhydrosugar Alcohols*, WO 2002036598, to Archer-Daniels-Midland Company.
114. H. Huang, D. Yu, M. Gu, 2009. *Method for Preparation of Hydronol with Tetravalent Metal Phosphate as Catalyst*, CN101492457a; M. Gu, D. Yu, Z. Hongman, P. Sun, H. Huang, 2009. Metal (iv) phosphates as solid catalysts for selective dehydration of sorbitol to isosorbide. *Catal. Lett.*, 133(1–2), 214.
115. H. Huang, P. Sun, D. Yu, Z. Tang, 2010. *Method for Preparing Isosorbide Taking Supported Heteropoly Acid as Catalyst*, CN101691376 (A).
116. H. Li, D. Yu, Y. Hu, P. Sun, J. Xia, H. Huang, 2010. Effect of preparation method on the structure and catalytic property of activated carbon supported nickel oxide catalysts. *Carbon*, 48(15), 4547.
117. J. Xia, D. Yu, Y. Hu, B. Zou, P. Sun, H. Li, H. Huang, 2011. Sulfated copper oxide: An efficient catalyst for dehydration of sorbitol to isosorbide. *Catal. Comm.*, 12(6), 544.
118. Z. Tang, D. Yu, P. Sun, H. Li, H. Huang, 2010. Phosphoric acid modified Nb2o5: A selective and reusable catalyst for dehydration of sorbitol to isosorbide. *Bull. Korean Chem. Soc.*, 31(12), 3679.
119. K.K. Bhatia, 2005. Process innovations for simultaneous reaction, separation, and purification of isosorbide. World Congress of Chemical Engineering, 7th, Glasgow, United Kingdom, p. 10.

10 Synthesis of Fine Chemicals Using Catalytic Nanomaterials

Structure Sensitivity

Dmitry Yu. Murzin, Yuliya Demidova,
Benjamin Hasse, Bastian Etzold,
and Irina L. Simakova

CONTENTS

10.1 INTRODUCTION

Conventional industrial synthesis methods, i.e., utilization of homogeneous catalysts, stoichiometric oxidants, strong acids and bases as well as metal salts and batch operations are nowadays undergoing a transition toward the use of solid catalysts and continuous processes. The changes are needed to allow sustainable product quality and diminish the so-called E factor (the ratio of waste to product), which is too high for fine chemicals [1], not being in accord with the green chemistry principles [2]. Heterogeneous catalysts are often cheaper than corresponding homogeneous catalysts composed of, for example, complex ligands, allowing, moreover, easy separation and reuse of the catalyst and in many cases also catalyst regeneration.

More widespread utilization of heterogeneous catalysis for production of fine chemicals is usually restricted by a limited knowledge of catalytic materials by synthetic organic chemists and few commercially available catalysts, which are seldom tailor-made to suit a particular reaction. For example, quite often 10% Pd/C or 10% Pt/C catalysts are used for hydrogenations, although these catalysts most probably do not possess optimum activity in the best case and are just inactive in the worst case.

Because the ratio between the surface and volume decreases when the size of nanoparticles increases, the fraction of surface atoms decreases as well [3]. Therefore, catalytic activity calculated per total amount of catalytic phase usually declines with the increase of cluster size as fewer exposed sites are available for catalysis. This could be compensated in batch operations, typical for production of fine chemicals, by extra amounts of catalysts introduced in a reactor. Such an approach, however, can lead to a situation where the rate of the process is becoming limited by mass transfer, for example, in three phase hydrogenations and oxidations.

Changes in the metal particle size lead to alterations of the relative ratio among edges, corners, and terrace atoms, which might exhibit different intrinsic reactivity leading to dependence of turnover frequency (TOF), defined as the activity per unit of exposed surface, on the cluster size [4–10]. Such dependence of TOF in heterogeneous catalytic reactions over metals or metal oxides was named structure sensitivity and is currently under intensive investigation.

In the current contribution two reactions, catalyzed by supported metals, were chosen in order to give examples, how the metal cluster sizes can influence activity, selectivity, and stability of catalysts.

10.2 CASE STUDIES

Two reactions were considered in the current work, both related to catalytic transformations of terpenes, namely isomerization of α-pinene to camphene and hydrogenation of thymol with menthol as a target product.

Turpentine oil, which contains pinenes, is produced either via chemical pulping from the Kraft process or directly via distillation of the resin extrudates from living trees. The most common monoterpenes α- and β-pinene are produced annually in the quantities 18,000 and 12,000 t, respectively [11]. *Picea abies*, common in Europe, contains 58% α-pinene and 24% β-pinene in the monoterpene fraction [12].

Isomerization of α-pinene to camphene [13–18] is an important reaction because the latter is used as an intermediate for production of camphor, isobornyl acetate, and isoborneol [19]. Camphene is either esterified or etherified to isobornylacetate or alkoxycamphane, and the formed product is further transformed to camphor. Isoborneol and its acetate are used as fragrances, whereas camphor is used in pharmaceutical applications.

Conventionally, camphene is produced via the transformation of α-pinene to (+)-bornyl chloride by reacting it with dry HCl followed by base-catalyzed dehalogenation of bornyl chloride to camphene [20]. An environmentally more benign method currently used for camphene production is the application of an acid-treated TiO_2 as a catalyst [21,22], but the drawback of this method is its low rate and poor selectivity [23]. A variety of products are generated in this reaction as could be seen from Figure 10.1.

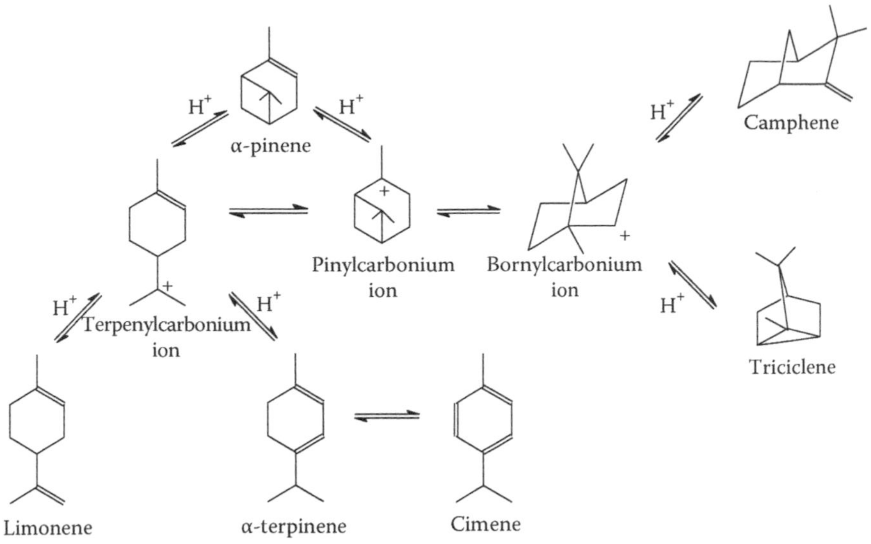

FIGURE 10.1 Pinene isomerization.

Several heterogeneous catalysts, other than amorphous TiO_2, have been applied as catalysts for isomerization of α-pinene to camphene in the liquid phase, such as various zeolites [24–30].

Contrary to a conventional way of α-pinene to camphene via liquid phase transformations over acid-hydrated TiO_2, Au/Al_2O_3 catalyst was found to afford conversion up to 99% in continuous α-pinene vapor phase isomerization and selectivity of 60%–80%, making this catalyst very promising from an industrial viewpoint [13,16].

Pinenes belong to terpenes, e.g., they are hydrocarbons resulting from the combination of several isoprene units. Terpenoids can be considered as modified terpenes, wherein methyl groups have been moved or removed or oxygen atoms added.

One of the most important terpenoids is menthol. After vanillin, it is the most widely used aroma chemical worldwide, finding use in the cosmetic, flavor, and fragrance, pharmaceutical, tobacco, and oral hygiene industries. Menthol is the main constituent of the essential oil of peppermint.

Menthol is synthesized by thymol hydrogenation with subsequent separation of menthols because hydrogenation of the aromatic ring results in stereoisomers of alkylated cyclohexanol (Figure 10.2). Hydrogenation represents a system of parallel–consecutive reactions for which selectivity strongly depends on the catalyst used, reaction conditions (especially pH, and solvent), and the reactant structure. Reaction mechanisms, catalysts, and their effects on stereoselectivity and products distribution, reaction kinetics, and the thermodynamics of hydrogenation of alkylated phenols are described in [31–37].

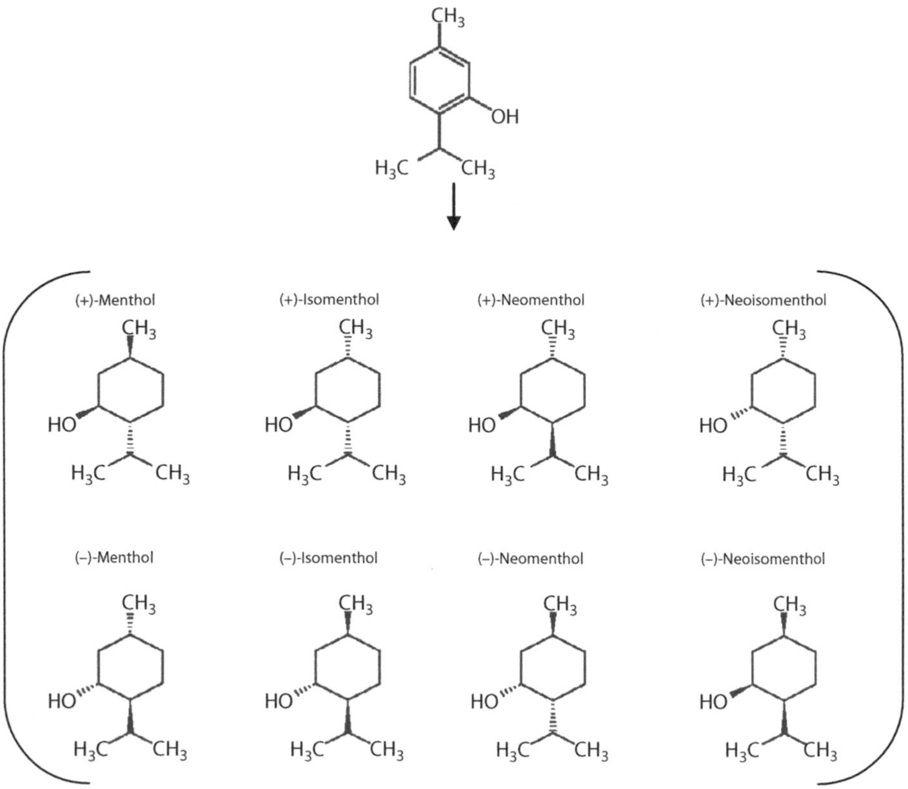

FIGURE 10.2 Thymol hydrogenation products.

10.3 CATALYST PREPARATION

Synthesis of supported metal catalysts bearing nanoclusters of a desired size and distribution is a challenging task. Among the most widely used wet methods to deposit nanoparticles on a support, incipient wetness impregnation, ion exchange, and deposition precipitation should be mentioned [38]. These methods being rather simple in implementation do not always provide satisfactory control over particle size.

One of the parameters that can influence the size of metal particles is related to electrostatic interactions between the active complex in the solution and the support during ion-exchange deposition. Matching complimentary charges on the support surface and metal complex could be used to optimize metal dispersion in impregnation.

During another method, namely deposition–precipitation, deposition of a particular compound is steered by a chemical reaction occurring in the liquid phase in a way that precipitation happens exclusively on the support, but not in the liquid bulk. In this method, along with impregnation, the final size of nanoparticles could be related with the preparation parameters, such as for instance the number of hydroxyl or carboxyl groups on the support.

It is known that surface hydroxyls or carboxyls on support particles in aqueous suspensions ionize. As a result, their surfaces develop a charge, which is dependent on the pH in the contacting solution. The difference in the zeta potential (charge at the interface) and the *pzc* (the zero point of charge, i.e., pH at which zeta potential is zero) can be considered as a descriptor, reflecting the number of hydroxyls.

As demonstrated experimentally, deposition–precipitation [39–42], deposition of colloidal nanoparticles [43], and impregnation [44] depend on the pH and the type of support, more precisely, on the point zero charge. Special attention in the literature was devoted to carbon supported metal catalysts, clarifying the effect of pH dependent oxygen surface groups on dispersion of platinum group metals [45–48].

Thermodynamic analysis of pH dependent cluster size was performed in [39]. The following expression was derived for the thermodynamically stable cluster size

$$r^3 - \frac{p_2}{\Delta z} r - \frac{p_2}{\Delta z} = 0 \qquad (10.1)$$

with

$$p_1 = 3 \frac{\left(\beta RT - \gamma_{ss}^0 \right)}{\gamma_{ss}^0 \pi \lambda}, p_2 = -2 \frac{4\alpha\gamma\Omega}{\gamma_{ss}^0 \pi \lambda}, \qquad (10.2)$$

where α is a dimensionless coefficient of an order of unity (α can be equal to 1 only for spherical particles), γ is the surface free energy, Ω is the atomic volume of the bulk metal, γ_{ss} is nonelectrostatic interfacial energy with the support, R is the gas constant, λ is the cluster size–independent proportionality coefficient between electrostatic and nonelectrostatic interfacial energy, β is a dimensionless constant ($\beta \approx 5 - 8$), $\Delta z = (\text{pH}-pzc)/pzc$ and could be either positive or negative (for example, for supports with high *pzc*, such as magnesia).

The size of several gold catalysts prepared by deposition–precipitation with urea was compared with theoretical predictions [39] showing a possibility to utilize Equation 10.1 to predict the cluster size.

Experimental data on controlled deposition of silver nanoparticles on an alumina support and catalytic results in 4-nitrophenol hydrogenation are summarized in [43]. Due to the experimental procedure applied in [43], the theoretical analysis done in [39] should be reformulated. Silver nanoparticles were synthesized by adjusting pH of a colloidal Ag aqueous solution with HCl and quickly mixing it with water. The surface of silver nanoparticles in the solution was negatively charged, and the size of nanoparticles assembly is rather controlled by *pzc* of the nanoparticles and solution pH. In Equation 10.1, *pzc* thus corresponds to the point of zero charge of the nanoparticles. The calculation results for experimental data of [43] are given in Figure 10.3, confirming good correspondence between theory and experiments. A positive value of p_2 means that the total interfacial energy of nanoparticles during deposition is lower than only nonelectrostatic interfacial energy, which might be understood as negatively charged silver particles are deposited on positively charged alumina. The

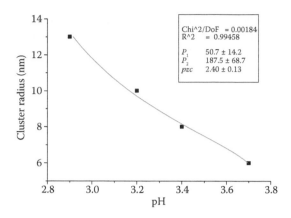

FIGURE 10.3 Average radius of silver nanoparticles as a function of pH in the preparation sequence. (Data from Mori, K., Kumami, A., Tomonari, M. and H. Yamashita, *J. Phys. Chem. C*, 113: 16850–54, 2009. With permission.)

value of *pzc* of silver nanoparticles calculated by the model is very similar to the one experimentally observed. The surface of silver nanoparticles in the solution becomes zero charged at pH = 2.3 [43], which is in close agreement with calculations.

10.4   ACTIVITY AND SELECTIVITY

Comparison of experimental data showing structure sensitivity in transformations of bulk and fine chemicals as a function of cluster size is presented in [49–53] covering among other reactions decarboxylation of fatty acids, allylic isomerization of allylbenzene to *trans*-β-methylstyrene, oxidation, oxidation of glycerol and cyclohexane, oxidative dehydrogenation of lignan hydroxymatairesinol (HMR), extracted from Norway spruce knots, as well as enantioselective hydrogenation of diketones.

The theoretical analysis in [52] took into account differences in the activation energy between edges and terraces, leading to different activities of edges and terraces in terms of reaction rates and selectivity.

The cubo-octahedral shape of nanoparticles was discussed in [52], showing that the fraction of edges can be described in a simplified way as $f_{edges} \approx 1/d_{cluster}$ when d is given in nm. In case of the two-step mechanism with two kinetically significant steps [54,55] and one most abundant surface intermediates

$$1.\, Z + A_1 \leftrightarrow ZI + B_1$$

$$2.\, ZI + A_2 \leftrightarrow Z + B_2 \qquad (10.3)$$

$$A_1 + A_2 \leftrightarrow B_1 + B_2,$$

where A_1 and A_2 are reactants, B_1 and B_2 are products, Z is the surface site, and I is an adsorbed intermediate; the expression for TOF is given by

$$v(d) = \frac{(k_1 k_2 P_{A_1} P_{A_2} - k_{-1} k_{-2} P_{B_1} P_{B_2}) e^{(1-2\alpha)\chi/d_{cluster}}}{(k_1 P_{A_1} + k_{-2} P_{B_2}) e^{-\alpha\chi/d_{cluster}} + (k_2 P_{A_2} + k_{-1} P_{B_1}) e^{(1-\alpha)\chi/d_{cluster}}} \qquad (10.4)$$

In Equation 10.4, α is the Polanyi parameter, P_{A_1} and others are partial pressures (for gas-phase reactions) or concentrations (for liquid-phase reactions), k_i is the kinetic constants, and parameter χ reflects the difference between the Gibbs energy of adsorption on edges and terraces in the following way:

$$\chi = (\Delta G_{ads,\ edges} - \Delta G_{ads,\ terraces})/RT.$$

Equation 10.4 can be rearranged in case of an irreversible reaction ($k_{-2} \approx 0$), giving

$$\text{conversion} = \frac{1}{r} \frac{p_1' e^{(1-\alpha)\chi/d_{cluster}}}{1 + p_2' e^{\chi/d_{cluster}}} \qquad (10.5)$$

where p_1' and p_2' are lumped combinations of constants and partial pressures of reactants.

Herein experimental data on thymol hydrogenation in cyclohexane over 2.5 wt% platinum on carbon-derived carbon support as a function of the metal cluster size [56] are compared with Equation 10.5. The size of platinum clusters was adjusted during preparation from 2.5 to 5.5 nm by the pH of the impregnation as described above. Larger clusters were obtained by varying the subsequent H_2 reduction temperature from 300°C to 600°C. Hydrogenation experiments demonstrated that reaction rate passes through a maximum between 7 and 9 nm. Such behavior in fact means that TOF increases with increasing mean pore size. Comparison with experimental data (Figure 10.4) illustrates a possibility to describe a maximum in the values of

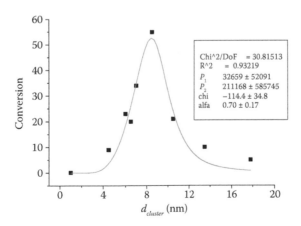

FIGURE 10.4 Modeled (Equation 10.5) conversion vs. measured one for thymol hydrogenation.

conversion in thymol hydrogenation. It is also visible that the values of parameters are not very well defined from a statistical viewpoint, thus making interpretation of parameters somewhat challenging.

Not only activity but also structure sensitive selectivity was observed for several reactions with rather bulky organic molecules.

Enantioselectivity, and regioselectivity in 1-phenyl-1,2-propanedione hydrogenation were influenced by the cluster size of platinum [57]. Phenylacetylene hydrogenation over gold catalysts [58,59] exhibited strong dependence of selectivity toward the intermediate product, styrene, on the cluster size. At a similar conversion level, an increase of the size from 2.5 to 30 nm resulted in a substantial decrease of styrene selectivity from 8% to 0.7%. Structure sensitive selectivity was observed for the carbon–nitrogen ring opening in pyrrole hydrogenation for platinum nanoparticles smaller than 2 nm [60].

In isomerization of allylbenzene to *cis* and *trans* β-methylstyrenes over gold catalysts [61], as the size of gold nanoparticles increased from 2 to 40 nm, the specific activity per surface gold decreased more than two orders of magnitude, while the *trans/cis* ratio was almost constant in the region 13–34 nm, being equal to 4.7, although decreasing to 2.5 at smaller particles sizes.

The isomerization of α-pinene conducted over gold catalysts with different mean particle size selectivity to *p*-cymene seen to be dependent on the cluster size at a similar conversion level. Thus, the selectivity to *p*-cymene was 17% for the catalyst with Au cluster size 2.7 nm, while it decreased to 9% for the catalyst with the mean cluster size of ca. 4 nm.

A mechanistic explanation for such dependence is that edges, corners, and terraces in metal clusters exhibit different selectivity.

10.5   STABILITY

Although structure sensitivity is a topic currently under intense investigation, cluster size–dependent deactivation aspects are seldom discussed. Several studies are available, however, on nanocluster catalyst poisoning [62,63] mainly aimed at elucidation that sites are preferentially involved in catalytic reactions. For example, selective poisoning by CS_2 confirmed [63] that in the hydrogenation of styrene, terrace sites are not involved. Another well-known reason for deactivation could be sintering [64], which could be also related to the cluster size. Thus, deactivation in Fischer Tropsch synthesis over Ru promoted cobalt catalysts due to sintering was related to cobalt cluster size [65].

Regarding catalyst coking, it can be stated that very few papers are available. For instance, DFT analysis of Pd_n clusters, with n ranging from 2 to 7, showed that coking is favored by increasing cluster size [66].

While deactivation by poisons present in the feed is sometimes considered as a reason for catalyst deactivation in reactions relevant for synthesis of fine chemicals, such as in hydrogenation of sitosterol [67,68], possible formation of coke as well as catalyst sintering is often neglected.

Many studies involving application of heterogeneous catalysts to synthetic organic chemistry problems often report only a value of conversion at a certain reaction time and corresponding selectivity for a particular catalyst, compare catalysts, and draw mechanistic conclusions based on such data. Interestingly enough, deactivation as a topic is not even mentioned in otherwise comprehensive books of Augustine [69] and Nishimura [70] devoted to heterogeneous catalysis in organic chemistry.

Herein we would like to address cluster size–dependent deactivation in case of α-pinene isomerization [16]. For this reaction, a "separable" deactivation model was applied, and the following time on stream-dependent activity function was obtained [16].

$$a = \frac{1}{1 + k_d K_p^2 C_o^2 t}, \tag{10.6}$$

where k_d is the deactivation constant, K_p and C_o are the adsorption constant of pinene and its initial concentration, respectively.

In the case of α-pinene isomerization, the following scheme was applied [16] to explain the catalyst deactivation from a mechanistic point of view where * is the surface site. In this scheme, two adsorbed pinene molecules form coke deposits. This proposal was required to explain experimentally observed activity decline with time on stream.

Scheme 10.1 could be analyzed using the theoretical approach developed in [52]. To this end, the scheme is rearranged in the following manner:

$$A + * \leftrightarrow A* \tag{10.7}$$

$$A* \pm A* \leftrightarrow coke \pm 2* \tag{10.8}$$

$$2A \rightarrow coke \tag{10.9}$$

The equilibrium constant for the first step of adsorption is defined according to [50]:

$$K_g = K_g^0 e^{-\chi/d_{cluster}}, \tag{10.10}$$

where K_g^0 is the cluster size–independent adsorption constant.

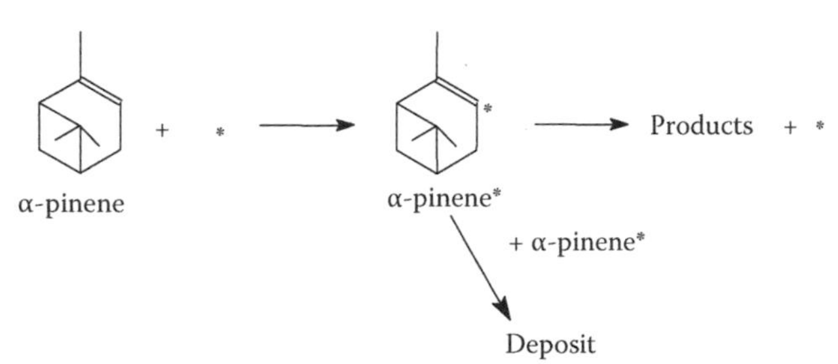

SCHEME 10.1

The overall equilibrium constant of coke formation,

$$K_{10} = K_8(K_9)^2, \tag{10.11}$$

is cluster size–independent, leading to

$$K_9 = K_9^0 e^{2\chi/d_{cluster}}. \tag{10.12}$$

The rate constant of the deactivation step could be expressed making use of the linear free energy (or Brønsted-Evans-Polanyi) relationship between reaction constants k and equilibrium constants K in a series of analogous elementary reactions $k = gK^\alpha$, where g and α (Polanyi parameter, $0 < \alpha < 1$) are constants, leading to

$$k_9 = gK_9^0 e^{2\alpha_9\chi/d_{cluster}}. \tag{10.13}$$

Finally, an expression for the activity function equation (10.6) takes the form

$$a = \frac{1}{1 + k_d K_p^2 C_o^2 t} = \frac{1}{1 + k_9 K_8^2 C_o^2 t} = \frac{1}{1 + k_d' C_o^2 t}, \tag{10.14}$$

with apparent cluster size–dependent deactivation constant k_d':

$$k_d' = gK_9^0 (K_8^0)^2 e^{2(\alpha_9-1)\chi/d_{cluster}}. \tag{10.15}$$

The analysis of Equation 10.15 shows that because $\alpha < 1$, with the increase in cluster size, the deactivation constant is becoming more prominent when parameter χ is positive, i.e., that Gibbs energy of adsorption on edges prevails over Gibbs energy of adsorption on terraces. A similar type of reasoning could be applied for the case when coke is formed just from one adsorbed molecule of the reactants leading, however, to the same qualitative conclusions. A detailed recent study of Somorjai and coworkers demonstrated that in hydrogenation of benzene and toluene on platinum catalysts deactivation due to formation of polymeric surface carbon was more pronounced with the larger Pt sizes [71].

Figure 10.5 demonstrates the deactivation data for three gold catalysts that were subjected to treatment in hydrogen for 4 h at 673 K. Additionally, catalyst A was treated for 4 h in oxygen at the same temperature, while catalyst B was exposed to oxygen for the same time at 873. Finally, catalyst C underwent treatment in oxygen at 873 K for 21 h.

As could be seen from Figure 10.5, deactivation was more severe for catalysts with somewhat larger mean particle size. At the same time, catalysts B and C, with the same mean particle size of 3.9 nm, differ noticeably with α-pinene conversion profiles probably because they possess different Au particle size distribution (Figure 10.6) with catalyst C exhibiting higher amounts of larger particles.

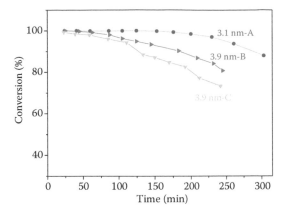

FIGURE 10.5 Time on stream behavior in pinene isomerization over gold catalysts. Reaction conditions: temperature, 473 K; catalyst, 2.1 wt% Au/Al_2O_3, 200 mg; carrier gas, H_2; residence time, 0.33 s; 10 vol% α-pinene in octane.

It should be also noted that in Figure 10.5 initial conversions were 100% for all catalysts and gradually decreased with time on stream. Such type of behavior could in fact correspond to the situation when not the whole catalyst bed is active and just a fraction of the catalyst is sufficient to afford complete conversion. Numerical analysis of data in Figure 10.5 is therefore extremely challenging. Detailed quantitative

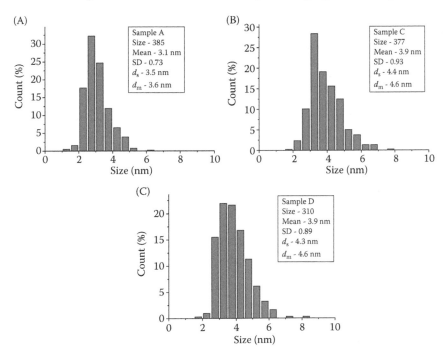

FIGURE 10.6 (**See color insert.**) Au particle size distribution of catalysts A, B, and C.

analysis using the theoretical approach described above would require in the future more experimental data for catalysts deactivation acquired with different particle sizes in the domain when the initial activity is well below 100%.

10.6 CONCLUSIONS

In the field of fine chemicals, despite the increasing significance of heterogeneous catalysis, kinetic studies, addressing structure sensitivity issues, i.e., the metal particles size, are rather sparse. In the current contribution, two reactions were mainly considered, both related to catalytic transformations of terpenes, namely isomerization of α-pinene to camphene and hydrogenation of thymol with menthol as a target product.

It was demonstrated that the metal size variations have a profound effect on activity, selectivity, and stability. Theoretical models were applied to describe experimental data on cluster size evolution depending on the relative point zero charge (difference between *pzc* and deposition–precipitation or impregnation pH in relation to *pzc*) as well as to account for TOF, selectivity, and stability dependence as a function of cluster size, showing a possibility to utilize these theoretical approaches.

REFERENCES

1. Sheldon, R.A. 1997. Catalysis and pollution prevention. *Chem. Ind.* 1: 12–5.
2. Anastas, P.T. and J.C. Warner. 1998. *Green Chemistry: Theory and Practice*. Oxford University Press: Oxford.
3. van Hardeveld, R. and F. Hartog. 1969. The statistics of surface atoms and surface sites on metal crystals. *Surf. Sci.* 15: 189–230.
4. Boudart, M. 1969. Catalysis by supported metals. *Adv. Catal.* 20: 153–66.
5. Klasovsky F. and P. Claus. 2008. The issue of size control. In *Metal Nanoclusters in Catalysis and Materials Science*, eds. B. Corain, G. Schmid and N. Toshima. Elsevier: Amsterdam, pp. 167–81.
6. van Santen, R.A. 2009. Complementary structure sensitive and insensitive catalytic relationships. *Acc. Chem. Res.* 42: 57–66.
7. van Santen, R.A., Neurock, M. and S.G. Shetty. 2010. Reactivity theory of transition-metal surfaces: a Brønsted–Evans–Polanyi linear activation energy-free-energy analysis. *Chem. Rev.* 110: 2005–48.
8. Schlögl, R. and S.B. Abd Hamid. 2004. Nanocatalysis: mature science revisited or something really new? *Angew. Chemie Int. Ed.* 43: 1628–37.
9. Henry, C.R. 2000. Catalytic activity of supported nanometer-sized metal clusters. *Appl. Surf. Sci.* 164: 252–9.
10. Somorjai, G.A. and J.Y. Park. 2008. Colloid science of metal nanoparticle catalysts in 2D and 3D structures. Challenges of nucleation, growth, composition, particle shape, size control and their influence on activity and selectivity. *Top. Catal.* 49: 126–35.
11. Fahlbusch, K.-G., Hammerschmidt, F.J., Panten, J., Pickenhagen, W. and D. Swatkowski. 1997. *Ullmann's Encyclopedia of Industrial Chemistry*, Vol. 14. Wiley-VCH: New York, p. 103.
12. Bauer, K., Garbe, D. and H. Surburg. 2001. *Common Fragrance and Flavor Materials. Preparation, Properties and Uses*. Wiley-VCH: Weinheim.

13. Simakova, I.L., Solkina, Yu.S., Moroz, B.L. et al. 2010. Selective vapour phase α-pinene isomerization to camphene over gold-on-alumina catalyst. *Appl. Catal. A: Gen.* 385: 136–43.
14. Roman-Aguirre, M., Gochi, Y.P., Sanchez, A.R., de la Torre, L. and A. Aguilar-Elguezabal. 2008. Synthesis of camphene from α-pinene using SO_3^{2-} functionalized MCM-41 as catalyst. *Appl. Catal. A: Gen.* 334: 59–64.
15. Gscheidmeier, M., Häberlein, H., Häberlein, H., Häberlein, J. and M. Häberlein. Process for the preparation of camphene by the rearrangement of α-pinene. US Patent 5,826,202, Oct 20, 1998.
16. Solkina, Yu.S., Reshetnikov, S., Estrada, M. et al. 2011. Evaluation of gold on alumina catalyst deactivation dynamics during α-pinene isomerization. *Chem. Eng. J.* 176–77: 42–8.
17. Chimal-Valencia, O., Robau-Sanchez, A., Collins-Martinez, V. and A. Aguilar-Elguezabal. 2004. Ion exchange resins as catalyst for the isomerization of α-pinene to camphene. *Biores. Technol.* 93: 119–23.
18. Mäki-Arvela, P., Holmbom, B., Salmi, T. and D.Yu. Murzin. 2007. Recent progress in synthesis of fine and specialty chemicals from wood and other biomass by heterogeneous catalytic processes. *Catal. Rev. Sci. Eng.* 49: 197–340.
19. Gscheidmeier, M., Gutmann, R., Wiesmuller, J. and A. Riedel. Process for the continuous preparation of terpene esters. US Patent 5,596,127, Jan 21, 1997.
20. Beri, M.L. and J.L. Sarin. 1936. Production of synthetic camphor from Indian turpentine. *Chem. Ind. (London, UK)* 605–7.
21. Etzel, G. Preparation of camphene from pinene. US Patent 2,551,795, May 8, 1951.
22. Rudakov, G.A. 1976. *Chemistry and Technology of Camphor*, Lesnaya promyshlennost: Moscow, 208 pp.
23. Corma, A., Iborra, S. and A. Velty. 2007. Chemical routes for the transformation of biomass into chemicals. *Chem. Rev.* 107: 2411–502.
24. Rachwalik, R., Olejniczak, Z., Jiao, J., Huang, J., Hunger, M. and B. Sulikowski. 2007. Isomerization of α-pinene over dealuminated ferrierite-type zeolites. *J. Catal.* 252: 161–70.
25. Ünveren, E., Gündüz, G. and F. Cakicioğlu-Özkanc. 2005. Isomerization of α-pinene over acid treated natural zeolite. *Chem. Eng. Commun.* 192: 386–404.
26. Allahverdiev, A.I., Gündüz, G. and D.Yu. Murzin. 1998. Kinetics of α-pinene isomerization. *Ind. Eng. Chem. Res.* 37: 2373–77.
27. Allahverdiev, A.I., Irandoust, S. and D.Yu. Murzin. 1999. Isomerization of α-pinene over clinoptilolite. *J. Catal.* 185: 352–62.
28. Allahverdiev, A., Irandoust, S., Andersson, B. and D.Yu. Murzin. 2000. Kinetics of α-pinene enantiomeric isomerization over clinoptilolite. *Appl. Catal. A: Gen.* 198: 197–206.
29. Özkan, F., Gündüz, G., Akpolat, O., Besun, N. and D.Yu. Murzin. 2003. Isomerization of α-pinene over ion-exchanged natural zeolites. *Chem. Eng. J.* 91: 257–75.
30. Zou, J.-J., Chang, N., Zhang, X. and L. Wang. 2012. Isomerization and dimerization of pinene using Al-incorporated MCM-41 mesoporous materials. *ChemCatChem* doi: 10.1002/cctc.201200106.
31. Dudas, J. and J. Hanika. 2009. Design, scale up and safe piloting of thymol hydrogenation and menthol racemisation. *Chem. Eng. Res. Des.* 87(1): 83–90.
32. Etzold, B., Jess, A. and M. Nobis. 2009 Epimerisation of menthol-stereoisomers: Kinetic studies of the heterogeneously catalysed menthol production. *Catal. Today* 140: 30–6.
33. Allakhverdiev, A.I., Kulkova, N.V. and D.Yu. Murzin. 1994. Liquid-phase stereoselective thymol hydrogenation over supported nickel-catalysts. *Catal. Lett.* 29: 57–67.
34. Allakhverdiev, A.I., Kulkova, N.V. and D.Yu. Murzin. 1995. Kinetics of thymol hydrogenation over a Ni-Cr_2O_3 catalyst. *Ind. Eng. Chem. Res.* 34: 1539–47.

35. Besson, M., Bullivant, L., Nicolaus-de Champ, N. and P. Gallezot. 1993. Kinetics of thymol hydrogenation on charcoal-supported platinum catalysts. *J. Catal.* 140: 30–40.
36. Dudas, J., Hanika, J., Lepuru, J. and M. Barkhuysen. 2005. Thymol hydrogenation in bench scale trickle-bed reactor. *Chem Biochem. Eng. Q.* 19: 255–62.
37. Tobicik, J. and L. Cerveny. 2003. Hydrogenation of alkyl-substituted phenols over nickel and palladium catalysts. *J. Mol. Catal. A: Chem.* 194: 249–54.
38. de Jong, K. 2009. *Synthesis of Solid Catalysts.* Wiley-VCH: Weinheim.
39. Murzin, D.Yu., Simakova, O.A., Simakova, I.L. and V.N. Parmon. 2011. Thermodynamic analysis of the cluster size evolution in catalyst preparation by deposition–precipitation. *React. Kinet. Mech. Catal.* 104: 259–66.
40. Louis, C. 2007. Deposition–precipitation syntheses of supported metal catalysts. In *Catalyst Preparation: Science and Engineering*, ed. J. Regalbuto, 319–39. CRC Press: Boca Raton.
41. Zanella, R., Delannoy, L. and C. Louis. 2005. Mechanism of deposition of gold precursors onto TiO_2 during the preparation by cation adsorption and deposition–precipitation with NaOH and urea. *Appl. Catal. A: Gen.* 291: 62–72.
42. Geus, J.W. and A.J. van Dillen. 1997. *Handbook of heterogeneous catalysis*, eds. G. Ertl, H. Knözinger and J. Weikamp, 240–57. Wiley-VCH: Weinheim.
43. Mori, K., Kumami, A., Tomonari, M. and H. Yamashita. 2009. A pH-induced size controlled deposition of colloidal Ag nanoparticles on alumina support for catalytic application. *J. Phys. Chem. C.* 113: 16850–54.
44. Soled, S., Wachter, W. and H. Wo. 2010. The use of zeta potential measurements in catalyst preparation. In *Studies in Surface Science and Catalysis, 10th International Symposium "Scientific Bases for the Preparation of Heterogeneous Catalysts,"* eds. E.M. Gagneaux, M. Devillers, S. Hermans, P. Jacobs, J. Materns and P. Ruiz, 101. Elsevier: Amsterdam.
45. Rodriguez-Reinoso, F. 1998. The role of carbon materials in heterogeneous catalysis. *Carbon.* 36: 159–175.
46. Prado-Burguette, C., Linares-Solano, A., Rodriguez-Reinoso, F. and C. Salimas-Martinez de Lecea. 1991. The effect of oxygen surface groups of the support on platinum dispersion in Pt/carbon catalysts. *J. Catal.* 115: 98–106.
47. Leon y Leon, C.A., Solar, J.M., Calemna, V. and L.R. Radovic. 1992. Evidence for the protonation of basal plane sites on carbon. *Carbon* 30: 797–811.
48. Coloma, F., Sepulveda-Escribano, A., Fierro, J.L.G. and F. Rodriguez-Reinoso. 1997. Gas phase hydrogenation of croton aldehydes over Pt/activated carbon catalysts. Influence of the oxygen surface groups on the support. *Appl. Catal. A. Gen.* 150: 165–83.
49. Murzin, D.Yu. 2009. Thermodynamic analysis of nanoparticle size effect on catalytic kinetics. *Chem. Eng. Sci.* 64: 1046–52.
50. Murzin, D.Yu. 2010. Size dependent heterogeneous catalytic kinetics. *J. Mol. Catal. A: Chem.* 315: 226–30.
51. Murzin, D.Yu. and V.N. Parmon. 2011. Quantification of cluster size effect (structure sensitivity) in heterogeneous catalysis. *Catal. Spec. Period. Rep. RSC* 23: 179–203.
52. Murzin, D.Yu. 2010. Kinetic analysis of cluster size dependent activity and selectivity. *J. Catal.* 276: 85–91.
53. Sotoodeh, F. and K.J. Smith. 2011. Structure sensitivity of dodecahydro-N-ethylcarbazole dehydrogenation over Pd catalysts. *J. Catal.* 279: 36–47.
54. Boudart, M. 1968. *Kinetics of Chemical Processes.* Prentice-Hall: Englewood Cliffs, NJ.
55. Temkin, M.I. 1979. The kinetics of some industrial heterogeneous catalytic reactions. *Adv. Catal.* 28: 173–291.
56. Hasse, B., Reissner, F., Dicenta, D., Hausmann, P. and B.J.M. Etzold. 2012. Carbide derived carbon (CDC) as model catalyst support—an exemplary study. *Book of Abstracts.* International Congress on Catalysis.

57. Murzin, D.Yu. and E. Toukoniitty. 2007. Nanocatalysis in asymmetric hydrogenation. *React. Kinet. Catal. Lett.* 90: 19–25.
58. Nikolaev, S.A. and V.V. Smirnov. 2009. Selective hydrogenation of phenylacetylene on gold nanoparticles. *Gold Bull.* 42: 182–9.
59. Nikolaev, S.A. and V.V. Smirnov. 2009. Synergistic and size effects in selective hydrogenation of alkynes on gold nanocomposites. *Catal. Today* 147S: S336–41.
60. Kuhn, J.N., Huang, W., Tsung, C.-K., Zhang, Y. and G.A. Somorjai. 2008. Structure sensitivity of carbon–nitrogen ring opening: Impact of platinum particle size from below 1 to 5 nm upon pyrrole hydrogenation product selectivity over monodisperse platinum nanoparticles loaded onto mesoporous silica. *J. Am. Chem. Soc.* 130: 14026–27.
61. Smirnov, V.V., Nikolaev, S.A., Murav'eva, G.P., Tyurina, L.A. and A.Yu. Vasil'kov. 2007. Allylic isomerization of allylbenzene on nanosized gold particles. *Kinet. Catal.* 48: 265–70.
62. Hornstein, B.J., Alken, J.D. and R.G. Finke. 2002. Nanoclusters in catalysis: A comparison of CS_2 catalyst poisoning of polyoxoanion- and tetrabutylammonium-stabilized 40 ± 6 Å Rh(0) nanoclusters to 5 Rh/Al_2O_3, including an analysis of the literature related to the CS_2 to metal stoichiometry issue. *Inorg. Chem.* 41: 1625–38.
63. Kiraly, Z., Veisz, B. and A. Malistar. 2004. CS_2 poisoning of size-selective cubooctahedral Pd particles in styrene hydrogenation. *Catal. Lett.* 95: 57–9.
64. Moulijn, J.A., van Diepen, A.E. and F. Kapteijn. 2001. Catalyst deactivation: Is it predictable? What to do? *Appl. Catal. A: Gen.* 212: 3–16.
65. Tavasoli, A., Malek Abasslou, R.M. and A.K. Dalai. 2008. Deactivation behavior of ruthenium promoted Co/γ-Al_2O_3 catalysts in Fischer–Tropsch synthesis. *Appl. Catal. A: Gen.* 346: 58–64.
66. Bertani, V., Cavallotti, C., Masi, M. and S. Carrà. 2000. Density functional study of the interaction of palladium clusters with hydrogen and CH_x species. *J. Phys. Chem.* 104: 11390–97.
67. Lindroos, M., Mäki-Arvela, P., Kumar, N., Salmi, T. and D.Yu. Murzin. 2003. Catalyst deactivation in selective hydrogenation of β-sitosterol to β-sitostanol over palladium. *Catal. Org. React.* 587–94.
68. Mäki-Arvela, P., Martin, G., Simakova, I. et al. 2009. Kinetics, catalyst deactivation and modeling in the hydrogenation of β-sitosterol to β-sitostanol over micro- and mesoporous carbon supported Pd catalysts. *Chem. Eng. J.* 154: 45–51.
69. Augustine, R.L. 1996. *Heterogeneous Catalysis for the Synthetic Chemist*. Marcel Dekker: New York.
70. Nishimura, S. 2001. *Handbook of Heterogeneous Catalytic Hydrogenation for Organic Synthesis*. Wiley-VCH: Amsterdam.
71. Pushkarev, V.V., An, K., Alayoglu, S., Beaumont, S.K. and G.A. Somorjai. 2012. Hydrogenation of benzene and toluene over size controlled Pt/SBA-15 catalysts: Elucidation of the Pt particle size effect on reaction kinetics. *J. Catal.* doi: 10.1016/j.jcat.2012.04.022.

11 Tunable Biomass Transformations by Means of Photocatalytic Nanomaterials

Juan Carlos Colmenares Quintero

CONTENTS

11.1 INTRODUCTION

The field of photocatalysis can be traced back a century ago to early observations of the chalking of titania-based paints and to studies of the darkening of metal oxides in contact with organic compounds in sunlight. Since the mid-1990s, it has become an extremely well-researched field due to the practical interest. The utilization of solar irradiation to supply energy or to initiate chemical reactions is already an established idea. If a wide-band gap semiconductor such as titanium dioxide (TiO_2) is irradiated with light, excited electron–hole pairs result that can be applied in solar cells to generate electricity or in chemical processes to create or degrade specific compounds. In this chapter, I will try to put together an overview of the more fundamental aspects of heterogeneous photocatalysis, which are in their own right extremely scientifically interesting and which also need to be better understood in order to make significant progress with applications.

The chapter will be concentrated mostly on titanium dioxide, as it is one of the most important and most widely used oxides in all application areas. The first part of this chapter will be devoted to a brief introduction of the principles and properties of catalytically photoinduced processes and materials, respectively; after which I will mention shortly the methodologies to prepare nanophotocatalysts with emphasis on the widely used sol–gel procedure. The last part will describe research, performed on the application of photoactive solid nanomaterials, in the selective photocatalytic transformation of biomass-derived molecules to high-value chemicals.

In conclusion, a critical evaluation of the work performed will be given, in which I will emphasize the questions that remained open until now and what kind of research is desired to further develop this field of science. If the field of photoinduced processes of such materials as TiO_2 is explored successfully, the effective utilization of sun energy (which is clean, safe, and abundant) in the photocatalytic valorization of biomass for solar chemicals production will, in the future, be able to provide energy and solve environmental pollution. Until that time, there is much work to be done, but not without a promising future.

11.2   BASICS OF HETEROGENEOUS PHOTOCATALYSIS

Heterogeneous photocatalysis is a discipline that includes a large variety of reactions: organic synthesis, water splitting, photoreduction, hydrogen transfer, $O_2^{18} - O_2^{16}$ and deuterium–alkane isotopic exchange, metal deposition, disinfection and anticancer therapy, water detoxification, removal of gaseous pollutants, etc. [1–3]. Among them, titania-assisted heterogeneous photocatalytic oxidation has received more attention for many years as an alternative method for purification of both air and water streams. The basic photophysical and photochemical principles underlying photocatalysis are already established and have been extensively reported [4,5].

A photocatalytic reaction is initiated when a photoexcited electron is promoted from the filled valence band (VB) of a semiconductor photocatalyst (e.g., TiO_2) to the empty conduction band (CB) as the absorbed photon energy, $h\upsilon$, equals or exceeds the band gap of the semiconductor photocatalyst, leaving behind a hole in the VB. In concert, electron and hole pair $(e^- - h^+)$ are generated. Equations 11.1 through 11.4 show the widely accepted chain reactions.

$$\text{Photoexcitation: } TiO_2 + h\upsilon \rightarrow e^- + h^+ \tag{11.1}$$

$$\text{Oxygen ionosorption: } (O_2)_{ads} + e^- \rightarrow O_2^{\cdot-} \tag{11.2}$$

$$\text{Ionization of water: } H_2O \rightarrow OH^- + H^+ \tag{11.3}$$

$$\text{Protonation of superoxides: } O_2^{\cdot-} + H^+ \rightarrow HOO\bullet \tag{11.4}$$

The hydroperoxyl radical formed in Equation 11.4 also has scavenging properties similar to O_2, thus doubly prolonging the lifetime of the photohole (Equations 11. 5 and 11.6).

$$HOO \cdot + e^- \rightarrow HO_2^- \qquad (11.5)$$

$$HOO^- + H^+ \rightarrow H_2O_2 \qquad (11.6)$$

Both the oxidation and reduction can take place at the surface of the photoexcited semiconductor photocatalyst. Recombination between electron and hole occurs unless oxygen is available to scavenge the electrons to form superoxides ($O_2^{\cdot-}$), its protonated form the hydroperoxyl radical ($HO_2 \cdot$), and subsequently H_2O_2.

11.2.1 NANOTITANIA-ASSISTED PHOTOCATALYSIS MECHANISM

Titania has been widely used as a photocatalyst for generating charge carriers, thereby inducing reductive and oxidative processes, respectively [6]. Generally, ΔG is negative for titania-assisted aerobic photocatalytic reactions, as opposed to a photosynthetic reaction [1]. The corresponding acid A of the nonmetal substituent is formed as by-product (Equation 11.7).

$$\text{Organic wastes} \Rightarrow TiO_2/O_2/h\upsilon \geq E_g \Rightarrow \text{Intermediate(s)} \Rightarrow CO_2 + H_2O + A \quad (11.7)$$

The [$>Ti^{IV}OH^{\cdot+}$] and [$>Ti^{III}OH$] represent the surface-trapped VB and CB electrons, respectively. The surface-bound OH radical represented by [$>Ti^{IV}OH^{\cdot+}$] is chemically equivalent to the surface-trapped hole, allowing the use of the former and latter terms interchangeably [7]. According to Lawless and Serpone [8], the trapped hole and a surface-bound OH radical are indistinguishable species. A good correlation occurs among charge carrier dynamics, their surface densities, and the efficiency of the photocatalytic degradation over TiO_2. In the last two decades, aqueous suspensions of TiO_2 have been probed by picosecond and, more recently, femtosecond absorption spectroscopies [9,10]. Traditionally, an electron scavenger has been used in such a study. A femtosecond spectroscopic study of TiO_2/SCN^- aqueous system by Colombo and Bowman [9] indicated dramatic increase in the population of trapped charge carriers within the first few picoseconds. The results also confirmed that for species adsorbed to TiO_2, the hole-transfer reaction can successfully compete with the picosecond electron–hole recombination process. In Figure 11.1, the interfacial photochemical reactions on TiO_2 surface are described.

Photoholes have a great potential to oxidize organic species directly (although mechanism not proven conclusively [11]) or indirectly via the combination with •OH predominant in aqueous solution (Equations 11.8 through 11.10) [4,12].

$$H_2O + h^+ \rightarrow OH \cdot + H^+ \qquad (11.8)$$

$$R - H + OH \cdot \rightarrow R \cdot + H_2O \qquad (11.9)$$

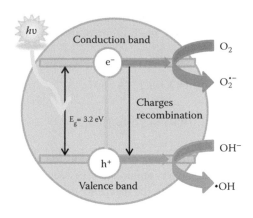

Photoexcitation: $TiO_2 + h\upsilon \rightarrow e^-_{CB} + h^+_{VB}$
Charge carrier trapping: $e^-_{CB} \rightarrow e^-_{TR}$
Charge carrier trapping: $h^+_{VB} \rightarrow h^+_{TR}$
Electron-hole recombination: $e^-_{TR} + h^+_{VB}(h^+_{TR}) \rightarrow e^-_{CB} + heat$

FIGURE 11.1   **(See color insert.)** Photocatalytic process over nano-TiO_2.

$$R\cdot + h^+ \rightarrow R^{+\cdot} \rightarrow Degradation\ products \tag{11.10}$$

Mediation of radical oxidative species in photooxidation was evidenced by photoluminescence and electroluminescence spectra of TiO_2 electrodes in aqueous solutions measured as functions of the electrode potential and the solution pH [13]. The originally absent radical oxidative species were found to accumulate after illumination under anodic bias. The primary photoreactions (Equations 11.1 through 11.7 and Figure 11.1) indicate the critical role of charge carriers (electron–hole pair) in photooxidative degradation. Essentially, hydroxyl radicals (•OH), holes (h^+), superoxide ions ($O_2^{\cdot-}$), and hydroperoxyl radicals (•OOH) are highly reactive intermediates that will act concomitantly to oxidize a large variety of organic pollutants. It is, however, argued experimentally that the oxidative reaction on the titania photocatalyst surface occurs mainly via the formation of holes (with quantum yield of 5.7×10^{-2}) not hydroxyl radicals formation (quantum yield 7×10^{-5}) [14]. The photoinduced phenomenon is affected by quantum size. Anpo et al. [15] observed a blue shift and increase in reaction yield and photocatalytic activity as the diameter of the TiO_2 particles become smaller, especially below 10 nm. This observation was attributed to the suppression of radiationless transfer and the concurrent enhancement of the activities of the charge carriers.

11.3   PROPERTIES AND CHARACTERISTICS OF NANOPHOTOCATALYSTS

Titanium metal is strong, light, and corrosion-resistant, and for this reason, it is used in metal alloys, extending to them its qualities. The most common compound of titanium is titanium dioxide, which for a long time has been used as white pigment. However, the most impressive and without precedent attention has been paid to the semiconducting properties of titanium dioxide and to its photocatalytic applications. Thus, the number of studies and scientific publications dealing with photocatalytic applications of titania has followed an exponential growth in the last 40 years. There are several reasons for this impressive growth:

1. Photocatalytic applications of titania have been associated with efforts toward environmental remediation and for the expansion of the use of renewable energy resources.
2. The use of "soft chemistry" techniques, such as the so-called sol–gel method, has allowed all laboratories, even those with limited knowledge and equipment in inorganic synthesis, to easily synthesize titania, either in amorphous or nanocrystalline form, in addition to easily depositing it in all kinds of shapes and geometries.
3. Titania, like titanium metal, is corrosion-resistant and is considered a non-toxic material, and for this reason, there are no prohibitions against its use.

Photocatalytic applications of titania soon marked four main interconnected research streams:

(a) Photodecomposition of organic pollutants for water and air cleaning
(b) Splitting of water and hydrogen production utilizing solar radiation
(c) Employing titania in photoelectrochemical processes
(d) Application in photoselective catalytic transformations in organic synthesis

An ideal photocatalyst for photocatalytic oxidation is characterized by the following attributes [1]:

Photostability
Chemically and biologically inert nature
Availability and low cost
Capability to adsorb reactants under efficient photonic activation ($h\upsilon \geq E_g$).

Titania is the most widely employed material in photocatalytic processes, but there are several materials currently considered as photocatalysts and/or supports for photocatalysis aside from titania. These include related metal oxides, metal chalcogenides, zeolites (as supports), etc. [3,16].

The anatase form of titania is reported to give the best combination of photoactivity and photostability [11]. Nearly all studies have focused on the crystalline forms of titania, namely anatase and rutile.

The minimum band gap energy required for photon to cause photogeneration of charge carriers over TiO_2 semiconductor (anatase form) is 3.2 eV corresponding to a wavelength of approx. 388 nm [17]. Practically, TiO_2 photoactivation takes place in the range of 300–388 nm. The photoinduced transfer of electrons occurring with adsorbed species on semiconductor photocatalysts depends on the band-edge position of the semiconductor and the redox potentials of the adsorbates [12].

11.4 TiO$_2$ NANOPHOTOCATALYST PREPARATION

Various methods are available for the preparation of TiO$_2$-based photocatalysts, such as electrochemical [18–20], multigelation [21], supercritical carbon dioxide [22], thin films and spin coating [23–25], and thin film by vacuum arc plasma evaporator [26], combining inverse micelle and plasma treatment [27,28], dip coating [29,30] and SILAR [31], two-step wet chemical [32], precipitation [33–35], thermal (ethanol thermal, hydrothermal, and solvothermal) [36–38], chemical solvent and chemical vapor deposition (CSD and CVD) [39–41], physical vapor deposition [42,43], microwave and ultrasonic irradiation [44–46], extremely low temperature [47,48], aerogel and xerogel [49,50], modified sol–gel [51,52], two-route sol–gel [53,54], and many other methods similar to sol–gel. Nevertheless, the benefits derived from preparing TiO$_2$ by sol–gel method, which include synthesis of nanosized crystallized powder of high purity at relatively low temperature, possibility of stoichiometry controlling process, preparation of composite materials, and production of homogeneous materials, have driven many researchers to the use of the method in preparing TiO$_2$-based nanophotocatalysts.

In summary, there are so many routes for the preparation of solid nanophotocatalysts. In this section of the chapter, I will focus on the most used and promising route of nanosized TiO$_2$ synthesis, which is the sol–gel method.

11.4.1 SOL–GEL PROCEDURE

Sol–gel technology has been in existence from as long ago as the mid-1800s [55], and was used almost a century later by Schott Glass Company (Jena, Germany). Sol–gel technology finds applications in the development of new materials for catalysis [56,57], chemical sensors [58], membranes [59,60], fibers [61,62], optical gain media [63], photochronic applications [64], and solid-state electrochemical devices [65], and in other areas of scientific and engineering fields. The sol–gel process is currently considered one of the most promising alternatives due to its inherent advantages including low sintering temperature and versatility of processing and homogeneity at the molecular level. This method allows the preparation of TiO$_2$-anatase at low temperature. Many studies revealed that different variants and modifications of the process have been used to produce pure thin films or nanopowders in large homogeneous concentration and under stoichiometry control [66–68].

One of the advantages of sol–gel synthesis of mesoporous materials is the possibility to form uniform films on a substrate. Using the method studied in detail by Sanchez et al. [69], uniform films of mesoporous titania on glass can be obtained dipping the glass slide into an acidic solution of titanium alcoxide in ethanol. The surfactant concentration is lower than the critical micellar concentration (CMC) immediately after dipping the glass, but CMC is reached when the glass is gradually removed from solution (together with ethanol evaporation) and the surfactant starts to template the formation of thin layers of a mesoporous titanium oxide.

(a)   (b)   (c)

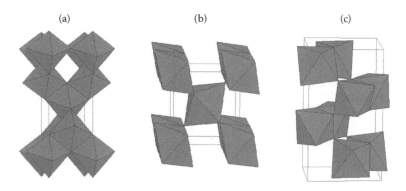

FIGURE 11.2 **(See color insert.)** Crystalline structure of anatase (a), rutile (b), and brookite (c). (Adapted from http://ruby.colorado.edu/~smyth/min/tio2.html.)

Applying the above methodology, Stucky et al. prepared highly structured materials constituted by anatase nanoparticles (5–10 nm) ordered forming a mesoporous film of TiO$_2$ perpendicular to the glass slide [70]. The as-synthesized TiO$_2$ material is initially structured in the 5- to 100-nm length scale forming mesopores, but the walls are formed by an amorphous TiO$_2$ phase. Calcination of the material produces crystallization of the as-synthetized amorphous titanium dioxide into anatase phase (Figure 11.2a) without destroying the mesoporous ordering of the film. Careful control of the calcination temperature (<500°C) is also crucial to avoid the formation of the significantly less photocatalytic active rutile and brookite phases (Figure 11.2b,c).

11.5 APPLICATION OF NANOSTRUCTURED PHOTOCATALYSTS IN BIOMASS VALORIZATION

By far, the main supply of energy on Earth is solar irradiation, a perennial source that exceeds every conceivable need of mankind in the future (25,000–75,000 kWh per day and hectare). Human beings use directly only a small fraction of such energy. Green plants use solar energy in the best photocatalytic process known, photosynthesis, a veritable chemical factory based on water splitting, where oxygen is liberated and reduced coenzymes are formed, which then reduce carbon dioxide to carbohydrates. The products serve both for building the plant structure (cellulose) and as an energy supply. These have been the (renewable) source of food, energy, and materials for mankind all along its history, either directly or after further elaboration by other living beings.

11.5.1 PHOTOCATALYTIC PRODUCTION OF HYDROGEN

Nanotechnology has boosted the modification of existing photocatalysts for the production of hydrogen and the discovery and development of new candidate materials. The rapidly increasing number of scientific publications on nanophotocatalytic H$_2$ production (1.7 times every year since 2004) provides clear evidence for the significance of this hot topic. Many papers studied the effect of different nanostructures and nanomaterials on the performance of photocatalysts because their energy conversion efficiency is principally influenced by nanoscale properties.

11.5.1.1 Water Splitting by Heterogeneous Photocatalysis

Fujishima and Honda [71] reported the splitting of water by the use of a semiconductor electrode of TiO_2 (rutile phase) connected through an electrical load to a platinum black counterelectrode. Irradiation of the TiO_2 electrode with near-UV light caused electrons to flow from it to the platinum counterelectrode via the external circuit.

The splitting of water into hydrogen and oxygen by solar light offers a promising method for the photochemical conversion and storage of solar energy due to the following:

1. Hydrogen is an environmentally valuable fuel, the combustion of which only produces water.
2. Hydrogen is easy to store and transport.
3. The raw material, water, is abundant and cheap.

For H_2 production to occur, the photocatalyst CB bottom-edge must be more negative than the reduction potential of H^+ to H_2 ($EH^+/H_2 = 0V$ vs. NHE at pH 0), whereas the photocatalyst VB top-edge should be more positive than the oxidation potential of H_2O to O_2 ($EO_2/H_2O = 1.23$ V vs. NHE at pH 0) for O_2 formation from water to occur. This theoretical minimum band gap for water splitting of 1.23 eV corresponds to light of about 1100 nm.

Up to now, over 140 materials and derivatives have been developed to photocatalyze the overall water splitting or produce hydrogen/oxygen in the presence of external redox agents [72–74]. A combinatorial method has been developed, and it has been demonstrated as a convenient way for quick selection of photocatalyst materials [75,76]. However, no semiconducting material has been found to be capable of catalyzing the overall water splitting under visible light with a quantum efficiency (QE) larger than the commercial application limit 30% at 600 nm [77].

11.5.1.2 Hydrogen Production by Photocatalytic Reforming of Biomass

Biomass has attracted much more attention for the substitution of the petroleum resource and the conversion to useful chemicals [78,79]. Fundamental reasons for biomass utilization are

1. Biomass is the only current sustainable source of pharmaceuticals, plastics, and biofuels. The real renewable source of carbon (production of around 1.0×10^{11} tons per year).
2. By 2030, 20% of the transportation fuel and 25% of the chemicals in the United States will be produced from biomass. The European Commission has set a goal that by 2020, 10% of the transportation fuels in EU will be based on biofuels.
3. It is expected that during the next 20 years, renewable energy (solar, wind, geothermal, and biofuels) will increase from 5% to 18% in the global scale.
4. Mankind has serious problems related to the lack of energy sources and huge amounts of organic wastes produced.

Due to its small relative molecular mass, the energy-storage capacity of H_2 is approximately 120 kJ/g—a figure that is almost three times that of the energy-storage capacity of oil.

Biomass sources have been utilized for the sustainable production of hydrogen [80,81]. A number of processes have been developed for this purpose (e.g., steam gasification [82], fast pyrolysis [83], supercritical conversion [84]). However, these processes require harsh reaction conditions including high temperatures and/ or pressures and consequently imply high costs. Compared with these energy-intensive thermochemical processes, photocatalytic reforming may be a good approach as this process can be driven by sunlight and performed at room temperature. Producing hydrogen by photocatalytic reforming of renewable biomass may also be more practical and valuable than that of photocatalytic water-splitting due to its potentially higher efficiency. Water-splitting processes are relatively low efficiency as limited by the recombination reaction between photogenerated electrons and holes [85]. The thermodynamics of photochemical water splitting were investigated in detail by Bolton et al. [86] who concluded that it is possible to store a maximum of 12% of the incident solar energy in the form of H_2, allowing for reasonable losses in the electron transfer steps and the catalytic reactions of water oxidation and reduction.

Pioneer studies in biomass photocatalytic reforming were conducted in 1980 [87]. Kawai and Sakata reported that hydrogen could be generated from carbohydrates on $RuO_2/TiO_2/Pt$ photocatalyst under 500-W Xe lamp irradiation. The same authors subsequently reported that hydrogen could also be generated under identical conditions from other biomass sources including cellulose, dead insects, and waste materials [88–90]. These studies demonstrate the feasibility of the photocatalytic production of hydrogen from biomass.

One of the major disadvantages of semiconductor photocatalytic systems, especially those involving photoinduced hydrogen production from water, is their relatively low efficiency, which is mainly limited by the recombination reaction between photogenerated electrons and holes. One of the ways to suppress the recombination hole rate is with the use of electron donors as sacrificial agents, the role of which is to react irreversibly with the photogenerated holes and/or oxygen thereby increasing the rate of hydrogen production. A large variety of organic compounds (most of them model compounds of biomass structure) have been used as electron donors for photocatalytic hydrogen production (Table 11.1), including alcohols, polyalcohols, sugars, and organic acids, as well as aliphatic and aromatic compounds. These compounds, especially alcohols, are satisfactory hole scavengers and undergo a relatively rapid and irreversible oxidation, which results in increased quantum yields and enhanced rates of photocatalytic hydrogen production.

Table 11.1 summarizes the latest achievements in hydrogen production from photocatalytic biomass-derived molecules wet reforming.

Photocatalytic wet reforming of biomass is relatively in the infancy stage, and at present, most investigations on it are still largely on the laboratory scale. There is not yet a single report about any pilot studies on photocatalytic reforming for commercial hydrogen production. Nevertheless, as the process is frequently carried out at room temperature and both the energy and feedstocks are from renewable sources,

TABLE 11.1

Summary of Literature Results on Photocatalytic Production of Hydrogen from Photoreforming of Biomass

Biomass-Derived Material	Photocatalysts	Reaction Conditions	Photocatalytic Activity and Concluding Remarks	Ref.
Glucose, sucrose, starch	Series of different 1.0 wt% noble metals (Pt, Pd, Au, Rh, Hg, Ru) loaded TiO_2 prepared by sol–gel technique and metal impregnation method	125-W high-pressure mercury lamp in anaerobic conditions Microwave pretreatment of starch	The rate of hydrogen evolution decreases in the order: $Pt/TiO_2 > Au/TiO_2 > Pd/TiO_2 > Rh/TiO_2 > Ag/TiO_2 > Ru/TiO_2$. Necessary conditions to reach a maximum rate of hydrogen production from glucose: pH 11, pure N_2 atmosphere, 1.0 wt% Pt loading, anatase TiO_2. The microwave irradiation pretreatment of starch (polysaccharide biomass) significantly enhances the hydrogen evolution rate.	[91]
Glucose	0.2 wt% Pt- or NiO-doped alkali tantalates ($MTaO_3$, M = Li, Na, K) obtained by solid-state reaction and metal impregnation followed by calcination (NiO–) and reduction (Pt–)	125-W high-pressure Hg lamp N_2 (>99.99%) atmosphere in a closed gas-recirculation system equipped with an inner irradiation-type quartz reaction cell	Among the studied samples, NiO loaded $NaTaO_3$ showed the highest activity for H_2 production from glucose solution. For $LiTaO_3$ and $KTaO_3$ samples, Pt co-catalyst was more effective for H_2 production than NiO.	[92]
Glucose	Series of oxide catalysts $Bi_xY_{1-x}VO_4$ (BYV) codoped with 1.0 wt% Pt prepared by solid-state reaction method and in situ photodeposition of Pt	350-W Xe lamp with a 430-nm cutoff filter; visible light	The highest activity for hydrogen production showed BYV with a B/Y ratio 1:1 at pH 3. After 2 or 3 h of reaction, the hydrogen production decreased, but when the gas product was replaced with N_2 atmosphere, hydrogen production recovered. It is supposed that the generated CO_2 might participate in the redox reaction, possibly forming CH_4, and in some way inhibit the hydrogen production.	[93]

Substrate	Catalyst/method	Light source	Results	Reference
Glucose, propanetriol and methanol	Pt/P25-x% R catalyst with tuned anatase–rutile structure obtained by thermal treatment of P25, 0.1 wt% Pt was loaded on TiO$_2$ by in situ photodeposition method	300-W Xe lamp with a shutter window filled with water to remove infrared light illumination	Maximum hydrogen production was achieved for Pt/P25–74% R and compared with P25 was enhanced up to three to five times. It is proposed that the anatase–rutile phase structure not only can enhance the charge separation and consequently the activity but also adjust the surface acid/base property, which suppresses the CO formation from several thousand parts per million to 5 ppm.	[94]
Glucose	Pt/TiO$_2$, Pd/TiO$_2$ systems prepared by sol–gel method with application of ultrasonic treatment and calcined in a different atmosphere	125-W Hg lamp, UV light (λ_{max} = 365 nm), under Ar atmosphere	Pt/TiO$_2$ calcined at 850°C gives the most effective system, because of strong metal–support interaction effect (SMSI), although oxidation/reduction treatment had led to a decrease in a surface area and transformation from anatase to rutile.	[95]
Glucose	Pt/ZnS-ZnIn$_2$S$_4$ prepared in methanol by solvothermal method, 0.50 wt% of Pt was deposited by in situ photoreduction	400-W metal halide lamp, visible light (λ > 420 nm)	ZnS-coated ZnIn$_2$S$_4$ has better activity for hydrogen evolution than pure ZnIn$_2$S$_4$. It may be attributed to enhancement of the adsorption of glucose by ZnS on the ZnIn$_2$S$_4$ surface. Maximum hydrogen generation is promoted over Pt/ZnS(17 mol%)-ZnIn$_2$S$_4$. Gluconic acid has been detected as an intermediate product of glucose degradation.	[96]
Olive mill wastewater	Mesoporous TiO$_2$ prepared by sol–gel method S_{BET} = 59 m^2/g and anatase crystal size of 19.5 nm	150-W Hg lamp, UV light (λ = 100–280 nm)	Maximum amount of hydrogen (38 mmol after 2 h of reaction) was evolved at TiO$_2$ dosage 2g/L, pH 3.	[97]

(continued)

TABLE 11.1 (Continued)
Summary of Literature Results on Photocatalytic Production of Hydrogen from Photoreforming of Biomass

Biomass-Derived Material	Photocatalysts	Reaction Conditions	Photocatalytic Activity and Concluding Remarks	Ref.
Sugars	Series of TiO_2 dispersed on SiO_2 TiO_2–SiO_2 and TiO_2 modified with metals (Cr^{2+}, Mn^{4+} and V^{5+}) and nonmetals N and S	200-W Ne lamp under N_2 atmosphere, UV–visible light	It was found that doping with nonmetals is more effective than doping with metals in the case of TiO_2-based photocatalysts and more effective for metals in the case of TiO_2–SiO_2–based photocatalysts. However, doping TiO_2 with nonmetals gives the most effective catalyst because the nonmetals are less effective in forming recombination center than metals.	[98]
	Au deposited on TiO_2–SiO_2 and TiO_2 Incipient wetness impregnation, impregnation with urea or thiourea, and deposition–precipitation were used as preparation methods			
Glycerol	Pt/TiO_2 prepared by wet impregnation method, metal loading varied 0.05–5.0 wt%	Xe-arc lamp with a water filter for the elimination of infrared radiation, UV–visible light	Optimal photocatalytic performance for hydrogen production is obtained for samples loaded with 0.1–0.5 wt% Pt. Results obtained show that the reaction proceeds with intermediate production of methanol and acetic acid and eventually results in complete conversion of glycerol to H_2 and CO_2. The reaction rate is higher in neutral and basic solutions and increases with increasing temperature from 40°C to 60°C–80°C.	[99]
Methanol	0.5 wt% Pt/TiO_2 P25 prepared by hydrothermal treatment of TiO_2 P25 and photochemical deposition of Pt	UV light adjusted by black band-pass filters to the range $\lambda = 300$–400 nm	The rate of hydrogen evolution increases with increasing methanol concentration. The highest activity in hydrogen evolution exhibits Pt-loaded TiO_2UV100. Detected reaction products were formaldehyde, formic acid, and carbon dioxide.	[100]

Methanol, Ethanol	TiO_2 films prepared by radio frequency magnetron sputtering RF–MS deposition method, thickness about 3 μm	500-W Xe arc lamp, UV light, and visible light (>450 nm). The concentration of alcohol was adjusted at 10 vol%	Hydrogen evolution from an aqueous solution of methanol proceeds on the Vis-TiO_2 thin films under visible light irradiation, but this reaction does not proceed on the UV-TiO_2 thin films. Adding 10 vol% of methanol improves the photocatalytic activity and increases the H_2 evolution rate from 0.08 to 0.52 μmol/h.	[101]
Acetic acid	20 wt% Fe/TiO_2 prepared from crude TiO_2 and impregnated with Fe	Medium-pressure Hg vapor lamp, UV light (λ = 365 nm), under N_2 atmosphere	The gaseous products identified in the reaction are CH_4, CO_2, C_2H_6, C_3H_8, and H_2. The most active photocatalyst toward all the identified gas production was A-Fe20N500 with 20 wt% Fe and calcined at 500°C. The maximum yield of hydrogen evolution (0.013 mmol) was lower than that of CH_4 and CO_2 probably due to acidic pH of the reaction environment. CH_4 formation follows the so-called photo-Kolbe reaction pathway.	[102]
Acetic acid	TiO_2 calcined at different temperatures: 600°C–800°C	Medium-pressure Hg vapor lamp, UV light (λ = 365 nm), under N_2 atmosphere	The gaseous products identified in the reaction were CH_4, CO_2, C_2H_6, C_3H_8, and H_2. The most active photocatalyst toward all the identified gas production was P700 (calcined at 700°C, composed 94% of anatase, crystallite size of 32 nm, S_{BET} = 30 m^2/g). The maximum yield of hydrogen evolution was 0.22 mmol/g of photocatalyst after 27 h of the process.	[103]
Acetic acid	CuO/SnO_2 photocatalyst prepared by co-precipitation method	300-W high-pressure Hg lamp, UV light	Photocatalytic H_2 production increases with the increment of CuO amount and the optimum CuO content is about 33.3 mol%.	[104]

(continued)

TABLE 11.1 (Continued)
Summary of Literature Results on Photocatalytic Production of Hydrogen from Photoreforming of Biomass

Biomass-Derived Material	Photocatalysts	Reaction Conditions	Photocatalytic Activity and Concluding Remarks	Ref.
Methanol	Series of noble metal (Ag, Au, Ag–Au alloy, and Pt)–modified TiO_2 synthesized by flame spray pyrolysis or by deposition of noble metal	Iron halogenide mercury arc lamp emitting in the $\lambda = 350$–450 nm	Methanol underwent oxidation up to CO_2 through the formation of formaldehyde and formic acid. Carbon monoxide, methane, methyl formate, acetaldehyde, and dimethyl ether were identified as side products. Hydrogen evolved at a constant rate, which significantly increased upon noble metal addition, Pt being the most effective co-catalyst, followed by gold and silver, according to their work function values.	[105]
Methanol	Au/TiO_2 synthesized by deposition–precipitation method on commercial Degussa P25	300-W Xe lamp with a glass filter to remove IR illumination from the lamp	The rate of H_2 production is greatly increased when the gold particle size is reduced from 10 to smaller than 3 nm. Both the rate of H_2 production and the CO selectivity increase with pH value up to the neutral value and decrease thereafter. The by-product CO is possibly formed via decomposition of the intermediate formic acid species.	[106]
Methanol	Series of $ALaTa_{1/3}Nb_{2-x/3}O_7$ ($A = K$, H; $x = 0, 2, 3, 4$, and 6) prepared by solid-state method and Pt co-catalyst prepared by deposition of Pt particles	100-W high-pressure Hg arc lamp, UV light	$HLaTa_{2/3}Nb_{4/3}O_7/Pt$ shows the best photocatalytic activity. The photocatalytic H_2 evolution rate reaches 136 cm^3 g^{-1} h^{-1}, which is 45.3 times larger than that of TiO_2(P-25) (ca. 3 cm^3 g^{-1} h^{-1}).	[107]

Oxygenated hydrocarbons, glycerol, glucose, sucrose	(B, N)-codoped TiO_2 prepared by hydrothermal synthesis and platinized by photodeposition method	300-W Xe lamp	The highest hydrogen production rate displayed Pt/(B, N)-TiO_2. The biomass-derived fuels serve as both hydrogen sources and electron donors, and they are oxidized into CO_2 as an ultimate product. The maximum hydrogen amount is achieved at 5 wt% glycerol concentration.	[108]
Glycerol and different biomass-derived compounds	0.5 wt% Pt/TiO_2 prepared by impregnation of TiO_2 powder (Degussa P25)	300- or 450-W Xe arc lamp with a water filter to remove infrared radiation, solar light source	The overall reaction is nonselective with respect to the organic substrate employed, and therefore, practically all biomass-derived compounds in solution or in suspension may be used as feedstock.	[109]
Glucose	Series of different metal/TiO_2 systems (metals: Pt, Rh, Ru, Ir, Au, Ni, Cu). Ni/TiO_2 and Cu/TiO_2 were prepared by impregnation method, whereas the other catalysts by in situ photodeposition method	300-W high-pressure Hg lamp	Rh/TiO_2 catalyst is found to be most active for H_2 production while with the most extremely low CO concentration. The rate of H_2 production increases in the following order: $Ir < Ru < Au < Ni \approx Cu \approx Pt < Rh$. The optimum loading of deposited metal is 0.3%.	[110]
Methanol	Series of mesoporous-assembled $SrTiO_3$ photocatalysts with different loaded metal co-catalyst: Au, Pt, Ag, Ni, Ce, and Fe, synthesized by sol–gel method with the aid of a structure-directing surfactant	176-W Hg lamp with the main emission $\lambda = 254$ nm (UV light) and 300-W Xe lamp with $\lambda > 400$ nm (visible light). Reaction temperature for UV light source = 45°C and for visible light = room temperature	The Au, Pt, Ag, and Ni loadings had a positive effect on the photocatalytic activity enhancement, whereas the Ce and Fe loadings did not. The best loaded metal was found to be Au due to its electrochemical properties compatible with the $SrTiO_3$-based photocatalyst and its visible light–harvesting enhancement. A 1 wt% Au-loaded $SrTiO_3$ photocatalyst exhibited the highest photocatalytic hydrogen production activity for both UV and visible light, and the rate was higher was under UV light irradiation.	[111]

(continued)

TABLE 11.1 (Continued)
Summary of Literature Results on Photocatalytic Production of Hydrogen from Photoreforming of Biomass

Biomass-Derived Material	Photocatalysts	Reaction Conditions	Photocatalytic Activity and Concluding Remarks	Ref.
Methanol	Series of mesoporous-assembled $0.93TiO_2$–$0.07ZrO_2$ loaded by nonprecious metals: Ag, Ni, and Cu, synthesized by a sol–gel process with the aid of a structure-directing surfactant and photochemical deposition method	A set of Hg lamps: 16 11-W UV light, 50 vol% aqueous solution of methanol	The most efficient loaded metal was found to be Cu due to its suitable physical, chemical, and electrochemical properties with the $0.93TiO_2$–$0.07ZrO_2$ mixed oxide-based photocatalyst. The 0.15 wt% Cu-loaded $0.93TiO_2$–$0.07ZrO_2$ mixed oxide photocatalyst exhibited the highest photocatalytic hydrogen production activity. UV light-harvesting ability is in the following order: Cu-loaded photocatalyst > Ni-loaded photocatalyst > Ag-loaded photocatalyst.	[112]
Methanol	A series of ZnO/TiO_2, SnO/TiO_2, CuO/TiO_2, Al_2O_3/TiO_2, and $CuO/Al_2O_3/TiO_2$ nanocomposites prepared by mechanical mixing	15-W black light, UV light ($\lambda = 352$ nm)	The maximal photocatalytic hydrogen production was observed with $CuO/Al_2O_3/TiO_2$ nanocomposites and the optimal component was 0.2 wt% CuO/0.3 wt% Al_2O_3/TiO_2.	[113]
Glycerol	Series of metal/TiO_2 catalysts (metal = Au, Pt, Pd)	400-W Xe arc lamp	The activity of Pd/TiO_2 was 1.7 times higher than that of Au/TiO_2 and was almost the same as that over Pt/TiO_2.	[114]
Ethanol, glycerol	CuO_x/TiO_2 photocatalyst prepared by water-in-oil microemulsion method. The nominal Cu loading was 2.5 wt%.	125-W medium-pressure Hg lamp, UV light ($\lambda = 365$ nm)	The study indicates appreciable advantages of the preparation method of photocatalyst for hydrogen production by photo-reforming of ethanol and glycerol water solutions with respect to conventional impregnated materials.	[115]

Glycerol	0.5 wt% Pt/TiO$_2$ modified with heteropoly blue (HPB) prepared by photoreduction method	250-W high-pressure Hg lamp for UV light and 300-W halogen lamp for visible light ($\lambda > 420$ nm)	The electrons formed by the excitation of HPB upon visible light irradiation would migrate to TiO$_2$, which would enable the hydrogen production.	[116]
Methane	Pt/TiO$_2$ prepared by impregnation or photodeposition of Pt on TiO$_2$	300-W Xe lamp, UV–vis light	Photocatalytic SRM to produce hydrogen could proceed on Pt/TiO$_2$ upon UV irradiation around room temperature, ca. 323 K. The catalyst prepared by photodeposition method exhibited the highest activity.	[117]
Methane	Pt/NaTaO$_3$:La prepared by solid-state reaction method	300-W Xe lamp, UV–vis light, $\lambda = 230$–280 nm and $\lambda = 310$–400 nm	The best photocatalyst, Pt/NaTaO$_3$:La(2%), exhibited more than two times higher activity than Pt/TiO$_2$ did. The photocatalytic reaction lasted for a long time without deactivation.	[118]
Methane	Pt/CaTiO$_3$ synthesized by three methods: co-precipitation, homogeneous precipitation and solid-state reaction methods	300-W Xe lamp, $\lambda = 230$–280 nm and $\lambda = 310$–400 nm	The highest activity for the PSRM was obtained over the sample prepared by the solid-state reaction method from anatase TiO$_2$. This would be originated from the large size of crystallites and few crystal defects.	[119]
Methane	Series of Ga$_2$O$_3$ doped with different metals: Pt, Rh, Au, Pd, and Ni as co-catalyst	300-W Xe lamp, $\lambda = 230$–280 nm and $\lambda = 310$–400 nm	The highest activity was obtained over Pt-loaded β-type Ga$_2$O$_3$.	[120]

photocatalytic reforming can be considered to be especially promising for sustainable large-scale production of green hydrogen.

11.5.2 High-Value Chemicals from Photocatalytic Transformations of Biomass

Lignocellulose is the fibrous material that builds the cell walls of plants. Due to its chemical structure, which consists of three major polymeric units, lignin, cellulose, and hemicellulose, it is characterized by rather recalcitrant nature [121]. Cellulose (40%–50%) that is a high molecular polymer of glucose, forms bundles that are additionaly attached together by hemicellulose (15%–20%). Moreover, cellulose and hemicellulose are surrounded by lignin polymer (15%–25%), and all these structures together provide an extra rigidity to the whole material. As a consequence, it is much more difficult to convert it than the first generation renewable feedstock (such as sugars, starches, vegetable oils) [122].

Lignocellulosic materials (i.e., forest wastes, agricultural residues, municipal paper wastes) are usually converted into a variety of useful products in multistep processes. This complexity regards to the necessary of transformation of this material in simpler fractions that are easier transformed into desired products. However, it also allows the simultaneous production of fuels, power, and chemicals to form lignocellulose in an integrated facility denoted as a biorefinery [123]. Basically, the conversion of lignocellulose into liquid hydrocarbon transportation fuels and chemicals is based on three major routes: gasification, pyrolysis, and pretreatment hydrolysis. After this first step, the obtained liquid or gaseous fractions are upgraded via various processes obtaining a plethora of chemicals.

As far as aqueous sugars are concerned, a variety of processes is proposed for their transformation into liquid chemicals. The main aqueous-phase routes to upgrade sugars and derivatives into liquid hydrocarbon transportation fuels are schematically presented by Serrano-Ruiz and Dumesic [124]. All of these routes are driven in milder reaction conditions that allows a better control of conversion and selectivity. Nevertheless, high costs must be employed for the pretreatment and hydrolysis steps as well as for the combustion of the isolated lignin fraction [124]. Because of the high number of these thermochemical and thermocatalytic processes (including the broad spectrum of selective conversion of biomass by thermal activation of the involved catalysts), I will focus my discussion on photonic activation of nanocatalysts and their activity/selectivity in biomass valorization.

The application of heterogeneous photocatalysis to environmental purification in gas and liquid phases is generally accepted and has been widely studied since the photoactivated semiconductors have proven their capability to mineralize various kinds of refractory, toxic, and nonbiodegradable organic pollutants under mild conditions. In this context, semiconductor photocatalysis has long been considered as "nonselective" processes, especially in the aqueous media.

Heterogeneous photocatalysis in selective organic synthesis is not frequently employed, although nowadays, the demands for replacement of traditional oxidation methods with cleaner ones are increasing. TiO_2-sensitized organic photosynthetic reactions include oxidation and oxidative cleavage, reduction, isomerization, substitution, and polymerization. These reactions can be carried out in oxidatively inert solvents. Photocatalytic transformations, based on titanium dioxide, of different compounds were reviewed accurately by Shiraishi et al. [125]. Herein, without the intention of minimizing the importance of other types of reactions such as isomerization, polimerization, and reduction, I will focus the discussion on selective photocatalytic oxidation of biomass-derived compounds in liquid phase.

Lignocellulosic substances (e.g., wood) undergo UV-induced degradative reactions. Early studies from Stillings and Van Nostrand on cellulose showed cotton fibers irradiated with UV light under nitrogen atmosphere underwent photochemical transformations that led to an increase in the number of reducing sugars with a corresponding evolution of carbon monoxide and carbon dioxide [126]. Twenty-five years later, Desai and Shields [127] also studied the photochemical degradation of cellulose filter paper working with rubber-stoppered static irradiation tubes that were initially filled with air (in contrast with the nitrogen-purged system of Stillings and Van Nostrand); they observed a spectrum of fully reduced and oxygenated hydrocarbons produced upon degradation. Such hydrocarbons were only observed after irradiation periods longer than 1–2 h, whereas no delay was observed in an initially oxygen-free atmosphere (vide infra). The photoconversion of biomass, the most versatile renewable resource, could therefore be turned into a wide range of oxygenated hydrocarbons (hydrogen carriers) by means of light-induced processes [128,129].

Catalytic selective photooxidation of biomass can provide a wide range of high added-value chemicals including some of the so-called top 12 sugar-derived platform molecules (e.g., 1,4-diacids [succinic, fumaric, and malic acids], 2,5-furandicarboxylic acid, 3-hydroxypropionic acid, aspartic acid, glucaric acid, glutamic acid, itaconic acid, levulinic acid, 3-hydroxybutyrolactone, glycerol, sorbitol, and xylitol/arabinitol) [130,131].

Selective oxidation of alcohols to carbonyls (Figure 11.3) is one of the most important chemical transformations in industrial chemistry. Carbonyl compounds such as ketones and aldehydes are precursors for many drugs, vitamins, and fragrances and are important intermediates in many complex syntheses [132,133]. Most of the employed reactions, however, use toxic, corrosive, or expensive oxidants and require stringent conditions such as high pressure, temperature, or strong mineral acids [134,135]. Additionally, only aldehydes and ketones that withstand high temperatures can be prepared by classical copper-catalyzed dehydrogenation in the gas phase. Catalytic photodehydrogenation reactions that take place at room temperature and atmospheric pressure offer an interesting route for aldehyde synthesis. The C_1–C_4 alcohols are easily converted in the liquid and gas phase (Figure 11.3a,b) and in the presence of oxygen into their corresponding aldehydes or ketones, which may be further transformed by noncatalytic processes into acids [136]. A higher reactivity is registered for primary alcohols [137], which opens up a promising selective synthesis method for hydroxycarboxylic acids.

Selective dehydrogenation of various primary and secondary alcohols is also observed [138]. If the alcohol is unsaturated, isomerization may occur, yielding the corresponding saturated aldehyde [139]. The photooxidation reaction is found to be very dependent on the nature of the used alcohols. Generally, the conversion per pass of primary alcohols is low (with a slightly higher value for secondary alcohols), but the selectivity is high (>95%) [140,141]. The initial step of the proposed mechanism

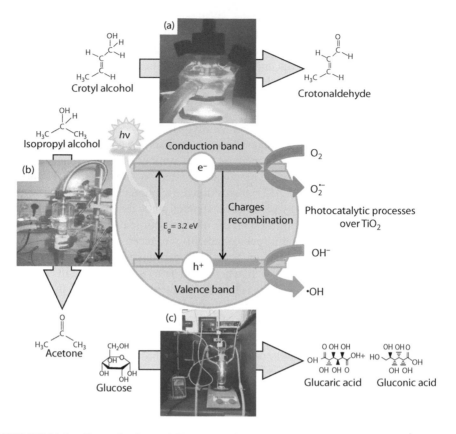

FIGURE 11.3 **(See color insert.)** Representative scheme of a variety of selective catalytic photooxidation of alcohols and sugars to aldehydes/ketones/carboxylic acids in gas (b) and liquid phase (a,c).

is the interaction of a surface hole with the hydroxyl group of the alcohol, forming a metal-oxo species with proton removal. This proton removal step becomes easier with increasing carbon chain length and branching. The higher the number of adjacent hydrogen atoms presents, the easier the removal and the greater the conversion.

In the absence of oxygen and in aqueous or mixed aqueous/organic solutions, aliphatic carboxylic acids are decarboxylated to the corresponding reduced hydrocarbons or hydrocarbon dimers [142,143]. Because CH_3COOH represents a common product of biological digestion, the photo-Kolbe reaction could be combined with biological waste treatments to generate combustible fuels.

Colmenares et al. [144] have recently determined the efficiency of heterogeneous nano-TiO_2 catalysts in the selective photocatalytic oxidation of glucose into high-valued organic compounds (Figure 11.3c and Scheme 11.1a). They have found that this reaction is highly selective (>70%) toward two organic carboxylic acids: glucaric and gluconic acids. Apart from these two products in the liquid phase, mostly CO_2 and traces of light hydrocarbons in the gas phase were detected, so the plausible reaction pathways can undergo as presented in Scheme 11.1.

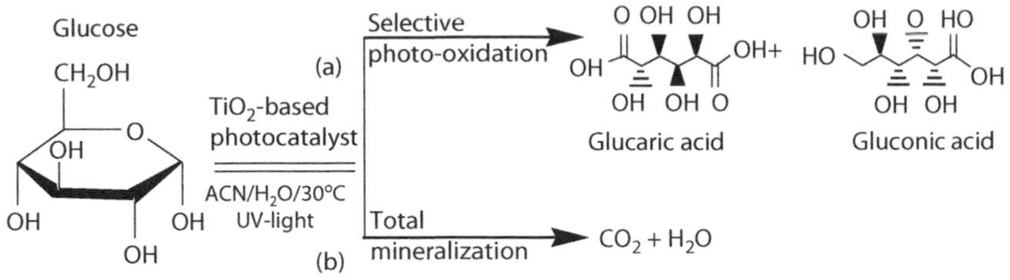

SCHEME 11.1 Reaction pathways in the liquid phase photocatalytic oxidation of glucose to carboxylic acids.

Among all photocatalytic systems tested, the best product selectivity was achieved with titania synthesized by the ultrasound-modified sol–gel method (TiO$_2$(US)) [144]. Colmenares et al. also observed that solvent composition and short illumination times have a considerable effect on photocatalysts activity/selectivity. The total organic compounds selectivity of 39.3% and 71.3% for 10% water/90% acetonitrile and 50% water/50% acetonitrile used as solvents, respectively, toward organic compounds in liquid phase was obtained with TiO$_2$(US). Such results suggest that synthesized nano-TiO$_2$ material can be used, for instance, in the decomposition of waste from the food industry and the simultaneous production of high-value chemicals when the wastes act as electron donors (here glucose) [16]. Balu and coworkers [145] reported the preparation of some very interesting magnetically separable TiO$_2$-guanidine-(Ni,Co)-Fe$_2$O$_4$ materials, which were subsequently investigated in the photocatalytic transformation of a biomass-derived platform molecule (malic acid, Scheme 11.2) in aqueous solution under visible light. An efficient separation of the photocatalyst after the reaction is one of the important issues of current photocatalytic research [1]. These materials combine the advantage of being easily recoverable using a simple magnet as well as offering the possibility to work under visible and sunlight irradiation due to the modification with guanidine, which remarkably decreases the band gap of the metal oxide.

Yasuda et al. [146] has recently reported the biological production of ethanol from lignocelluloses with TiO$_2$-photocatalytic (under UV irradiation) pretreatment. Enzymatic saccharification (SA) and subsequent fermentation (FE) are the mostly used processes for bioethanol production from cellulose materials [147], but for the efficient biological saccharification of lignocellulosic materials, some pretreatment

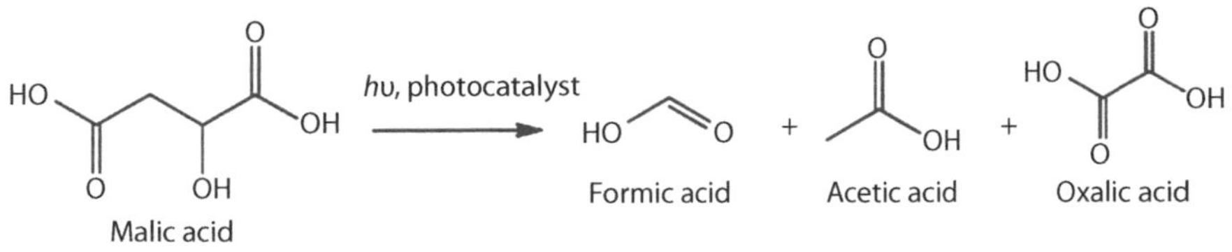

SCHEME 11.2 Photocatalytic transformation of malic acid into different chemicals. (Adapted from A.M. Balu, B. Baruwati, E. Serrano, J. Cot, J. Garcia-Martinez, R.S. Varma, and R. Luque, *Green Chemistry*, 13, 2750–2758, 2011. With permission.)

must be applied. There are several already applied pretreatment methods, such as dilute sulfuric acid, alkali, or pressured hot water. Photocatalytic pretreatment of cellulose materials has been scarcely applied and reported in the literaure. As a lignocellulosic source, Napier grass (*Pennisetum purpureum* Schumach) and silver grass (*Miscanthus sinensis* Anderss) were selected. They have observed that photocatalytic pretreatment did not affect the final product distribution, showing that TiO_2 did not disturb the biological reactions by the cellulase and yeast. However, the photocatalytic pretreatment was remarkably effective for the shortening of the reaction time in the SA and FE reactions compared with non-pretreatment and pretreatment with NaOH. Thus, the photocatalytic pretreatment is an environmentally conscious process without acid and alkali. Moreover, this is a first finding on the photocatalytic pretreatment, which includes an important inside for the bioethanol production process from soft cellulose. Another example of a combination of photocatalytic process with the one of biomass oxidation is given by Tian et al. [148]. Herein, the integration of the photochemical and electrochemical oxidation, for the modification and degradation of Kraft lignin was applied. Ta_2O_5–IrO_2 thin film was used as electrocatalyst and TiO_2 nanotube arrays were used as a photocatalyst. They discovered that the rate constant of the photochemical–electrochemical process is much larger than the sum of the rate constants of the photocatalytic oxidation and electrochemical oxidation, revealing that there is a strong synergetic effect resulting from the integration of the photocatalytic oxidation and electrochemical oxidation. The oxidation of lignin gave two products: vanillin and vanillic acid, which may well be of interest to the industries of food, aromas, and perfumes. Selective photooxidation of methanol to methyl formate in the gas phase was reported by Kominami et al. [149]. The reaction was carried out in a flow-type reactor to avoid deep oxidation of methanol in the presence of TiO_2 particles and under UV light irradiation. A high selectivity to methyl formate was observed (91 mol%) with no catalyst deactivation.

Selective photocatalytic biomass transformations hold significant promise for the development of economically and environmentally friendly synthesis processes to produce a significant number of important chemicals.

11.6   FUTURE DIRECTIONS AND OUTLOOK

Three main challenges existing in heterogeneous photocatalysis need therefore to be examined so as to understand what factors govern photochemical processes in heterogeneous systems.

First, identification of the factors determining the activity of photocatalysts and subsequent realization of how these factors influence their activity. Sometimes, the aim is to achieve the greatest photocatalytic activity possible, whereas in others, the desire may be to completely shut down the photochemical activity of the solids surface.

Another major and no less significant challenge is to discover how to govern the selectivity of photocatalysts and what factors manipulate this selectivity. For example, even in conventional applications of photocatalysis in water and air purification, one may often wish to achieve complete mineralization of organic pollutants without necessarily producing hazardous by-products. Of greater importance for heterogeneous photocatalysis, however, may be the photochemical synthesis of desired high-value chemicals.

The last challenge deals with efforts on how to improve the spectral sensitivity of solid metal-oxide photocatalysts so that they can absorb considerably more light energy, thus significantly improving the efficiencies in processes.

Several future trends for further development are also currently under investigation. Most of such research lines are presently at their infancies but they are envisaged to hold a great potential in the near future. These include the following:

- One of the biggest problems at this moment is that most articles are not comparable to each other due to differences in used TiO_2 and reaction parameters (e.g., light, amount of catalyst, reactor setup, concentration of substrates). Standardization of reactions is one of the most important steps that should be taken in this field to get beyond the present level
- The preparation of photocatalytic nanostructures capable of selective photocatalytic degradation of organic pollutants
- Novel preparation of ternary mixed oxide systems for photooxidative degradation
- Designing of more reliable photocatalyst that can be photoactivated by visible and/or solar light (e.g., photosensitizing TiO_2 in the visible by cation/anion doping)
- Exploring the possibilities to work with other materials than titania
- Advancing photocatalysis for preparative fine chemistry especially from biomass-derived intermediates
- Novel photocatalysts for the production of energy: gas and liquid solar fuels.

ACKNOWLEDGMENTS

This research was supported by a Marie Curie International Reintegration Grant within the 7th European Community Framework Programme. This scientific work was financed from the 2012–2014 Science Financial Resources, granted for the international co-financed project implementation (Project 473/7.PR/2012, Ministry of Science and Higher Education of Poland). This work was supported by the National Science Centre (NCN) within Research Project DEC-2011/01/B/ST5/03888.

REFERENCES

1. Carp, O.; Huisman, C.L.; Reller, A. Photoinduced reactivity of titanium dioxide. *Prog. Solid State Chem.* 32 (2004) 33–177.
2. Herrmann, J.M. Heterogeneous photocatalysis: fundamentals and applications to the removal of various types of aqueous pollutants. *Catal. Today* 53 (1999) 115–129.
3. Colmenares, J.C. In: Activation of heterogenous nanocatalysts by solar light: principles, synthesis and application. *Catalysis: Principles, Types and Applications.* Editor: Minsuh Song. Nova Science Publishers, Inc. New York, 2011. pp. 101–165.
4. Zhao, J.; Yang, X. Photocatalytic oxidation for indoor air purification: a literature review. *Build. Environ.* 38 (2003) 645–654.

5. Dung, N.T.; Khoa, N.V.; Herrmann, J.M. Photocatalytic degradation of reactive dye RED-3BA in aqueous TiO_2 suspension under UV-visible light. *Int. J. Photoenergy*, 7 (2005) 11–15.

6. Kim, H.; Lee, S.; Han, Y.; Park, J. Preparation of dip-coated TiO_2 photocatalyst on ceramic foam pellets. *J. Mater. Sci.* 40 (2005) 5295–5298.

7. Horvath, I.T. *Encyclopedia of Catalysis*; John Wiley & Sons: New York, USA, 2003.

8. Lawless, D.; Serpone, N.; Meisel, D. Role of hydroxyl radicals and trapped holes in photocatalysis. A pulse radiolysis study. *J. Phys. Chem.* 95 (1991) 5166–5170.

9. Colombo, D.P., Jr.; Bowman, R.M. Does interfacial charge transfer compete with charge carrier recombination? A femtosecond diffuse reflectance investigation of TiO_2 nanoparticles. *J. Phys. Chem.* 100 (1996) 18445–18449.

10. Yang, X.; Tamai, N. How fast is interfacial hole transfer? In situ monitoring of carrier dynamics in anatase TiO_2 nanoparticles by femtosecond laser spectroscopy. *Phys. Chem. Chem. Phys.* 3 (2001) 3393–3398.

11. Zeltner, W.A.; Tompkin, D.T. Shedding light on photocatalysis. *Ashrae Trans.* 111 (2005) 532–534.

12. Fujishima, A.; Rao, T.N.; Tryk, D.A. Titanium dioxide photocatalysis. *J. Photochem. Photobiol. C* 1 (2000) 1–21.

13. Nakato, Y.; Tsumura, A.; Tsubomura, H. Photo- and electroluminescence spectra from an n-titanium dioxide semiconductor electrode as related to the intermediates of the photooxidation reaction of water. *J. Phys. Chem.* 87 (1983) 2402–2405.

14. Ishibashi, K.; Fujishima, A.; Watanabe, T.; Hashimoto, K. Quantum yields of active oxidative species formed on TiO_2 photocatalyst. *J. Photochem. Photobiol. A* 134 (2000) 139–142.

15. Anpo, M.; Shima, T.; Kodama, S.; Kubokawa, Y. Photocatalytic hydrogenation of propyne with water on small-particle titania: size quantization effects and reaction intermediates. *J. Phys. Chem.* 91 (1987) 4305–4310.

16. Colmenares, J.C.; Luque, R.; Campelo, J.M.; Colmenares, F.; Karpiński, Z.; Romero, A.A. Nanostructured photocatalysts and their applications in the photocatalytic transformation of lignocellulosic biomass: an overview. *Materials* 2 (2009) 2228–2258.

17. Perkowski, J.; Bzdon, S.; Bulska, A.; Józwiak, W.K. Decomposition of detergents present in car-wash sewage by titania photo-assisted oxidation. *Polish J. Environ. Stud.* 15 (2006) 457–465.

18. Song, W.; Xiaohong, W.; Wei, Q.; Zhaohua, J. TiO_2 films prepared by micro-plasma oxidation method for dye-sensitized solar cell. *Electrochim. Acta* 53 (2007) 1883–1889.

19. Karuppuchamy, S.; Suzuki, N.; Ito, S.; Endo, T. A novel one-step electrochemical method to obtain crystalline titanium dioxide films at low temperature. *Curr. Appl. Phys.* 9 (2009) 243–248.

20. Fan, L.; Ichikuni, N.; Shimazu, S.; Uematsu, T. Preparation of Au/ TiO_2 catalysts by suspension spray reaction method and their catalytic property for CO oxidation. *Appl. Catal. A: Gen.* 246 (2003) 87–95.

21. Inoue, S.; Muto, A.; Kudou, H.; Ono, T. Preparation of novel titania support by applying the multi-gelation method for ultra-deep HDS of diesel oil. *Appl. Catal. A: Gen.* 269 (2004) 7–12.

22. Wu, C.-I.; Huang, J.-W.; Wen, Y.-L.; Wen, S.-B.; Shen, Y.-H.; Yeh, M.-Y. Preparation of TiO_2 nanoparticles by supercritical carbon dioxide. *Mater. Lett.* 62 (2008) 1923–1926.

23. Patil, K.R.; Sathaye, S.D.; Khollam, Y.B.; Deshpande, S.B.; Pawaskar, N.R.; Mandale, A.B. Preparation of TiO_2 thin films by modified spin-coating method using an aqueous precursor. *Mater. Lett.* 57 (2003) 1775–1780.

24. Cheng, F.; Peng, Z.; Liao, C.; Xu, Z.; Gao, S.; Yan, C.; Wang, D.; Wang, J. Chemical synthesis and magnetic study of nanocrystalline thin films of cobalt spinel ferrites. *Solid State Commun.* 107 (1998) 471–476.

25. Prasad, S.; Vijayalakshmi, A.; Gajbhiye, N.S. Synthesis of ultrafine cobalt ferrite by thermal decomposition of citrate precursor. *Therm. Anal. Calorim.* 52 (1998) 595–607.
26. Miyata, T.; Tsukada, S.; Minami, T. Preparation of anatase TiO_2 thin films by vacuum arc plasma evaporation. *Thin Solid Films* 496 (2006) 136–140.
27. Arimitsu, N.; Nakajima, A.; Kameshima, Y.; Shibayama, Y.; Ohsaki, H.; Okada, K. Preparation of cobalt-titanium dioxide nanocomposite films by combining inverse micelle method and plasma treatment. *Mater. Lett.* 61 (2007) 2173–2177.
28. Lin, X.M.; Sorensen, C.N.; Klabunde, K.J.; Hadjipanayis, G.C. Temperature dependence of morphology and magnetic properties of cobalt nanoparticles prepared by an inverse micelle technique. *Langmuir* 14 (1998) 7140–7146.
29. Kluson, P.; Luskova, H.; Cajthaml, T.; Solcov, O. Non thermal preparation of photoactive titanium (IV) oxide thin layers. *Thin Solid Films* 495 (2006) 18–23.
30. Sun, J.; Wang, X.; Sun, J.; Sun, R.; Sun, S.; Qiao, L. Photocatalytic degradation and kinetics of Orange G using nano-sized $Sn(IV)/TiO_2/AC$ photocatalyst *J. Mol. Catal. A: Chem.* 260 (2006) 241–246.
31. Pathan, H.M.; Min, S.-K.; Desai, J.D.; Jung, K.-D.; Joo, O.-S. Preparation and characterization of titanium dioxide thin films by SILAR method. *Mater. Chem. Phys.* 97 (2006) 5–9.
32. Gu, D.-E.; Yang, B.-C.; Hu, Y.-D. V and N co-doped nanocrystal anatase TiO_2 photocatalysts with enhanced photocatalytic activity under visible light irradiation. *Catal. Commun.* 9 (2008) 1472–1476.
33. Lee, A.-C.; Lin, R.-H.; Yang, C.-Y.; Lin, M.-H.; Wang, W.-Y. Preparations and characterization of novel photocatalysts with mesoporous titanium dioxide (TiO_2) via a sol–gel method. *Mater. Chem. Phys.* 109 (2007) 275–280.
34. Li, Y.; Demopoulos, G.P. Precipitation of nanosized titanium dioxide from aqueous titanium(IV) chloride solutions by neutralization with MgO. *Hydrometallurgy* 90 (2008) 26–33.
35. Sun, J.; Qiao, L.; Sun, S.; Wang, G. Photocatalytic degradation of Orange G on nitrogen-doped TiO_2 catalysts under visible light and sunlight irradiation. *J. Hazard. Mater.* 155 (2008) 312–319.
36. Zhu, J.; Deng, Z.; Chen, F.; Zhang, J.; Chen, H.; Anpo, M.; Huang, J.; Zhang, L. Hydrothermal doping method for preparation of Cr^{3+}-TiO_2 photocatalysts with concentration gradient distribution of Cr^{3+}. *Appl. Catal. B: Environ.* 62 (2006) 329–335.
37. Peng, F.; Cai, L.; Huang, L.; Yu, H.; Wang, H. Preparation of nitrogen-doped titanium dioxide with visible-light photocatalytic activity using a facile hydrothermal method. *J. Phys. Chem. Solids* 69 (2008) 1657–1664.
38. Zhao, X.; Liu, M.; Zhu, Y. Fabrication of porous TiO_2 film via hydrothermal method and its photocatalytic performances. *Thin Solid Films* 515 (2007) 7127–7134.
39. Kim, B.-H.; Lee, J.-Y.; Choa, Y.-H.; Higuchi, M.; Mizutani, N. Preparation of TiO_2 thin film by liquid sprayed mist CVD method. *Mater. Sci. Eng. B* 107 (2004) 289–294.
40. Ghorai, T.K.; Dhak, D.; Biswas, S.K.; Dalai, S.; Pramanik, P. Photocatalytic oxidation of organic dyes by nano-sized metal molybdate incorporated titanium dioxide ($M_xMo_xTi_{1-x}O_6$) (M = Ni, Cu, Zn) photocatalysts. *J. Mol. Catal. A: Chem.* 273 (2007) 224–229.
41. Zhang, X.; Lei, L. One step preparation of visible-light responsive Fe–TiO_2 coating photocatalysts by MOCVD. *Mater. Lett.* 62 (2008) 895–897.
42. Wu, J.M.; Shih, H.C.; Wu, W.T.; Tseng, Y.K.; Chen, I.C. Thermal evaporation growth and the luminescence property of TiO_2 nanowires. *J. Cryst. Growth*, 281 (2005) 384.
43. Xiang, B.; Zhang, Y.; Wang, Z.; Luo, X. H.; Zhu, Y. W.; Zhang, H.Z.; Yu, D. P. Field-emission properties of TiO_2 nanowire arrays. *J. Phys. D* 38 (2005) 1152.
44. Corradi, A.B.; Bondioli, F.; Focher, B.; Ferrari, A.M.; Grippo, C.; Mariani, E.; Villa, C. Conventional and microwave-hydrothermal synthesis of TiO_2 nanopowders. *J. Am. Ceram. Soc.* 88 (2005) 2639.

45. Peng, F.; Cai, L.; Yu, H.; Wang, H.; Yang, J. Synthesis and characterization of substitutional and interstitial nitrogen-doped titanium dioxides with visible light photocatalytic activity. *J. Solid State Chem.* 181 (2008) 130–136.

46. Colmenares, J.C.; Aramendía, M.A.; Marinas, A; Marinas, J.M.; Urbano, F.J. Titania nano-photocatalysts synthesized by ultrasound and microwave methodologies: application in depuration of water from 3-chloropyridine. *J. Mol. Catal. A: Chem.* 331 (2010) 58–63.

47. Colmenares, J.C.; Aramendía, M.A.; Marinas, A.; Marinas, J.M.; Urbano. F.J. Synthesis, characterization and photocatalytic activity of different metal-doped titania systems. *Appl. Catal. A* 306 (2006) 120–127.

48. Jeon, M.S.; Yoon, W.S.; Joo, H.; Lee, T.K.; Lee, H. Preparation and characterization of a nano-sized Mo/Ti mixed photocatalyst. *Appl. Surf. Sci.* 165 (2000) 209–216.

49. Horikawa, T.; Katoh, M.; Tomida, T. Preparation and characterization of nitrogen-doped mesoporous titania with high specific surface area. *Micropor. Mesopor. Mater.* 110 (2008) 397–404.

50. Zhou, M.; Yu, J. Preparation and enhanced daylight-induced photocatalytic activity of C,N,S-tridoped titanium dioxide powders. *J. Hazard. Mater.* 152 (2008) 1229–1236.

51. Andronic, L.; Duta, A. The influence of TiO_2 powder and film on the photodegradation of methyl orange. *Mater. Chem. Phys.* 112 (2008) 1078–1082.

52. Liu, C.; Tang, X.; Mo, C.; Qiang, Z. Characterization and activity of visible-light-driven TiO_2 photocatalyst co-doped with nitrogen and cerium. *J. Solid State Chem.* 181 (2008) 913–919.

53. Huang, M.; Xu, C.; Wu, Z.; Huang, Y.; Lin, J.; Wu, J. Photocatalytic discolorization of methyl orange solution by Pt modified TiO_2 loaded on natural zeolite. *Dyes Pigments* 77 (2008) 327–334.

54. Li, F.B.; Li, X.Z. The enhancement of photodegradation efficiency using $Pt–TiO_2$ catalyst. *Chemosphere* 48 (2002) 1103–1111.

55. Brinker, C.J.; Scherer, G.W. *Sol-Gel Science: The Physics and Chemistry of Sol-Gel Processing*, Academic Press, Boston, 1990.

56. Oye, G.; Glomm, W.R.; Vralstad, T.; Volden, S.; Magnusson, H.; Stocker, M.; Sjoblom, J. Synthesis, functionalisation and characterisation of mesoporous materials and sol–gel glasses for applications in catalysis, adsorption and photonics. *Adv. Colloid. Interface Sci.* 123–126 (2006) 17–32.

57. Arnelao, L.; Barreca, D.; Moraru, B. A molecular approach to RuO_2-based thin films: sol–gel synthesis and characterisation. *J. Non-Crystal. Solids* 316 (2003) 364–371.

58. Maduraiveeran, G.; Ramaraj, R. A facile electrochemical sensor designed from gold nanoparticles embedded in three-dimensional sol–gel network for concurrent detection of toxic chemicals. *Electrochem. Commun.* 9 (2007) 2051–2055.

59. Pakizeh, M.; Omidkhah, M.R.; Zarringhalam, A. Synthesis and characterization of new silica membranes using template–sol–gel technology. *Int. J. Hydrogen Energy* 32 (2007) 1825–1836.

60. Park, S.H.; Park, J.-S.; Yim, S.-D.; Park, S.-H.; Lee, Y.-M.; Kim, C.-S. Preparation of organic/inorganic composite membranes using two types of polymer matrix via a sol–gel process. *J. Power Sources* 181 (2008) 259–266.

61. Li, X.J.; Zeng, Z.R.; Gao, S.Z. Preparation and characteristics of sol–gel-coated calix[4] arene fiber for solid-phase microextraction. *J. Chromatogr. A* 1023 (2004) 15–25.

62. Yun, L. High extraction efficiency solid-phase microextraction fibers coated with open crown ether stationary phase using sol–gel technique. *Anal. Chim. Acta* 486 (2003) 63–72.

63. Jeronimo, P.C.A.; Araujo, A.N.; Conceica, M.; Montenegro, B.S.M. Optical sensors and biosensors based on sol–gel films. *Talanta* 72 (2007) 13–27.

64. Volkan, M.; Stokes, D.L.; Vo-Dinh, T. A sol–gel derived AgCl photochromic coating on glass for SERS chemical sensor application. *Sensors Actuators B: Chem.* 106 (2005) 660–667.

65. Dunn, B.; Farrington, G.C.; Katz, B. Sol-gel approaches for solid electrolytes and electrode materials. *Solid State Ionics* 70–71 (1994) 3–10.

66. Bu, S.; Jin, Z.; Liu, X.; Yin, T.; Cheng, Z. Preparation of nanocrystalline TiO_2 porous films from terpineol-ethanol-PEG system. *J. Mater. Sci.* 41 (2006) 2067–2073.

67. Campostrini, R.; Ischia, M.; Palmisano, L. Pyrolysis study of Sol-gel derived TiO_2 powders: Part I. TiO_2-anatase prepared by reacting titanium(IV) isopropoxide with formic acid. *J. Therm. Anal. Calorim.* 71 (2003) 997–1009.

68. Li, Y.; White, T.J.; Lim, S.H. Low-temperature synthesis and microstructural control of titania nano-particles. *J. Solid State Chem.* 177 (2004) 1372–1381.

69. Crepaldi, E.L.; De Soler-Illia, G.J.A.A.; Grosso, D.; Cagnol, F.; Ribot, F.; Sanchez, C. Controlled formation of highly organized mesoporous titania thin films: from mesostructured hybrids to mesoporous nanoanatase TiO_2. *J. Am. Chem. Soc.* 125 (2003) 9770–9786.

70. Bartl, M.H.; Boettcher, S.W.; Frindell, K.L.; Stucky, G.D. 3-D Molecular assembly of function in titania-based composite material systems. *Acc. Chem. Res.* 38 (2005) 263–271.

71. Fujishima, A.; Honda, K. Electrochemical photolysis of water at a semiconductor electrode. *Nature* 238 (1972) 37–38.

72. Kudo, A.; Miseki, Y. Heterogeneous photocatalyst materials for water splitting. *Chem. Soc. Rev.* 38 (2009) 253–278.

73. Navarro R.M.; Alvarez M.C.; del Valle, F.; Villoria de la Mano, J.A.; Fierro J.L.G. Water spliting on semiconductor catalysts under visible-light irradiation. *ChemSusChem* 2 (2009) 471–485.

74. Kitano, M.; Hara, M. Heterogeneous photocatalytic cleavage of water. *J. Mater. Chem.* 20 (2010) 627-641.

75. Lettmann, C.; Hinrichs, H.; Maier, W.F. Combinatorial discovery of new photocatalysts for water purification with visible light. *Angew Chem. Int. Ed.* 40 (2001) 3160–3164.

76. Seyler, M.; Stoewe, K.; Maier, W. New hydrogen-producing photocatalysts—a combinatorial search. *Appl. Catal. B* 76 (2007) 146–157.

77. Maeda, K.; Domen, K. New non-oxide photocatalysts designed for overall water splitting under visible light. *J. Phys. Chem. C* 111 (2007) 7851–7861.

78. Chareonlimkun, A.; Champreda, V.; Shotipruk, A.; Laosiripojana, N. Catalytic conversion of sugarcane bagasse, rice husk and corncob in the presence of TiO_2, ZrO_2 and mixed-330 oxide TiO_2–ZrO_2 under hot compressed water (HCW) condition. *Bioresour. Technol.* 101 (2010) 4179–4186.

79. Goh, C.S.; Lee, K.T.; Bhatia, S. Hot compressed water pretreatment of oil palm fronds to enhance glucose recovery for production of second generation bio-ethanol. *Bioresour. Technol.* 101 (2010) 7362–7367.

80. Ni, M.; Leung, D.Y.C.; Leung, M.K.H.; Sumathy, K. An overview of hydrogen production from biomass. *Fuel Process Technol.* 87 (2006) 461–472.

81. Iwasaki, W. A consideration of the economic efficiency of hydrogen production from biomass. *Int. J. Hydrogen Energy* 28 (2003) 939–944.

82. Rapagna, S.; Jand, N.; Foscolo, P.U. Catalytic gasification of biomass to produce hydrogen rich gas. *Int. J. Hydrogen Energy* 23 (1998) 551–557.

83. Li, S.G.; Xu, S.P.; Liu, S.Q.; Yang, C.; Lu, Q.H. Fast pyrolysis of biomass in free-fall reactor for hydrogen-rich gas. *Fuel Process Technol.* 85 (2004) 1201–1211.

84. Hao, X.H.; Guo, L.J.; Mao, X.; Zhang, X.M.; Chen, X.J. Hydrogen production from glucose used as a model compound of biomass gasified in supercritical water. *Int. J. Hydrogen Energy* 28 (2003) 55–64.

85. Ni, M.; Leung, M.K.H.; Leung, D.Y.C.; Sumathy, K. A review and recent developments in photocatalytic water-splitting using TiO_2 for hydrogen production. *Renewable Sust. Energy Rev.* 11 (2006) 401–425.

86. Bolton, J.R.; Strickler, S.J.; Connolly, J.S. Limiting and realizable efficiencies of solar photolysis of water. *Nature* 316 (1985) 495.

87. Kawai, T.; Sakata, T. Conversion of carbohydrate into hydrogen fuel by a photocatalytic process. *Nature* 286 (1980) 474–476.

88. Kawai, M.; Kawai, T.; Tamaru, K. Production of hydrogen and hydrocarbon from cellulose and water. *Chem. Lett.* 8 (1981) 1185–1188.

89. Kawai, T.; Sakata, T. Photocatalytic hydrogen production from water by the decomposition of poly(vinyl chloride), protein, algae, dead insects, and excrement. *Chem. Lett.* 1 (1981) 81–84.

90. Sakata, T.; Kawai, T. Hydrogen production from biomass and water by photocatalytic processes. *Nouv. J. Chem.* 5 (1981) 279–281.

91. Fu, X.; Long, J.; Wang, X.; Leung, D.Y.C.; Ding, Z.; Wu, L.; Zhang, Z.; Li, Z.; Fu, X. Photocatalytic reforming of biomass: A systematic study of hydrogen evolution from glucose solution. *Int. J. Hydrogen Energy* 33 (2008) 6484–6491.

92. Fu, X.; Wang, X.; Leung, D.Y.C.; Xue, W.; Ding, Z.; Huang, H.; Fu, X. Photocatalytic reforming of glucose over La doped alkali tantalate photocatalysts for H_2 production. *Catal. Commun.* 12 (2010) 184–187.

93. Jing, D.; Liu, M.; Shi, J.; Tang, W.; Guo, L. Hydrogen production under visible light by photocatalytic reforming of glucose over an oxide solid solution photocatalyst. *Catal. Commun.* 12 (2010) 264–267.

94. Xu, Q.; Ma, Y.; Zhang, J.; Wang, X.; Feng, Z.; Li, C. Enhancing hydrogen production activity and suppressing CO formation from photocatalytic biomass reforming on Pt/TiO_2 by optimizing anatase–rutile phase structure. *J. Catal.* 278 (2011) 329–335.

95. Colmenares, J.C.; Magdziarz, A.; Aramendia, M.A.; Marinas, A.; Marinas, J.M.; Urbano, F.J.; Navio, J.A. Influence of the strong metal support interaction effect (SMSI) of Pt/TiO_2 and Pd/TiO_2 systems in the photocatalytic biohydrogen production from glucose solution. *Catal. Commun.* 16 (2011) 1–6.

96. Li, Y.; Wang, J.; Peng, S.; Lu, G.; Li, S. Photocatalytic hydrogen generation in the presence of glucose over ZnS-coated $ZnIn_2S_4$ under visible light irradiation. *Int. J. Hydrogen Energy* 35 (2010) 7116–7126.

97. Badawy, M.I.; Ghaly, M.I.; Ali, M.E.M. Photocatalytic hydrogen production over nanostructured mesoporous titania from olive mill wastewater. *Deslination* 267 (2011) 250–255.

98. Ilie, M.; Cojocaru, B.; Parvulescu, V.I.; Garcia, H. Improving TiO_2 activity in photoproduction of hydrogen from sugar industry wastewaters. *Int. J. Hydrogen Energy* 36 (2011) 15509–15518.

99. Daskalaki, V.M.; Kondarides, D.I. Efficient production of hydrogen by photo-induced reforming of glycerol at ambient conditions. *Catal. Today* 144 (2009) 75–80.

100. Kandiel, T.A.; Dillert, R.; Robben, L.; Bahnemann, D.W. Photonic efficiency and mechanism of photocatalytic molecular hydrogen production over platinized titanium dioxide from aqueous methanol solutions. *Catal. Today* 161 (2011) 196–201.

101. Fukumoto, S.; Kitano, M.; Takeuchi, M.; Matsuoka, M.; Anpo, M. Photocatalytic hydrogen production from aqueous solutions of alcohol as model compounds of biomass using visible light-responsive TiO_2 thin films. *Catal. Lett.* 127 (2009) 39–43.

102. Mozia, S.; Heciak, A.; Morawski, A.W. Photocatalytic acetic acid decomposition leading to the production of hydrocarbons and hydrogen on Fe-modified TiO_2. *Catal. Today* 161 (2011) 189–195.

103. Mozia, S.; Heciak, A.; Morawski, A.W. The influence of physico-chemical properties of TiO_2 on photocatalytic generation of C_1–C_3 hydrocarbons and hydrogen from aqueous solution of acetic acid. *Appl. Catal. B* 104 (2011) 21–29.

104. Zheng, X.-J.; Wu, Y.-J.; Wei, L.-F.; Xie, B.; Wei, M.-B. Photocatalytic H_2 production from acetic acid solution over CuO/SnO_2 nanocomposites under UV irradiation. *Int. J. Hydrogen Energy* 35 (2010) 11709–11718.

105. Chiarello, G.L.; Aguirre, M.H.; Selli, E. Hydrogen production by photocatalytic steam reforming of methanol on noble metal-modified TiO_2. *J. Catal.* 273 (2010) 182–190.

106. Wu, G.; Chen, T.; Su, W.; Zhou, G.; Zong, X.; Lei, Z.; Li, C. H_2 production with ultra-low CO selectivity via photocatalytic reforming of methanol on Au/TiO_2 catalyst. *Int. J. Hydrogen Energy* 33 (2008) 1243–1251.

107. Li, Y.; Huang, Y.; Wu, J.; Huang, M.; Lin, J. Photocatalytic activities for hydrogen evolution of new layered compound series $HLaTa_{x/3}Nb_{2-x/3}O_7$/Pt (x = 0, 2, 3, 4, and 6). *J. Hazard. Mater.* 177 (2010) 458–464.

108. Luo, N.; Jiang, Z.; Shi, H.; Cao, F.; Xiao, T.; Edwards, P.P. Photo-catalytic conversion of oxygenated hydrocarbons to hydrogen over heteroatom-doped TiO_2 catalysts. *Int. J. Hydrogen Energy* 34 (2009) 125–129.

109. Kondarides, D.I.; Daskalaki, V.M.; Patsoura, A.; Verykios, X.E. Hydrogen production by photo-induced reforming of biomass components and derivatives at ambient conditions. *Catal. Lett.* 122 (2008) 26–32.

110. Wu, G.; Tao, Ch.; Zhou, G.; Zong, X.; Li, C. H_2 production with low CO selectivity from photocatalytic reforming of glucose on metal/TiO_2 catalysts. *Sci. China, Ser. B Chem.* 51 (2008) 97–100.

111. Puangpetch, T.; Sreethawong, T.; Chavadej, S. Hydrogen production over metal-loaded mesoporous-assembled $SrTiO_3$ nanocrystal photocatalysts: Effects of metal type and loading. *Int. J. Hydrogen Energy* 35 (2010) 6531–6540.

112. Onsuratoom, S.; Puangpetch, T.; Chavadej, S. Comparative investigation of hydrogen production over Ag-, Ni-, and Cu-loaded mesoporous-assembled TiO_2–ZrO_2 mixed oxide nanocrystal photocatalysts. *Chem. Eng. J.* 173 (2011) 667–675.

113. Miwa, T.; Kaneco, S.; Katsumata, H.; Suzuki, T.; Ohta, K.; Verma, S.Ch.; Sugihara, K. Photocatalytic hydrogen production from aqueous methanol solution with CuO/Al_2O_3/TiO_2 nanocomposite. *Int. J. Hydrogen Energy* 35 (2010) 6554–6560.

114. Bowker, M.; Davie, P.R.; Al-Mazroai, L.S. Photocatalytic reforming of glycerol over gold and palladium as an alternative fuel source. *Catal. Lett.* 128 (2009) 253–255.

115. Gombac, V.; Sordelli, L.; Montini, T.; Delgado, J.J.; Adamski, A.; Adami, G.; Cargnello, M.; Bernal S.; Fornasiero, P. CuO_x–TiO_2 photocatalysts for H_2 production from ethanol and glycerol solutions. *J. Phys. Chem. A* 114 (2010) 3916–3925.

116. Fu, N.; Lu, G. Hydrogen evolution over heteropoly blue-sensitized Pt/TiO_2 under visible light irradiation. *Catal. Lett.* 127 (2009) 319–322.

117. Yoshida, H.; Hirao, K.; Nishimoto, J.; Shimura, K.; Kato, S.; Itoh, H.; Hattori, T. Hydrogen production from methane and water on platinum loaded titanium oxide photocatalysts. *J. Phys. Chem. C* 112 (2008) 5542–5551.

118. Shimura, K.; Kato, S.; Yoshida, T.; Itoh, H.; Hattori, T.; Yoshida, H. Photocatalytic steam reforming of methane over sodium tantalate. *J. Phys. Chem. C* 114 (2010) 3493–3503.

119. Shimura, K.; Miyanaga, H.; Yoshida, H. Scientific bases for the preparation of heterogeneous catalysts. *Stud. Surf. Sci. Catal.* 175 (2010) 85–92.

120. Shimura, K., Yoshida, T., Yoshida, H. *J. Phys. Chem. C*, 114 (2010) 11466–11474.

121. Martin-Alonso, D.; Bond, J.Q.; Dumesic, J.A. Catalytic conversion of biomass to biofuels. *Green Chem.* 12 (2010) 1493.

122. Lenk, F.; Broring, S.; Herzog, P.; Leker, J. On the usage of agricultural raw materials – energy or food? An assessment from an economics perspective. *Biotechnology* 2 (2007) 1497–1504.

123. Kamm, B. Production of platform chemicals and synthesis gas from biomass. *Angew. Chem., Int. Ed.* 46 (2007) 5056.

124. Serrano-Ruiz, J.C.; Dumesic, J.A. Catalytic routes for the conversion of biomass into liquid hydrocarbon transportation fuels. *Energy Environ. Sci.* 4 (2011) 83.

125. Shiraishi, Y.; Hirai, T. Selective organic transformations on titanium oxide-based photocatalysts. *J. Photochem. Photobiol. C* 9 (2008) 157–170.

126. Stillings, R.A.; Van Nostrand, R.J. The action of ultraviolet light upon cellulose. I. Irradiation effects. II. Post-irradiation effects. *J. Am. Chem. Soc.* 1944, 66, 753–760.

127. Desai, R.L.; Shields, J.A. Photochemical degradation of cellulose material. *Makromol. Chem.* 122 (1969) 134–144.

128. Timpa, J.D.; Griffin, G.W. Photoinduced electron transfer cleavage of lignocellulosics. *Cellulose Chem. Technol.* 19 (1985) 279–289.

129. Greenbaum, E.; Tevault, C.V.; Ma, C.Y. New photosynthesis: Direct photoconversion of biomass to molecular oxygen and volatile hydrocarbons. *Energy Fuels* 9 (1995) 163–167.

130. Werpy, T.; Pedersen, G. Top Value Chemicals from Biomass. Volume I-Results of Screening for Potential Candidates from Sugars and Synthesis Gas. US Department of Energy report, August 2004.

131. Bozell, J.J.; Petersen, G.R. Technology development for the production of biobased products from biorefinery carbohydrates—the US Department of Energy's "Top 10" revisited. *Green Chem.* 12 (2010) 539–554.

132. Sheldon, R.A.; Kochi, J.K. *Metal-catalyzed oxidation of organic compounds.* New York: Academic Press; 1981.

133. Hudlicky, M. *Oxidation in organic chemistry.* Washington (DC): American Chemical Society; 1990.

134. Larock, R.C. *Comprehensive organic transformation.* New York: VCH; 1989.

135. Canelli, G.; Cardillo, G. *Chromium oxidations in organic chemistry.* Berlin: Springer; 1984.

136. Lopez-Tenllado, F.J.; Marinas, A.; Urbano, F.J.; Colmenares, J.C.; Hidalgo, M.C.; Marinas, J.M.; Moreno, J.M. Selective photooxidation of alcohols as test reaction for photocatalytic activity. *Appl. Catal. B* 128 (2012) 150–158.

137. Harwey, P.R.; Rudham, R.; Ward, S. Photocatalytic oxidation of liquid alcohols and binary alcohol mixtures by rutile. *J. Chem. Soc. Faraday Trans. I* 79 (1983) 2975.

138. Hussein, F.H.; Pattenden, G.; Rudham, R.; Russel, R. Photo-oxidation of alcohols catalysed by platinised titanium dioxide. *Tetrahedron Lett.* 25 (1984) 3363–3364.

139. Pichat, P.; Disdier, J.; Mozzanega, M.N.; Herrmann, J.M. *Proceedings of the 8th International Congress in Catal,* vol. III. Weinheim: Verlag-Chemie-Dechema; 1984, p. 487.

140. Aramendía, M.A.; Colmenares, J.C.; Marinas, A.; Marinas, J.M.; Moreno, J.M.; Navio, J.A.; Urbano, F.J. Effect of the redox treatment of Pt/TiO$_2$ system on its photocatalytic behaviour in the gas phase selective photooxidation of propan-2-ol. *Catal. Today* 128 (2007) 235–244.

141. Aramendía, M.A.; Colmenares, J.C.; López-Fernández, S.; Marinas, A.; Marinas, J.M.; Urbano, F.J. Screening of different zeolite-based catalysts for gas-phase selective photooxidation of propan-2-ol. *Catal. Today* 129 (2007) 102–109.

142. Kreutler, B.; Bard, A.J. Heterogeneous photocatalytic preparation of supported catalysts. Photodeposition of platinum on titanium dioxide powder and other substrates. *J. Am. Chem. Soc.* 100 (1978) 4317.

143. Sakata, T.; Kawai, T.; Hashimoto, K. Heterogeneous photocatalytic reactions of organic acids and water. New reaction paths besides the photo-Kolbe reaction. *J. Phys. Chem.* 88 (1984) 2344.

144. Colmenares, J.C.; Magdziarz, A.; Bielejewska, A. High-value chemicals obtained from selective photo-oxidation of glucose in the presence of nanostructured titanium photocatalysts. *Bioresour. Technol.* 102 (2011) 11254–11257.

145. Balu, A.M.; Baruwati, B.; Serrano, E.; Cot, J.; Garcia-Martinez, J.; Varma, R.S.; Luque, R. Magnetically separable nanocomposites with photocatalytic activity under visible light for the selective transformation of biomass-derived platform molecules. *Green Chem.* 13 (2011) 2750–2758.

146. Yasuda, M.; Miuraa, A.; Yukia, R.; Nakamuraa, Y.; Shiragamia, T.; Ishii, Y.; Yokoia, H. The effect of TiO2-photocatalytic pretreatment on the biological production of ethanol from lignocelluloses. *J. Photochem. Photobiol. A* 220 (2011) 195–199.

147. Balat, M.; Balat, H.; Öz, C. Progress in bioethanol processing. *Prog. Energy Combust. Sci.* 34 (2008) 551–573.

148. Tian, M.; Wen, J.; MacDonald, D.; Asmussen, R.M.; Chen, A. A novel approach for lignin modification and degradation. *Electrochem. Commun.* 12 (2010) 527–530.

149. Kominami, H.; Sugahara, H.; Hashimoto, K. Photocatalytic selective oxidation of methanol to methyl formate in gas phase over titanium(IV) oxide in a flow-type reactor. *Catal. Commun.* 11 (2010) 426–429.

Index

Page numbers followed by f and t indicate figures and tables, respectively.